Recent Progress in Function Theory and Operator Theory

CONTEMPORARY MATHEMATICS

799

Recent Progress in Function Theory and Operator Theory

AMS Special Session
Recent Progress in Function Theory and Operator Theory
April 6, 2022
Virtual

Alberto A. Condori
Elodie Pozzi
William T. Ross
Alan A. Sola
Editors

AMERICAN MATHEMATICAL SOCIETY

2000 *Mathematics Subject Classification.* Primary 47B38, 47B37, 47B35, 47B32, 30J10, 30J05, 30H35, 30H10.

Library of Congress Cataloging-in-Publication Data

Names: AMS Special Session on Recent Progress in Function Theory and Operator Theory (2022 : Online) | Condori, Alberto Ademir, 1980- editor. | Pozzi, Elodie, 1984- editor. | Ross, William T., 1964- editor. | Sola, Alan Albert, 1981- editor.

Title: Recent progress in function theory and operator theory : AMS Special Session on Recent Progress in Function Theory and Operator Theory, April 6, 2022, virtual / Alberto A. Condori, Elodie Pozzi, William T. Ross, Alan A. Sola, editors.

Description: Providence, Rhode Island : American Mathematical Society, [2024] | Series: Contemporary mathematics, 0271-4132 ; 799 | Includes bibliographical references.

Identifiers: LCCN 2023053403 | ISBN 9781470472467 (paperback) | ISBN 9781470476120 (ebook)

Subjects: LCSH: Operator theory. | Functions of complex variables. | AMS: Operator theory – Special classes of linear operators – Operators on function spaces (general). | Operator theory – Special classes of linear operators – Operators on special spaces (weighted shifts, operators on sequence spaces, etc.). | Operator theory – Special classes of linear operators – Toeplitz operators, Hankel operators, Wiener-Hopf operators. | Operator theory – Special classes of linear operators – Operators in reproducing-kernel Hilbert spaces (including de Branges, de Branges-Rovnyak, and other structured spaces). | Functions of a complex variable – Function theory on the disc – Blaschke products. | Functions of a complex variable – Function theory on the disc – Inner functions. | Functions of a complex variable – Spaces and algebras of analytic functions – BMO-spaces. | Functions of a complex variable – Spaces and algebras of analytic functions – Hardy spaces.

Classification: LCC QA329 .A498 2024 | DDC 515/.724–dc23/eng/20231129

LC record available at https://lccn.loc.gov/2023053403

Contemporary Mathematics ISSN: 0271-4132 (print); ISSN: 1098-3627 (online)

DOI: https://doi.org/10.1090/conm/799

∞ The paper used in this book is acid-free and falls within the guidelines
established to ensure permanence and durability.
Visit the AMS home page at `https://www.ams.org/`

10 9 8 7 6 5 4 3 2 1   24 23 22 21 20 19

Contents

Preface

Function theory is a classical subject that examines the properties of individual elements in a function space or class, while operator theory usually deals with concrete operators acting on such spaces or collections of functions. These topics occupy a central position in analysis, with important connections to partial differential equations, spectral theory, approximation theory, and several complex variables. For instance, some inverse problems for PDEs in two dimensions are often restated as involving complex derivatives; this facilitates the use of complex function theory to address such questions. In addition, with the aid of certain canonical representations or "models," the study of general operators can often be reduced to that of the operator of multiplication by the independent variable or variables acting on spaces of analytic functions or its compressions to co-invariant subspaces. Thus, a detailed understanding of operators is often connected with questions concerning analytic functions (e.g., zero sets, interpolation, multiplicative structure granted by factorizations in spaces of analytic functions, and so forth). On the other hand, problems concerning best approximation by analytic functions are often solved using concrete operators (e.g., compressed shifts or Hankel operators) acting on other spaces of analytic functions too. This fascinating interplay between all of these subjects drives the field, which has continued to evolve in recent years.

Accordingly, this conference-proceedings volume contains recent progress on the subject, such as Crouzeix's conjecture, Sendov's conjecture, the invariant subspace problem, maximal valence of logharmonic polynomials, cross ratios for several complex variables, Bergman–projection estimates in pseudoconvex domains and with matrix weights, boundedness of the commutator of inhomogeneous singular integral operators with bmo functions, interpolating sequences for spaces of analytic functions on the unit disk with p-summable Maclaurin coefficients, regularity of kernel-summability methods using vector-valued integrals, and Cesàro and Rhaly operators.

The editors thank Christine Thivierge and Teresa McClure at the AMS for their kind assistance with the production of this volume.

Contemporary Mathematics
Volume **799**, 2024
https://doi.org/10.1090/conm/799/16017

Three open problems in function theoretic operator theory

Pamela Gorkin and Elodie Pozzi

ABSTRACT. We discuss three open problems in operator and function theory: Crouzeix's conjecture, Sendov's conjecture, and the invariant subspace problem. We include the full conjecture, prior work, and related questions.

1. Introduction

When one thinks of operator theory, the first thing that may come to mind is a linear operator on a function or Hilbert space. Important examples can be found in differential or integral operators. Or, perhaps matrices come to mind. While function theory may not seem to be operator theoretic, many operators, or matrices, are naturally associated with functions or offer insight into functional behavior. Indeed, some of the operators we consider here are defined by a functional symbol. Thus, the field of operator theory is both broad and deep, encompassing the study of matrices, operators, and related functional behavior. The open questions that we consider here reflect this.

We begin with Crouzeix's conjecture (see Section 2), a question that was originally stated for matrices and extends naturally to bounded operators. In Section 2.2, we turn to a class of operators that is associated with certain bounded analytic functions. One of the most important operators we can study are the shift operators (one-sided or bilateral) and we will consider these in this paper. But we also focus on their relatives: model space operators or compressions of the shift operator on spaces defined in terms of finite Blaschke products—functions analytic on an open set containing the closed unit disk $\overline{\mathbb{D}}$, mapping the open unit disk to itself, and the unit circle $\mathbb{T}$ to itself. In Section 3, we view Blaschke products as hyperbolic polynomials and compare and contrast these with polynomials. As we study these connections, it will become clear that the location of the zeros of these functions as well as the location of the critical points yield information about these functions. This leads us to Section 4, where we consider the Sendov conjecture. This is a problem about the location of critical points of polynomials.

Motivation for studying these questions comes from finite dimensional spaces. Crouzeix's conjecture, for example, requires knowledge of a set called the numerical range of a matrix or, more generally, an operator. This set is useful to study, in

---

2020 *Mathematics Subject Classification.* Primary 47A15; Secondary 47A12, 47A16, 30C10.
Key words and phrases. Compression of the shift operator, numerical range, critical point, invariant subspace, composition operator.

part, because it contains the eigenvalues of the matrix. The study of eigenvalues and eigenvectors is important in both theoretical and applied settings.

On a finite-dimensional (complex, nonzero) space, every nonzero operator T on a vector space V has an eigenvalue. A beautiful proof of this can be found in [**2**] and goes like this: If n is the dimension of V, let v be a nonzero vector and consider the set of $n+1$ vectors $\{v, Tv, T^2v, \ldots, T^nv\}$. These cannot all be linearly independent and so there exist constants α_j, not all zero, λ_j and a constant $\alpha \neq 0$ such that

$$\alpha_0 v + \alpha_1 Tv + \cdots \alpha_n T^n v = \alpha(T - \lambda_1 I)(T - \lambda_2 I) \cdots (T - \lambda_n I)v = 0.$$

Therefore, one of the factors is not injective. Consequently, T has an eigenvalue. One consequence of this is that every nonzero operator on a (complex) finite dimensional vector space has an invariant subspace; that is, a proper closed subspace $\mathcal{M}$ such that $T\mathcal{M} \subseteq \mathcal{M}$. In Section 5, we consider the situation in infinite-dimensional spaces and, perhaps, the most well-known question in operator theory: the invariant subspace problem.

2. Crouzeix's conjecture

Let A denote an $n \times n$ matrix. The numerical range of A, also called the field of values, is denoted by $W(A)$ and defined by

$$W(A) = \{\langle Ax, x\rangle : x \in \mathbb{C}^n, \|x\| = 1\}.$$

The numerical range is always convex and, in the finite-dimensional setting, it is always compact. It follows from the definition that the numerical range is a unitary invariant; that is, if A and B are two matrices and there is a unitary matrix U such that $A = U^\star BU$, then $W(A) = W(B)$. It is possible for two matrices to have the same numerical range without being unitarily equivalent (they need not even have the same size!). In some cases, we can describe the numerical range in detail. In other cases, it is very difficult to do so; in fact, in most cases it is very difficult. However, for 2×2 matrices with complex entries, we know precisely what the numerical range is.

The elliptical range theorem [**52**, p. 3] says that the numerical range of a 2×2 matrix A with eigenvalues λ_1 and λ_2 is an elliptical disk with foci at λ_1 and λ_2 and minor axis of length $\sqrt{\operatorname{tr}(A^\star A) - |\lambda_1|^2 - |\lambda_2|^2}$. One reason for studying the numerical range is that it gives us more information about a matrix than the eigenvalues do. For example, the elliptical range theorem tells us that the 2×2 Jordan block

$$J_2 = \begin{pmatrix} 0 & 1 \\ 0 & 0 \end{pmatrix}$$

has $W(J_2) = \{z \in \mathbb{C} : |z| \leq 1/2\}$. Thus, the zero matrix and J_2 have the same spectrum, so the spectrum cannot distinguish between the two matrices, but the numerical range of the matrices is different. It should be noted that the spectrum of A, denoted $\sigma(A)$, is always contained in the numerical range.

When we have a matrix A and a polynomial p, we are often interested in the norm of $p(A)$. One important result in this regard is von Neumann's inequality, which says that for a contraction T and an arbitrary complex polynomial p,

$$\|p(T)\|_{op} \leq \sup_{z \in \mathbb{D}} |p(z)|.$$

It's natural to ask whether or not we need the entire set $\mathbb{D}$. For example, is it perhaps true that

$$\|p(T)\|_{op} \leq \sup_{z \in \sigma(T)} |p(z)|?$$

A simple example shows us that this is not the case: take the matrix J_2 and the polynomial $p(z) = z$. In 2004, Michel Crouzeix [**29**] proved the following theorem.

THEOREM 2.1. *Let A be a 2×2 matrix with entries in $\mathbb{C}$. Then for all polynomials p,*

$$\|p(A)\|_{op} \leq 2 \max_{W(A)} |p(z)|.$$

If $\|A\| < 1$, the numerical range is contained in the open unit disk and Theorem 2.1 implies that we can obtain a bound on the norm of $p(A)$ on a smaller set than $\mathbb{D}$. Of course, the price we have to pay is the appearance of the constant 2.

Crouzeix notes that in a personal communication, C. Badea showed that the result holds true when $W(A)$ is a disk. We provide this proof in the next section. In this same paper, Crouzeix notes that he has been unable to answer the following question: "In the two dimensional case we have proved that the disk corresponds to the worst constant; does this result remain true in any dimension?" Thus, the Crouzeix conjecture was born.

CONJECTURE 2.2 (Crouzeix's conjecture). *Let A be a $n \times n$ matrix and let p be a complex polynomial. Then*

$$\|p(A)\|_{op} \leq 2 \max_{W(A)} |p(z)|. \tag{2.1}$$

In what follows, we will write $\|T\|$ in place of $\|T\|_{op}$.

If Crouzeix's conjecture holds, it will also hold for bounded operators on Hilbert space. If it is established for polynomials, it will also hold for functions that can be approximated by polynomials in the appropriate norm.

In some cases, not only does the conjecture hold, but stronger results hold. For example, if A is a diagonal matrix, the result clearly holds with the constant $C = 2$ replaced by $C = 1$. From this and the facts that every normal matrix is unitarily equivalent to a diagonal matrix and the numerical range is invariant under unitary matrices, it is clear that Crouzeix's conjecture hold for every normal matrix and the constant C may be chosen to be equal to 1. Looking at the Jordan block J_2 again, we see that if we take the polynomial p to be the identity function $p(z) = z$ and the matrix to be J_2, then

$$\|p(J_2)\|_{op} = 1 = 2 \max_{W(J_2)} |p(z)|.$$

So no constant smaller than 2 can work, in general. Indeed, at first glance, it is not clear that there should be any constant that works, let alone the constant 2. However, in [**30**], Crouzeix showed that the conjecture holds with 11.08 in place of the 2. In fact, in [**32**], Crouzeix and Palencia showed that (2.1) holds with the constant 2 replaced by $1 + \sqrt{2}$. At the time of this writing this remains the best constant. Crouzeix and Palencia's proof is quite accessible and depends on a key lemma. Another proof of the key lemma can be found in [**71**].

Like any really good question, this one is not only intrinsically interesting, it has generated a great deal of mathematics. In Section 2.1 we discuss work on the conjecture itself, and in Section 4 we discuss other problems that grew out of work on this conjecture.

2.1. Work on the Crouzeix conjecture. We begin by showing that Crouzeix's conjecture holds for all matrices with circular disks as the numerical range. This can be found in [**29**, Theorem 3.4], where it is attributed to C. Badea, and [**22**, Theorem 14]. This proof relies on the following result of Ando. Recall that the numerical radius of a matrix is the largest absolute value of the numbers in the numerical range.

THEOREM 2.3. [**1**] *Let A be an $n \times n$ matrix. Then the numerical radius of A is at most one if and only if there exists a Hermitian contraction M and a contraction C such that $A = (I+M)^{1/2}C(I-M)^{1/2}$. Moreover, it is possible to choose C and M so that C is isometric on the range of $I - M$.*

If we let B be a Hermitian matrix such that

$$\cos^2(B) = \frac{1}{2}(I+M),$$

then we see that we can write A as $A = 2\cos(B)C\sin(B)$. (See the discussion in [**22**, p. 15] for details.) In this proof we see one useful idea that comes up frequently: If we can write $A = X^{-1}CX$, where C is a contraction, then $p(A) = X^{-1}p(C)X$ for all polynomials p. If we take norms, then, we have $\|p(A)\| \leq \|X^{-1}\|\,\|p(C)\|\,\|X\|$. The objective turns to bounding $\|X^{-1}\|\,\|X\|$. Sometimes we do obtain a bound that is at most 2, but sometimes we cannot. Then we need to develop other methods to attack this conjecture.

THEOREM 2.4. *Let $\mathcal{D}$ be a disk containing $W(A)$. Then for every complex polynomial p,*

$$\|p(A)\| \leq 2\|p\|_{\mathcal{D}}.$$

PROOF. For every $\alpha, \beta \in \mathbb{C}$, the numerical range satisfies $W(\alpha A + \beta I) = \alpha W(A) + \beta$. This allows us to assume that $W(A) \subseteq \mathbb{D}$. In this case, the discussion above implies that there is a Hermitian matrix B and a contraction C such that $A = 2\cos(B)C\sin(B)$. Let $g(x)$ be the continuous function defined by

$$g(x) = \max\{1, 2|\cos(x)|\},$$

and let $H = g(B)$. Using the functional calculus for self-adjoint operators, we may let $T = H^{-1}AH$. Now, $\|\sin(B)H\| \leq 1$ and $2\|H^{-1}\cos(B)\| \leq 1$, so T is a contraction. Now, $\|H\| \leq 2$ and $\|H^{-1}\| \leq 1$, so we may apply von Neumann's inequality to $p(T)$ and we obtain

$$\|p(A)\| \leq \|H\|\,\|p(T)\|\,\|H^{-1}\| \leq 2\|p\|_{\mathbb{D}}.$$

In particular, if $W(A)$ is a disk, then Crouzeix's conjecture holds. □

The numerical range of a Jordan block is always a disk and, therefore, from the previous discussion, the conjecture is true in that case. A natural class of matrices to consider, then, is the class of perturbations of Jordan blocks. In [**50**], Greenbaum and Choi proved the conjecture for perturbed Jordan blocks of the form

$$\begin{pmatrix} \lambda & 1 & & \\ & \ddots & \ddots & \\ & & \ddots & 1 \\ a & & & \lambda \end{pmatrix},$$

where $a, \lambda \in \mathbb{C}$. Using similar techniques, Choi [**21**] extended this result to matrices of the form

$$\begin{pmatrix} \lambda & \alpha_1 & & \\ & \ddots & \ddots & \\ & & \ddots & \alpha_{n-1} \\ \alpha_n & & & \lambda \end{pmatrix},$$

where $\lambda, \alpha_j \in \mathbb{C}$. Following this, Glader, Kurula, and Lindström [**46**] showed that the conjecture holds for all matrices that can be written as $aI + DP$ or $aI + PD$ where $a \in \mathbb{C}$, the matrix D is diagonal, and the matrix P is a permutation matrix. The proof proceeds by reducing the problem to the main result in Choi's paper. But this is not all that they do. They also show that Crouzeix's conjecture holds for all 3×3 matrices with complex entries of the form

$$\begin{pmatrix} a & b_1 & 0 \\ c_1 & a & b_2 \\ 0 & c_2 & a \end{pmatrix}.$$

They note that this is the same class as the 3×3 matrices with complex entries that have elliptic numerical range and one eigenvalue at the center of the ellipse. In [**31**], Crouzeix focused on 3×3 nilpotent matrices and showed that the conjecture held there as well.

We turn to another setting in which one can obtain results pertaining to Crouzeix's conjecture.

2.2. Crouzeix's conjecture and compressions of the shift. Let Θ denote a degree-n finite Blaschke product; that is,

$$\Theta(z) = \lambda \prod_{j=1}^{n} \frac{z - a_j}{1 - \bar{a}_j z}, \qquad \text{for } a_1, \ldots, a_n \in \mathbb{D} \text{ and } \lambda \in \mathbb{T}.$$

Finite Blaschke products are precisely the functions that are holomorphic on a neighborhood of $\overline{\mathbb{D}}$ and map the unit circle to itself. For Θ a finite Blaschke product, we let $H^2 = H^2(\mathbb{D})$ denote the standard Hardy space on the unit disk $\mathbb{D}$ and let $K_\Theta = H^2 \ominus \Theta H^2$ denote the model space associated to Θ. If M_z denotes multiplication by the independent variable z, then the compression of the shift operator with symbol Θ is denoted by S_Θ and defined by

$$S_\Theta = P_\Theta M_z|_{K_\Theta},$$

where P_Θ is the orthogonal projection from H^2 onto K_Θ. Since $\deg B = n$, the space K_Θ has dimension n and so, we can interpret S_Θ as an $n \times n$ matrix by writing down its representation with respect to an orthonormal basis of K_Θ. For more on this representation see, for example, [**34, 45**].

There are other reasons for focusing on this class. These operators S_Θ are particularly important because Sz.-Nagy and Foias [**86**] showed that every completely nonunitary $n \times n$ contraction of class C_0 with defect index 1 is unitarily equivalent to S_Θ for some finite Blaschke product Θ. So, if we can show that Crouzeix's conjecture holds for this class of operators, we would obtain the results for a larger class of operators.

We now consider Crouzeix's conjecture for such compressions of the shift. Note that this includes several classes of matrices for which we know the conjecture is true; for example, Jordan blocks and perturbed Jordan blocks. Jordan blocks correspond to compressions of the shift with symbol $\Theta = z^n$ and the perturbed Jordan blocks correspond to those with symbol $\Theta = \frac{z^n - a}{1 - \bar{a}z^n}$. Much is known about these operators and their numerical ranges, [**34**,**41**,**43**]. In fact, two operators in the class of compressions of the shift operator on spaces of the same (finite) dimension with the same numerical range are unitarily equivalent, [**43**,**44**]. Related work from a geometric perspective can be found in [**33**,**61**].

In [**10**], the authors pose a conjecture related to Crouzeix's conjecture that should be easier to solve, given its more specific nature. To state it, we note that in [**77**], Sarason proved that if f is bounded and holomorphic on $\mathbb{D}$, then

$$f(S_\Theta) = P_\Theta M_f|_{K_\Theta},$$

where M_f is multiplication by f. It follows from this that if B is a finite Blaschke product with $\deg B < \deg \Theta$, then $\|B(S_\Theta)\| = 1$ and details are provided in [**42**, p. 512]. If we put this together with Crouzeix's conjecture, we obtain a function theoretic conjecture: If we are to have $1 = \|B(S_\Theta)\| \leq 2 \max_{z \in W(S_\Theta)} |B(z)|$, then the following must be true.

CONJECTURE 2.5 (Level set Crouzeix conjecture). *Let Θ, B be finite Blaschke products with* $\deg B < \deg \Theta$. *Then*

$$\max\{|B(z)| : z \in W(S_\Theta)\} \geq \tfrac{1}{2}.$$

Level sets of inner functions have been studied in other contexts; see [**4**, **7**, **23**–**25**, **63**, **85**], yet it appears that much remains to be done here.

In [**10**], the authors proved several results that might be helpful. In particular, they show that if B is a finite Blaschke product and we let

$$\Omega_r^B = \{z \in \mathbb{C} : |B(z)| < r\},$$

then the following holds.

THEOREM 2.6. *Let B and C be finite Blaschke products with* $\deg B = \deg C$. *If there is some $r \in (0,1)$ with $\Omega_r^B \subseteq \Omega_r^C$, then there exists $\lambda \in \mathbb{T}$ such that $B = \lambda C$.*

In addition, (in their Corollary 4.2) they show that if Θ and B are Blaschke products with the degree of B strictly less than Θ and $B(z) = \lambda \left(\frac{z - z_0}{1 - \overline{z_0} z}\right)^n$ for some $\lambda \in \mathbb{T}$ and $z_0 \in \mathbb{D}$, then the level set conjecture holds. Crouzeix's work [**31**] shows that when we consider $\Theta = \lambda \left(\frac{z - z_0}{1 - \overline{z_0} z}\right)^3$, then the full level set holds, while [**10**] adds additional information for $n = 4, 5$.

When a problem is difficult, it's natural to consider a simpler version of it. We pose the following question here.

QUESTION 2.7. Let A be a matrix with elliptical numerical range. Does Crouzeix's conjecture hold in this case?

We have already discussed the fact that every 2×2 matrix has (possibly degenerate) elliptical numerical range and we noted that Crouzeix established the result in this case. The matrices considered in [**46**] also have elliptical numerical range and the conjecture holds in this case as well. This may be a good problem to focus on because a lot is known about matrices with elliptical numerical range in some special cases. For example, 3×3 matrices with elliptical numerical range have been analyzed in [**56**] and the follow-up paper [**72**]. Compressions of the shift with Blaschke symbol of degree 3 with elliptical numerical range were analyzed in [**49**], following related work of Fujimura, [**40**].

We noted that one useful technique is to write the matrix $A = XDX^{-1}$. When A has distinct eigenvalues we can take the matrix D to be diagonal (which is why we have chosen D to represent this matrix). Therefore, the verification of the conjecture often comes down to finding $\|X\|\,\|X^{-1}\|$. For A with distinct eigenvalues we can define the minimal condition number by

$$\inf\{\|X\|\,\|X^{-1}\| : A = XDX^{-1}\}.$$

The following question was posed in [**9**], along with other related questions.

QUESTION 2.8. Let Θ be a finite Blaschke product with distinct zeros. What is the minimal condition number of an eigenvector matrix X for the matrix representing S_Θ?

In 2018, Greenbaum and Overton [**51**] considered the Crouzeix ratio,

$$\|p\|_{W(A)}/\|p(A)\|.$$

If the conjecture is true, the minimal value over all polynomials should be 1/2. They performed "extensive numerical experiments" using Chebfun to find the minimal value. They conclude that "overall, our results strongly support Crouzeix's conjecture."

3. From Blaschke products to polynomials

Blaschke products are rational functions and therefore an obvious connection to polynomials exists. But we can say more: If we let $p(z) = (z - a_1)\cdots(z - a_n)$ with $a_1, \dots, a_n \in \mathbb{D}$ and consider $\overline{p\left(\frac{1}{\overline{z}}\right)}$, we see that

$$p^*(z) = \overline{p\left(\frac{1}{\overline{z}}\right)} = (1 - \overline{a_1}z)\cdots(1 - \overline{a_n}z)/z^n.$$

Thus,

$$B(z) = \frac{p(z)}{z^n p^*(z)}.$$

Polynomials of degree n have n zeros (counted according to multiplicity) in the complex plane $\mathbb{C}$, while Blaschke products of degree n have n zeros (counted according to multiplicity) in $\mathbb{D}$. We measure the Euclidean distance in $\mathbb{C}$ using $|z - a_j|$ for the distance from z to a_j, but in $\mathbb{D}$ there are points near the unit circle that are, in the Euclidean sense, close together, but a Blaschke product (which has sup norm 1 on the disk) may be zero at one point and very close to one at another. In this sense, then, they would be far apart. For this reason, a more useful measure is often the pseudohyperbolic distance: For $z, w, \in \mathbb{D}$ we define the pseudohyperbolic distance between z and w to be $\rho(z,w) = \left|\frac{z-w}{1-\overline{w}z}\right|$. Thus, we might think about polynomials as "Euclidean polynomials" while Blaschke products may be thought of as "hyperbolic polynomials." For this point of view, the reader might consult [**35, 84**]. In practice, the results about each may reflect the distinction between the Euclidean point of view and the pseudohyperbolic viewpoint. For example, consider the following two classical results.

The Gauss-Lucas theorem tells us that the critical points of a polynomial always lie in the convex hull of the zeros of the polynomial.

THEOREM 3.1 (Gauss-Lucas). *The critical points of a polynomial lie in the convex hull of its zeros.*

An 1836 note of Gauss contains this as an implication, and an explicit statement and proof appears in an 1874 paper of Lucas, [**58**]. Here is a simple proof, see [**83**].

Let $p(z) = \lambda \prod_{j=1}^{n}(z - z_j)$ and

$$f(z) = \frac{p'(z)}{p(z)} = \sum_{j=1}^{n} \frac{1}{z - z_j}.$$

Note that if $p'(w) = 0$ then either $w = z_j$ for some j or $f(w) = 0$. If $w = z_j$, the result is clear. If $f(w) = 0$, then the complex conjugate satisfies $\overline{f(w)} = 0$. Thus,

$$\sum \frac{1}{\overline{w - z_j}} = \sum \frac{w - z_j}{|w - z_j|^2} = 0.$$

Solving for w yields

$$w = \sum_{j=1}^{n} a_j z_j, \text{ where } a_j := \frac{|w - z_j|^{-2}}{\sum_{k=1}^{n} |w - z_k|^{-2}},$$

completing the proof.

On the other hand, we have the following theorem, due to J. L. Walsh, [**88**].

THEOREM 3.2 (J. Walsh). *The critical points of a finite Blaschke product are contained in the hyperbolic convex hull of the zeros.*

4. Critical points and Sendov's conjecture

What else can we say about the location of the critical points of polynomials? There is much more that we know and, surprisingly, much more to know about the location. For example, in 1864, Siebeck proved the following theorem.

THEOREM 4.1 ([**82**]). *Let $z_1, z_2, z_3 \in \mathbb{C}$ be non-collinear points. Let m_1, m_2, $m_3 > 0$. Then the zeros of the function*

$$F(z) = \frac{m_1}{z - z_1} + \frac{m_2}{z - z_2} + \frac{m_3}{z - z_3}$$

are the foci of the ellipse that touches the line segments $\overline{z_1 z_2}$, $\overline{z_1 z_3}$, and $\overline{z_2 z_3}$ at the points that $\zeta_1, \zeta_2, \zeta_3$ at the points that divide these segments in ratios $m_1 : m_2$, $m_2 : m_3$, and $m_3 : m_1$.

Marden [**59**] includes a version of this result in which the m_j may be smaller than 0 as well as a statement for more than three points, but for the purposes of this paper we restrict ourselves to this case. This theorem is a starting point for many interesting results.

Let p be a complex polynomial of degree three with distinct zeros z_1, z_2, z_3 and p' its derivative. Using partial fractions

$$\frac{p'(z)}{3p(z)} = \frac{m_1}{z - z_1} + \frac{m_2}{z - z_2} + \frac{m_3}{z - z_3}.$$

Thus, Siebeck's theorem tells us that the critical points of p are the foci of an ellipse that is inscribed in the triangle with vertices z_1, z_2, z_3. There are two very interesting things about this: first, it suggests that given a polynomial, the location of its critical points may provide interesting information about the geometry of the polynomial. Second, and perhaps surprisingly, this is connected to the numerical range.

Let's begin with the first observation. In 1958, Blagovest Sendov wondered where the critical points of a polynomial were relative to the zeros and he made the following conjecture.

CONJECTURE 4.2 (Sendov conjecture). *Assume that all zeros of a complex polynomial p lie inside the closed unit disk. Then every disk of radius one about a zero of p contains a critical point of p.*

Walter Hayman included this conjecture in his 1967 book of problems [**54**] and attributed it to Iliev, who had communicated the problem. As a consequence, the early literature refers to this under (variations of the spelling of) Iliev's conjecture.

It should be noted that the polynomial $q(z) = z^n - 1$ shows that we cannot expect that a disk smaller than radius one would have this property.

There are several proofs of Sendov's conjecture for polynomials of degree 3 (see [**26, 76, 78**]). A simple proof for degree four appears in, for example, [**27**] and one for degree five can be found in [**60**]. It is more difficult to prove the result for degree 6 (see [**15**] for a restricted case and [**12**] for the general case), degree 7 ([**13, 16**]) and degree 8 [**17**]. When geometric conditions are imposed on the zeros of the polynomials the proofs are often more accessible. For example, the result has been proven when all zeros are real [**70**], all zeros on the unit circle [**47**], one of the zeros lies at the center of the disk [**11**], and [**78**] when the convex hull of the zeros form a triangle. In a different direction, still assuming that the zeros of the polynomial p are contained in the closed unit disk, Dégot [**36**] showed that if a is a zero of p, then there exists an integer N such that the disk $|\zeta - a| \leq 1$ contains a critical point of p when the degree of p is larger than N. More recently, Tao [**87**] showed that there exists a constant n_0 such that Sendov's conjecture holds for $n \geq n_0$. In [**81**, p. 206], T. Sheil-Small notes that "odds of 4 - 1 in favour of the conjecture are cautious odds in the circumstances."

That's a brief summary of the history and an incomplete list of references, but enough to serve as a guide to prior results. We turn now to our claim that there is a connection between this result and the numerical range. To illustrate this, we show how the elliptical range theorem can be used to establish Siebeck's theorem.

NUMERICAL RANGE PROOF OF SIEBECK'S THEOREM, [**48**]. Let z_1, z_2, z_3 denote the zeros of the polynomial and let us assume that they are not all colinear. As before, we use partial fractions to write

$$\frac{p'(z)}{3p(z)} = \frac{m_1}{z - z_1} + \frac{m_2}{z - z_2} + \frac{m_3}{z - z_3}.$$

Let $D = diag\{z_1, z_2, z_3\}$ denote the diagonal matrix with diagonal entries z_1, z_2, z_3. Let $v_1 = \begin{bmatrix} 1/\sqrt{3} \\ 1/\sqrt{3} \\ 1/\sqrt{3} \end{bmatrix}$. We choose an orthonormal basis for $\mathbb{C}^3$ with v_1 as first vector and we let

$$v_2 = \begin{bmatrix} 1/\sqrt{2} \\ -1/\sqrt{2} \\ 0 \end{bmatrix} \text{ and } v_3 = \begin{bmatrix} 1/\sqrt{6} \\ 1/\sqrt{6} \\ -2/\sqrt{6} \end{bmatrix}.$$

Let

$$P = \begin{pmatrix} 1/\sqrt{2} & 1/\sqrt{6} \\ -1/\sqrt{2} & 1/\sqrt{6} \\ 0 & -2/\sqrt{6} \end{pmatrix}.$$

Since the columns of P are orthonormal $P^\star P = I_2$. We let $A = P^\star DP$ denote the corresponding compression of D to a 2×2 matrix. A computation shows that

$$A = \begin{pmatrix} (z_1 + z_2)/2 & (z_1 - z_2)/\sqrt{12} \\ (z_1 - z_2)/\sqrt{12} & (z_1 + z_2 + 4z_3)/6. \end{pmatrix}$$

We are now ready to apply the elliptical range theorem, which implies that the numerical range of A is an elliptical disk. Now if a and b denote the eigenvalues of A, then the trace of A is just $\operatorname{tr}(A) = a + b$ and the determinant is $\det(A) = ab$. Computing,

$$\operatorname{tr}(A) = \frac{2}{3}(z_1 + z_2 + z_3),$$

and the determinant of A satisfies

$$\det(A) = \frac{1}{3}(z_1 z_3 + z_1 z_2 + z_2 z_3).$$

Now, the derivative of

$$p(z) = (z - z_1)(z - z_2)(z - z_3)$$

is

$$\begin{aligned} p'(z) &= 3\left(z^2 - \frac{2}{3}(z_1 + z_2 + z_3)z + \frac{1}{3}(z_1 z_2 + z_1 z_3 + z_2 z_3)\right) \\ &= 3(z^2 - \operatorname{tr}(A)z + \det(A)) \\ &= 3(z^2 - (a+b)z + ab) \\ &= 3(z-a)(z-b). \end{aligned}$$

So Siebeck's theorem and the computations above tell us that the foci of the ellipse are the critical points of the polynomial p. What about the points of tangency? Well, if $\|x\| = 1$, then $\|Px\|^2 = \langle Px, Px\rangle = \langle x, P^\star Px\rangle = \|x\|^2 = 1$ and

$$\langle P^\star DPx, x\rangle = \langle DPx, Px\rangle.$$

Therefore, $\langle DPx, Px\rangle \in W(D)$, so the elliptical disk that is the numerical range of A is contained inside the numerical range of D. But D is diagonal and, in particular, normal. So $W(D)$ is the convex hull of its eigenvalues, z_1, z_2, z_3. Therefore, the elliptical disk is contained inside the convex hull of these three points. If we let $x = [1\,, 0]^t$, then $\|x\| = 1$ and

$$\langle Ax, x\rangle = \frac{z_1 + z_2}{2}.$$

Thus, the midpoint of the line joining z_1 and z_2 lies on the ellipse and since the ellipse lies inside the triangle $\Delta_{z_1 z_2 z_3}$, that must be the point of tangency. Reordering the points produces a unitarily equivalent matrix, so the same argument can be made for the other sides. (See also the original proof in [**48**] for another way to look at this problem.) □

In a paper dedicated to Julius Borcea, [**55**], the authors considered several conjectures on the critical points of polynomials. This paper also makes connections between such problems and matrix theoretic ones and, in Proposition 7.1, it gives a short proof of Siebeck's theorem along the lines of the one above.

Let $p(z) = \prod_{k=1}^{n}(z - z_k)$ and

$$\sigma_\infty(p) = \min_{c} \max_{k=1,\ldots,n} |z_k - c|.$$

Recall that given two sets, S and T, the Hausdorff distance is defined by

$$H(S,T) = \max\{h(S,T), h(T,S)\},$$

where

$$h(S,T) = \sup_{s\in S} \inf_{t\in T} \|s - t\|.$$

When the sets are compact,

$$h(S,T) = \max_{s\in S} \min_{t\in T} \|s - t\|.$$

If $h(S,T) = d$, then each point of s has a nearest neighbor in T within a distance d and there is also a point of S that is a distance d from the nearest point of T.

CONJECTURE 4.3 (Schmeisser). *Let p be a degree n polynomial, with $n \geq 2$. Let $\zeta \in \mathbb{C}$ be a point in the convex hull of the zeros of p. Then the closed disk centered at ζ of radius $\sigma_\infty(p)$ contains a critical point of p.*

Borcea formulated a generalization of this conjecture in a matrix theory context. Letting $\sigma(A)$ denote the spectrum of the matrix, and A_k the $(n-1)\times(n-1)$ matrix obtained from A by deleting the kth row and columns, the conjecture takes this form.

CONJECTURE 4.4. *Let A be an $n \times n$ normal matrix. Then*

$$H(W(A), \bigcup_{k=1}^{n} \sigma(A_k)) \leq \min_{c\in\mathbb{C}} \|A - cI\|.$$

The interested reader may find it useful to consult Sendov's 2012 paper for other results and conjectures, [**80**]. Several attempts to prove the Sendov conjecture have been posted, but as of this writing, the conjecture remains open.

5. The invariant subspace problem

5.1. Some partial answers to the invariant subspace problem. The invariant subspace problem is an important open problem in functional analysis that appeared in the 1900s and can be stated as follows.

QUESTION 5.1. For every bounded linear operator T on an infinite-dimensional complex Banach space X, does there exist a proper closed subspace $\mathcal{M}$ such that $T\mathcal{M} \subseteq \mathcal{M}$?

In what follows, we let

$$\sigma_p(T) = \{\lambda \in \mathbb{C} : \ker(T - \lambda I) \neq \{0\}\}$$

denote the point spectrum of T. Note that if T is a bounded linear operator on X such that $\sigma_p(T) \neq \emptyset$, then for $\lambda \in \sigma_p(T)$, the eigenspace E_λ is a nontrivial closed invariant subspace for T. In what follows, we use $n.t.i.s$ in place of "nontrivial closed invariant subspace."

Much progress has been made on Question 5.1 in special settings—so much so, that we will not be able to give a complete list here. However, we note that there are some partial answers to Question 5.1 depending on properties of the space X or on the properties of the operator T.

- The topology of the space X plays an important role. Indeed, if X is a non-separable infinite dimensional space, then for $x \in X\backslash\{0\}$, it follows that
$$\mathcal{M}_x := \overline{\text{span}}\{T^n(x) : n \in \mathbb{N}\},$$
is a $n.t.i.s$; the non-separability of X guarantees that $\mathcal{M}_x \neq X$. When nonempty, the subspace $\mathcal{M}_x$ is the smallest closed invariant subspace containing x. When $\mathcal{M}_x = X$, the vector x is said to be cyclic.
- For a separable Banach space X with dimension greater than 2 and for a Hilbert space, a result obtained by Aronszajn and Smith in 1954 (see [**3**]) says that every compact operator has an invariant subspace. More generally, in 1966, Bernstein and Robinson showed that a polynomially compact operator T (that is, $p(T)$ is compact for some nonzero polynomial p) has a $n.t.i.s$ [**8**]. Their proof uses non-standard analysis. In 1966, Halmos proved the same result, avoiding arguments from non-standard analysis [**53**].

In [**57**], Lomonosov generalized the results from [**3,8,53**] by proving that every nonzero bounded linear compact operator T has a $n.t.i.s.$ The proof relies on a fixed point argument. Lomonosov notes that the arguments he used in his proof also show the existence of a $n.t.i.s$ for a nonscalar bounded linear operator R on a complex Banach space that commutes with a nonzero compact operator. Pearcy and Shields describe Lomonosov's proof and discuss the consequences of Lomonosov's result in [**67**].

In 1976 and 1987, in the Banach space setting, Per Enflo [**38,39**] showed that there exists an operator on a Banach space with no nontrivial invariant subspaces. The example was simplified by Beauzamy in 1985, [**6**]. Charles Read, in 1984, discovered a counterexample in a classical Banach space, describing an operator on ℓ_1 with no nontrivial invariant subspaces, see [**73**]. Read later [**74**] constructed an operator on ℓ_1 without a nontrivial closed invariant subset; in particular, for every vector x the set $\{T^n(x) : n \geq 0\}$ is dense.

In what follows, $\mathcal{L}(\mathcal{H})$ denotes the space of bounded linear operators on a Hilbert space $\mathcal{H}$. We write codim N for the dimension of M, where M is a subspace such that $M \oplus N = \mathcal{H}$ with M orthogonal to N. We use the following well-known fact from operator theory: If T is a bounded linear operator on a separable Hilbert space with closed range, the range of T^* is closed. An operator $T \in \mathcal{L}(\mathcal{H})$ is Fredholm if $\dim(\ker T) < \infty$ and $\operatorname{Im} T$ has finite codimension. The index of $T \in \mathcal{L}(\mathcal{H})$ is given by
$$\operatorname{Ind}(T) = \dim(\ker T) - \dim(\ker T^*).$$
If T is a Fredholm operator, $\operatorname{Ind}(T) < \infty$ and $\operatorname{Ind}(T^*) = -\operatorname{Ind}(T)$. An operator $T \in \mathcal{L}(\mathcal{H})$ is a Kato operator if $\operatorname{Im} T$ is closed and $\ker T \subseteq \cap_{k=1}^{\infty} T^k(\mathcal{H})$, see [**62**, Ch. 12, Def. 3]. The space $\cap_{k=1}^{\infty} T^k(\mathcal{H})$ is denoted by $\mathcal{R}^{\infty}(T)$.

5.2. The invariant subspace problem via universal operators. The existing results do not provide an answer to the invariant subspace problem (Question 5.1) on a complex separable Hilbert space of infinite dimension. One point of view consists in reformulating the problem by focusing on one specific operator called a universal operator.

DEFINITION 5.2 (Universal operator, [**75**]). Let $\mathcal{H}$ be a complex separable Hilbert space of infinite dimension. An operator $U \in \mathcal{L}(\mathcal{H})$ is universal (in the sense of Rota) if for every $T \in \mathcal{L}(\mathcal{H})$, there exists $\lambda \in \mathbb{C}$, $\lambda \neq 0$ and an invariant subspace $\mathcal{M}$ such that

$$\lambda JT = UJ,$$

where $J : \mathcal{H} \longrightarrow \mathcal{M}$ is a linear isomorphism.

In other words, every such T is similar to $U_{|\mathcal{M}}$. Definition 5.2 was introduced by Caradus in [**18**]. The link between a universal operator and the invariant subspace problem is given in Theorem 5.3.

THEOREM 5.3. *Let $U \in \mathcal{L}(\mathcal{H})$ be a universal operator. The following assertions are equivalent:*

(1) *Every operator $T \in \mathcal{L}(\mathcal{H})$ has a n.t.i.s;*
(2) *Every minimal n.t.i.s of U has dimension* 1.

Here, the minimality of a *n.t.i.s* is taken to be in the sense of the inclusion.

The importance of Theorem 5.3 is that it states that the description of all nontrivial invariant subspaces for a universal operator will provide an answer to the invariant subspace problem. In [**18**], Caradus gave a sufficient condition for an operator to be universal.

THEOREM 5.4 (The Caradus Theorem, [**18**]). *Let $U \in \mathcal{L}(\mathcal{H})$. If* $\dim(\ker U) = \infty$ *and $Im(U) = \mathcal{H}$, then U is a universal operator.*

Observe that if $\mathcal{M}$ is a *n.t.i.s* of T and the restriction of T to $\mathcal{M}$ is universal then T is universal. Using Theorem 5.4 we can now present more examples of universal operators on a complex separable Hilbert space with infinite dimension.

EXAMPLE 5.5. Let $\|.\|_{\mathcal{H}}$ be a norm on $\mathcal{H}$. Let

$$\ell^2(\mathbb{N}, \mathcal{H}) = \left\{ (x_n)_{n\in\mathbb{N}},\ x_n \in \mathcal{H} \text{ with } \sum_{n=1}^{+\infty} \|x_n\|_{\mathcal{H}}^2 < \infty \right\}.$$

Rather than viewing the shift operator on the Hardy space $H^2(\mathbb{D})$ as we did in Section 2, we consider it on the space of square summable sequences $\ell^2(\mathbb{N}, \mathcal{H})$. The shift operator S on $\ell^2(\mathbb{N}, \mathcal{H})$ is defined as

$$\begin{aligned} S\colon \ell^2(\mathbb{N}, \mathcal{H}) &\longrightarrow \ell^2(\mathbb{N}, \mathcal{H}) \\ (x_1, x_2, \ldots) &\longmapsto (0, x_1, x_2, \ldots). \end{aligned}$$

Its adjoint S^* is given by

$$\begin{aligned} S^*\colon \ell^2(\mathbb{N}, \mathcal{H}) &\longrightarrow \ell^2(\mathbb{N}, \mathcal{H}) \\ (x_1, x_2, \ldots) &\longmapsto (x_2, x_3 \ldots) \end{aligned}$$

is surjective and its kernel is isomorphic to $\mathcal{H}$. By Theorem 5.4, S^* is a universal operator.

EXAMPLE 5.6. A bilateral shift is an operator $T\colon \ell^2(\mathbb{Z},\mathcal{H}) \longrightarrow \ell^2(\mathbb{Z},\mathcal{H})$ such that

$$(Tx)_n = K_{n-1}(x_{n-1}), \quad n \in \mathbb{Z},$$

where $(K_n)_{n\in\mathbb{N}}$ is a sequence of invertible elements in $\mathcal{L}(\mathcal{H})$ such that

$$\sup_{n\in\mathbb{Z}} \|K_n\| < \infty \text{ and } \sup_{n\in\mathbb{Z}} \|K_n^{-1}\| < \infty.$$

In [**66**], the authors established the following result.

THEOREM 5.7. *Let T be a bilateral shift defined in Example 5.6. Assume that $0 < a < b$ and*

$$K_n \longrightarrow \begin{cases} a\,I & \text{if } n \to -\infty \\ b\,I & \text{if } n \to +\infty \end{cases}$$

in the norm on $\mathcal{L}(\mathcal{H})$. Then

$$\sigma_p(T) = \{\lambda \in \mathbb{C} : a < |\lambda| < b\},$$

and each $\lambda \in \sigma_p(T)$ is an eigenvalue of T^* of infinite multiplicity, but $T - \lambda I$ is bounded below. Moreover, for $\lambda \in \sigma_p(T)$, $T^* - \overline{\lambda} I$ is a universal operator.

Indeed, under the same hypothesis as in Theorem 5.7, for $a < |\lambda| < b$, the operator $T - \lambda I$ is bounded below, which implies that $T - \lambda I$ is injective with closed range. Thus, $T^* - \overline{\lambda} I$ is surjective. For such a $\overline{\lambda}$, the kernel of $T^* - \overline{\lambda} I$ is of infinite dimension. By the Caradus theorem (Theorem 5.4), it follows that $T^* - \overline{\lambda} I$ is universal.

As a consequence, for $a, b, c, d \in \mathbb{R}$, $a < b$ and $c < d$, the weighted shift $S : \ell^2(\mathbb{Z}, L^2([c,d])) \longrightarrow \ell^2(\mathbb{Z}, L^2([c,d]))$ such that

$$(Sx)_n = k_{n-1}x_{n-1}, \quad n \in \mathbb{Z},$$

where $(k_n)_{n\in\mathbb{N}}$ is a sequence of positive and continuous functions on $[c,d]$ with

$$k_n \longrightarrow \begin{cases} b & \text{if } n \to -\infty \\ a & \text{if } n \to +\infty \end{cases}, \text{ uniformly on } [c,d],$$

has the property that $S - \lambda I$ is universal for $a < |\lambda| < b$.

Many examples of universal operators are given via composition operators on separable Hilbert spaces, where the composition operator with symbol ϕ is denoted by C_ϕ. For example, when $\mathcal{H}$ is the Hardy space $H^2(\mathbb{D})$ defined by

$$\mathcal{H} = H^2(\mathbb{D}) = \left\{ f : \mathbb{D} \longrightarrow \mathbb{C} : \ f(z) = \sum_{n=0}^{+\infty} a_n z^n, \ \|f\|_{H^2(\mathbb{D})} := \left(\sum_{n=0}^{+\infty} |a_n|^2 \right)^{1/2} < \infty \right\},$$

if $\phi : \mathbb{D} \longrightarrow \mathbb{D}$ is analytic, by the Littlewood subordination theorem the composition operator $C_\phi : f \longmapsto f \circ \phi$ is linear and bounded on $H^2(\mathbb{D})$. When $\phi \in H^\infty(\mathbb{D})$ is such that $|\phi| = 1$ almost everywhere on $\mathbb{T}$, the function ϕ is said to be an inner function. In [**64**], Nordgren showed that if $\phi : \mathbb{D} \longrightarrow \mathbb{D}$ is an inner function with $\phi(0) = 0$, then C_ϕ is an isometry on $H^2(\mathbb{D})$. As a consequence, we obtain the following result.

THEOREM 5.8. *Let $\phi : \mathbb{D} \longrightarrow \mathbb{D}$ be an analytic inner function with $\phi(0) = 0$ and ϕ not an automorphism. Then C_ϕ^* is a universal operator.*

When ϕ satisfies the hypothesis of Theorem 5.8, by the previous discussion, C_ϕ is an isometry and, consequently, C_ϕ is injective and has closed image. This implies that $\operatorname{Im}(C_\phi^*)$ is closed and C_ϕ^* is surjective. Since ϕ is not an automorphism of $\mathbb{D}$, the operator C_ϕ is not a Fredholm operator, see [**14**]. This further implies that $\operatorname{codim}(\operatorname{Im} C_\phi) = \dim(\ker C_\phi^*) = \infty$. By Theorem 5.4, C_ϕ^* is universal.

Nordgren, Rosenthal and Wintrobe obtained the following result on composition operators on the Hardy space $H^2(\mathbb{D})$.

THEOREM 5.9 (Theorem 6.2, [**65**]). *If ϕ is a hyperbolic disk automorphism and λ is in the interior of the spectrum of C_ϕ, then $C_\phi - \lambda I$ is universal.*

Cowen and Gallardo gave a new proof of this result in [**28**]. They observed that it is sufficient to prove the universality of $C_\phi - \lambda I$ for a specific $\phi = \phi_{t_0}$, $t_0 > 0$, where ϕ_{t_0} is a hyperbolic disk automorphism with fixed points at ± 1. Then the authors use the similarity of $C_{\phi_{z_0}} - \lambda$ to the Toeplitz operator $T^*_{\psi_{t_0} - \overline{\lambda}}$, where

$$\psi_{t_0}(z) = \left(\frac{1-z}{1+z}\right)^{it_0/\pi},$$

and properties of Toeplitz operators (in particular, the point spectrum and existence of a left-inverse).

The hypothesis of Theorem 5.4 relies on the dimension of the kernel of T and T^*. In [**68**], the author used semi-Fredholm theory to generalize Theorem 5.4.

THEOREM 5.10 ([**68**]). *Let $U \in \mathcal{L}(\mathcal{H})$. If*

(1) $\dim(\ker U) = \infty$*;*
(2) $\operatorname{codim}(\operatorname{Im} U) < \infty$*,*

then U is a universal operator.

This theorem can be rephrased in terms of semi-Fredholm operators. Recall that

- an upper semi-Fredholm $T \in \mathcal{L}(\mathcal{H})$ is an operator such that the dimension of $\ker T$ is finite and $\operatorname{Im} T$ is closed;
- a lower semi-Fredholm $T \in \mathcal{L}(\mathcal{H})$ is an operator such that $\operatorname{Im} T$ has finite codimension.

REMARK 5.11. Let T be a lower semi-Fredholm operator. Then its adjoint T^* is an upper semi-Fredholm operator and

$$\operatorname{Ind}(T^*) = -\operatorname{Ind}(T).$$

Theorem 5.10 states that a lower semi-Fredholm operator with infinite index is universal. This theorem relies on the following fact about Kato operators:

LEMMA 5.12 ([**68**]). *If $T \in \mathcal{L}(\mathcal{H})$ is a Kato operator with $\dim(\ker T) = \infty$, then T is a universal operator.*

Indeed, if $T_1 = T_{|\mathcal{R}^\infty(T)}$ (the restriction of T to $\mathcal{R}^\infty(T)$), then

$$\ker T_1 = \ker T \cap \mathcal{R}^\infty(T) = \ker T.$$

By [**62**, Ch.12, Thm. 15], $T(\mathcal{R}^\infty(T)) = \mathcal{R}^\infty(T)$, which shows that $T_1 \in \mathcal{L}(\mathcal{R}^\infty(T))$ is surjective. By Theorem 5.4, the operator T_1 is universal and, consequently, T is also a universal operator.

A semi-Fredholm operator has a Kato decomposition in the following sense, (see [**62**, Thm. 21]):

PROPOSITION 5.13. *Let T be a semi-Fredholm operator. There are two closed invariant subspaces $\mathcal{N}$ and $\mathcal{D}$ with $\dim \mathcal{N} < \infty$ such that $\mathcal{H} = \mathcal{N} \oplus \mathcal{D}$, $T_{|\mathcal{N}}$ is a Kato operator and $T_{|\mathcal{D}}$ is nilpotent.*

It follows that lower semi-Fredholm operators with an infinite index are universal operators. In [**69**, Thm. 4.3.3] and using Theorem 5.10, the author shows the following results on the Hardy spaces $H^2(\mathbb{D})$ and the Hardy space of the upper half-plane $H^2(\mathbb{C}+)$, which is defined as the space of analytic functions on $\mathbb{C}+$ such that

$$\|f\|_{H^2(\mathbb{C}_+)} := \left(\sup_{y>0} \int_{-\infty}^{+\infty} |f(x+iy)|^2 dx\right)^{1/2} < \infty.$$

PROPOSITION 5.14. *Let $\phi : \mathbb{D} \longrightarrow \mathbb{D}$ be an inner function with one fixed point in $\mathbb{D}$. Assume that ϕ is not an automorphism. Then C_ϕ^* is universal.*

The proof relies on a theorem from Bayart [**5**] stating that under the hypothesis of Proposition 5.14, the composition operator C_ϕ is similar to an isometry.

The same kind of result is obtained in [**69**, Thm. 4.3.5] for the composition operator C_ϕ on $H^2(\mathbb{C}_+)$. First, recall that if $\phi : \mathbb{C}_+ \longrightarrow \mathbb{C}_+$ is analytic, the composition operator C_ϕ is bounded $H^2(\mathbb{C}_+)$ if and only if ϕ has a finite angular derivative at infinity, see [**37**, Thm. 3.1].

PROPOSITION 5.15. *Let $\phi : \mathbb{C}_+ \longrightarrow \mathbb{C}_+$ be an analytic function with a finite angular derivative at infinity. Assume that ϕ is not an automorphism on $\mathbb{C}^+$. If C_ϕ is similar to an isometry on $H^2(\mathbb{C}^+)$, then C_ϕ^* is universal. In particular, if ϕ is a rational function defined by*

$$\phi(z) = z + \alpha + \sum_{j=1}^{n} \frac{\mu_j}{z - \gamma_j}, \ \alpha \in \mathbb{R}, \ \gamma_j \in \mathbb{R}, \ \mu_j < 0 \text{ with } n \geq 1,$$

then C_ϕ^* is universal.

PROOF. Let $V \in \mathcal{L}(\mathcal{H})$ be invertible and $\mathcal{J}$ be an isometry such that $V^{-1}C_\phi V = \mathcal{J}$. Since $\mathcal{J}$ is an isometry, C_ϕ is injective and has closed range. This implies that C_ϕ^* is surjective. To apply the Caradus Theorem, it remains to show that $\ker(C_\phi^*)$ has infinite dimension. Since ϕ is not an automorphism, for $N \geq 1$ fixed, it is possible to find $2N$ distinct points $w_1, \dots, w_N$ and $w_1', \dots, w_N'$ such that $\phi(w_j) = \phi(w_j')$ with $j \in \{1, \dots, N\}$. Denote by k_{w_j} the reproducing kernel at w_j on $H^2(\mathbb{C}_+)$. Then for all $f \in H^2(\mathbb{C}_+)$, the function $h_j = k_{w_j} - k_{w_j'}$ satisfies

$$\begin{aligned}
\langle f, C_\phi^*(h_j)\rangle_{H^2(\mathbb{C}_+)} &= \langle C_\phi(f), k_{w_j} - k_{w_j'}\rangle_{H^2(\mathbb{C}_+)} \\
&= f(\phi(w_j)) - f(\phi(w_j')) \\
&= 0,
\end{aligned}$$

which shows that $h_j \in \ker(C_\phi^*)$. Moreover, the family $(h_j)_{1 \leq j \leq N}$ is linearly independent, as it is an orthogonal family for the inner product $\langle \cdot, \cdot \rangle$ defined by

$$\langle f, g\rangle = \langle V^{-1^*} \mathcal{J} V^* f, V^{-1^*} \mathcal{J} V^* g\rangle_{H^2(\mathbb{C}_+)},$$

proving that $\dim(\ker C_\phi^*) \geq N$. As N was chosen arbitrarily, it follows that $\ker C_\phi^*$ has infinite dimension. □

In [**79**, Prop. 2.1], the authors give a new version of Theorem 5.10.

PROPOSITION 5.16. *Let $\mathcal{H} = M \oplus M^{\perp}$, where $M \subset \mathcal{H}$ is an infinite-dimensional subspace of $\mathcal{H}$ and let*

$$V = \begin{pmatrix} U & A \\ 0 & B \end{pmatrix} \in \mathcal{L}(\mathcal{H}).$$

Assume that U is a universal operator on M. Then for every A and B, the operator V is universal.

The authors [**79**, Cor. 2.3] also give a necessary condition for an element in $\mathcal{L}(\mathcal{H})$ to be universal.

PROPOSITION 5.17. *Let $T \in \mathcal{L}(\mathcal{H})$ be universal. Then*

(1) *the interior of $\sigma_p(T)$ is not empty;*
(2) *the interior of the essential spectrum $\sigma_e(T)$ is not empty;*
(3) *there is a nonzero eigenvalue with infinite multiplicity.*

Recently in [**19**], Carmo and Waleed Noor give a characterization for the composition operator C_ϕ on $H^2(\mathbb{C}_+)$ for an affine self-map of $\mathbb{C}_+$ to be universal. Note that the authors work in the right half-plane $\{z = x + iy : x > 0\}$. More precisely, they obtained the following result.

THEOREM 5.18. *Let ϕ be a linear fractional transformation, $a \in (0, 1) \cup (1, \infty)$ and $\mathrm{Re}(b) > 0$. The composition operator $C_\phi - \lambda I$ is universal for some $\lambda \in \mathbb{C}$ if and only if $\phi(z) = az + b$ is hyperbolic and not an automorphism.*

The authors of [**19**] also obtained a version for the composition operator C_ϕ on $H^2(\mathbb{D})$, where $\phi : \mathbb{D} \longrightarrow \mathbb{D}$ is a linear fractional $\phi(z) = \frac{az+b}{cz+d}$ with $ad - bc \neq 0$.

THEOREM 5.19. *If ϕ is a linear fractional self-map of $\mathbb{D}$, then $C_\phi - \lambda I$ is universal on $H^2(\mathbb{D})$ for some $\lambda \in \mathbb{C}$ if and only if ϕ is hyperbolic.*

A natural question arising from Theorems 5.18 and 5.19 is the following.

QUESTION 5.20. If $\phi : \mathbb{C}_+ \longrightarrow \mathbb{C}_+$ is an affine map, is $C_\phi^* - \lambda I$ universal on $H^2(\mathbb{C}_+)$ for some $\lambda \in \mathbb{C}$ if and only if ϕ is elliptic and not an automorphism?

We conclude this paper with the following results in $H^2(\mathbb{D})$ on the multiplication operator $M_\psi : H^2(\mathbb{D}) \longrightarrow H^2(\mathbb{D})$ with $\psi \in H^\infty(\mathbb{D})$ that appear in [**20**, Thm. 8.1.6].

THEOREM 5.21. *Let $\phi \in H^\infty(\mathbb{D})$ be an inner function and K_ϕ be the model space defined by ϕ. Then, M_ψ is unitarily equivalent to a shift on $\ell^2(K_\psi)$ of multiplicity equal to the degree of ψ.*

It is easy to see that M_ψ is an isometry on $H^2(\mathbb{D})$ when ψ is inner. The proof of Theorem 5.21 relies on the following decomposition of $H^2(\mathbb{D})$:

$$H^2(\mathbb{D}) = \bigoplus_{n=0}^{+\infty} \psi^n K_\psi,$$

where the dimension of K_ψ is equal to the degree of ψ. Theorem 5.21 combined with Theorem 5.4 leads to the following corollary:

COROLLARY 5.22. *Let $\psi \in H^\infty(\mathbb{D})$ be an inner function that is not rational. Then M_ψ^* is a universal operator.*

References

[1] T. Ando, *Structure of operators with numerical radius one*, Acta Sci. Math. (Szeged) **34** (1973), 11–15. MR318920

[2] S. Axler, *Linear algebra done right*, 3rd ed., Undergraduate Texts in Mathematics, Springer, Cham, 2015, DOI 10.1007/978-3-319-11080-6. MR3308468

[3] N. Aronszajn and K. T. Smith, *Invariant subspaces of completely continuous operators*, Ann. of Math. (2) **60** (1954), 345–350, DOI 10.2307/1969637. MR65807

[4] A. D. Baranov, *Weighted Bernstein inequalities and embedding theorems for model subspaces* (Russian, with Russian summary), Algebra i Analiz **15** (2003), no. 5, 138–168, DOI 10.1090/S1061-0022-04-00829-5; English transl., St. Petersburg Math. J. **15** (2004), no. 5, 733–752. MR2068792

[5] F. Bayart, *Similarity to an isometry of a composition operator*, Proc. Amer. Math. Soc. **131** (2003), no. 6, 1789–1791, DOI 10.1090/S0002-9939-02-06759-X. MR1955266

[6] B. Beauzamy, *Un opérateur, sur un espace de Banach, sans sous-espace invariant: simplification de l'exemple de P. Enflo* (French), Seminar on functional analysis, 1983/1984, Publ. Math. Univ. Paris VII, vol. 20, Univ. Paris VII, Paris, 1984, pp. 51–53. MR825304

[7] R. D. Berman, *The level sets of the moduli of functions of bounded characteristic*, Trans. Amer. Math. Soc. **281** (1984), no. 2, 725–744, DOI 10.2307/2000082. MR722771

[8] A. R. Bernstein and A. Robinson, *Solution of an invariant subspace problem of K. T. Smith and P. R. Halmos*, Pacific J. Math. **16** (1966), 421–431. MR193504

[9] K. Bickel, P. Gorkin, A. Greenbaum, T. Ransford, F. L. Schwenninger, and E. Wegert, *Crouzeix's conjecture and related problems*, Comput. Methods Funct. Theory **20** (2020), no. 3-4, 701–728, DOI 10.1007/s40315-020-00350-9. MR4175507

[10] K. Bickel and P. Gorkin, *Blaschke products, level sets, and Crouzeix's conjecture*, J. Anal. Math. (2023), `https://doi.org/10.1007/s11854-023-0295-y`.

[11] B. D. Bojanov, Q. I. Rahman, and J. Szynal, *On a conjecture of Sendov about the critical points of a polynomial*, Math. Z. **190** (1985), no. 2, 281–285, DOI 10.1007/BF01160464. MR797543

[12] I. Borcea, *On the Sendov conjecture for polynomials with at most six distinct roots*, J. Math. Anal. Appl. **200** (1996), no. 1, 182–206, DOI 10.1006/jmaa.1996.0198. MR1387976

[13] I. Borcea, *The Sendov conjecture for polynomials with at most seven distinct zeros*, Analysis **16** (1996), no. 2, 137–159, DOI 10.1524/anly.1996.16.2.137. MR1397576

[14] P. S. Bourdon, *Fredholm multiplication and composition operators on the Hardy space*, Integral Equations Operator Theory **13** (1990), no. 4, 607–610, DOI 10.1007/BF01210404. MR1058089

[15] J. E. Brown, *On the Sendov conjecture for sixth degree polynomials*, Proc. Amer. Math. Soc. **113** (1991), no. 4, 939–946, DOI 10.2307/2048768. MR1081693

[16] J. E. Brown, *A proof of the Sendov conjecture for polynomials of degree seven*, Complex Variables Theory Appl. **33** (1997), no. 1-4, 75–95, DOI 10.1080/17476939708815013. MR1624899

[17] J. E. Brown and G. Xiang, *Proof of the Sendov conjecture for polynomials of degree at most eight*, J. Math. Anal. Appl. **232** (1999), no. 2, 272–292, DOI 10.1006/jmaa.1999.6267. MR1683144

[18] S. R. Caradus, *Universal operators and invariant subspaces*, Proc. Amer. Math. Soc. **23** (1969), 526–527, DOI 10.2307/2036577. MR250104

[19] J. R. Carmo and S. Waleed Noor, *Universal composition operators*, J. Operator Theory **87** (2022), no. 1, 137–156, DOI 10.7900/jot. MR4373152

[20] I. Chalendar and J. R. Partington, *Modern approaches to the invariant-subspace problem*, Cambridge Tracts in Mathematics, vol. 188, Cambridge University Press, Cambridge, 2011, DOI 10.1017/CBO9780511862434. MR2841051

[21] D. Choi, *A proof of Crouzeix's conjecture for a class of matrices*, Linear Algebra Appl. **438** (2013), no. 8, 3247–3257, DOI 10.1016/j.laa.2012.12.045. MR3023274

[22] D. Choi, *Estimating Norms of Matrix Functions using Numerical Ranges*, ProQuest LLC, Ann Arbor, MI, 2013. Thesis (Ph.D.)–University of Washington. MR3193123

[23] J. Cima and R. Mortini, *One-component inner functions*, Complex Anal. Synerg. **3** (2017), no. 1, Paper No. 2, 15, DOI 10.1186/s40627-016-0008-8. MR3628449

[24] J. Cima and R. Mortini, *One-component inner functions II*, Advancements in complex analysis—from theory to practice, Springer, Cham, [2020] ©2020, pp. 39–49, DOI 10.1007/978-3-030-40120-7_2. MR4264037

[25] B. Cohn, *Carleson measures for functions orthogonal to invariant subspaces*, Pacific J. Math. **103** (1982), no. 2, 347–364. MR705235

[26] G. L. Cohen and G. H. Smith, *A simple verification of Ilieff's conjecture for polynomials with three zeros*, Amer. Math. Monthly **95** (1988), no. 8, 734–737, DOI 10.2307/2322254. MR966243

[27] G. L. Cohen and G. H. Smith, *A proof of Ilieff's conjecture for polynomials with four zeros*, Elem. Math. **43** (1988), no. 1, 18–21. MR925540

[28] C. C. Cowen and E. A. Gallardo-Gutiérrez, *A new proof of a Nordgren, Rosenthal and Wintrobe theorem on universal operators*, Problems and recent methods in operator theory, Contemp. Math., vol. 687, Amer. Math. Soc., Providence, RI, 2017, pp. 97–102. MR3635755

[29] M. Crouzeix, *Bounds for analytical functions of matrices*, Integral Equations Operator Theory **48** (2004), no. 4, 461–477, DOI 10.1007/s00020-002-1188-6. MR2047592

[30] M. Crouzeix, *Numerical range and functional calculus in Hilbert space*, J. Funct. Anal. **244** (2007), no. 2, 668–690, DOI 10.1016/j.jfa.2006.10.013. MR2297040

[31] M. Crouzeix, *Spectral sets and* 3×3 *nilpotent matrices*, Topics in functional and harmonic analysis, Theta Ser. Adv. Math., vol. 14, Theta, Bucharest, 2013, pp. 27–42. MR3184340

[32] M. Crouzeix and C. Palencia, *The numerical range is a* $(1+\sqrt{2})$*-spectral set*, SIAM J. Matrix Anal. Appl. **38** (2017), no. 2, 649–655, DOI 10.1137/17M1116672. MR3666309

[33] U. Daepp, P. Gorkin, and R. Mortini, *Ellipses and finite Blaschke products*, Amer. Math. Monthly **109** (2002), no. 9, 785–795, DOI 10.2307/3072367. MR1933701

[34] U. Daepp, P. Gorkin, A. Shaffer, and K. Voss, *Finding ellipses*, Carus Mathematical Monographs, vol. 34, MAA Press, Providence, RI, 2018. What Blaschke products, Poncelet's theorem, and the numerical range know about each other, DOI 10.1090/car/034. MR3932079

[35] U. Daepp, P. Gorkin, G. Semmler, and E. Wegert, *The beauty of Blaschke products*, Handbook of the mathematics of the arts and sciences, Springer, Cham, [2021] ©2021, pp. 45–78, DOI 10.1007/978-3-319-57072-3_88. MR4432771

[36] J. Dégot, *Sendov conjecture for high degree polynomials*, Proc. Amer. Math. Soc. **142** (2014), no. 4, 1337–1349, DOI 10.1090/S0002-9939-2014-11888-0. MR3162254

[37] S. Elliott and M. T. Jury, *Composition operators on Hardy spaces of a half-plane*, Bull. Lond. Math. Soc. **44** (2012), no. 3, 489–495, DOI 10.1112/blms/bdr110. MR2966995

[38] P. Enflo, *On the invariant subspace problem in Banach spaces*, Séminaire Maurey-Schwartz (1975–1976): Espaces L^p, applications radonifiantes et géométrie des espaces de Banach, École Polytech., Palaiseau, 1976, pp. Exp. Nos. 14-15, 7. MR473871

[39] P. Enflo, *On the invariant subspace problem for Banach spaces*, Acta Math. **158** (1987), no. 3-4, 213–313, DOI 10.1007/BF02392260. MR892591

[40] M. Fujimura, *Inscribed ellipses and Blaschke products*, Comput. Methods Funct. Theory **13** (2013), no. 4, 557–573, DOI 10.1007/s40315-013-0037-8. MR3138353

[41] S. R. Garcia, J. Mashreghi, and W. T. Ross, *Introduction to model spaces and their operators*, Cambridge Studies in Advanced Mathematics, vol. 148, Cambridge University Press, Cambridge, 2016, DOI 10.1017/CBO9781316258231. MR3526203

[42] S. R. Garcia and W. T. Ross, *A non-linear extremal problem on the Hardy space*, Comput. Methods Funct. Theory **9** (2009), no. 2, 485–524, DOI 10.1007/BF03321742. MR2572653

[43] H.-L. Gau and P. Y. Wu, *Numerical range of* $S(\phi)$, Linear and Multilinear Algebra **45** (1998), no. 1, 49–73, DOI 10.1080/03081089808818577. MR1665615

[44] H.-L. Gau and P. Y. Wu, *Numerical range circumscribed by two polygons*, Linear Algebra Appl. **382** (2004), 155–170, DOI 10.1016/j.laa.2003.12.003. MR2050103

[45] H.-L. Gau and P. Y. Wu, *Numerical range and Poncelet property*, Taiwanese J. Math. **7** (2003), no. 2, 173–193, DOI 10.11650/twjm/1500575056. MR1978008

[46] C. Glader, M. Kurula, and M. Lindström, *Crouzeix's conjecture holds for tridiagonal* 3×3 *matrices with elliptic numerical range centered at an eigenvalue*, SIAM J. Matrix Anal. Appl. **39** (2018), no. 1, 346–364, DOI 10.1137/17M1110663. MR3769703

[47] A. W. Goodman, Q. I. Rahman, and J. S. Ratti, *On the zeros of a polynomial and its derivative*, Proc. Amer. Math. Soc. **21** (1969), 273–274, DOI 10.2307/2036982. MR239062

[48] P. Gorkin, *Four theorems with their foci on ellipses*, Amer. Math. Monthly **126** (2019), no. 2, 99–111, DOI 10.1080/00029890.2019.1537459. MR3910625

[49] P. Gorkin and N. Wagner, *Ellipses and compositions of finite Blaschke products*, J. Math. Anal. Appl. **445** (2017), no. 2, 1354–1366, DOI 10.1016/j.jmaa.2016.01.067. MR3545246

[50] A. Greenbaum and D. Choi, *Crouzeix's conjecture and perturbed Jordan blocks*, Linear Algebra Appl. **436** (2012), no. 7, 2342–2352, DOI 10.1016/j.laa.2011.12.005. MR2889994

[51] A. Greenbaum and M. L. Overton, *Numerical investigation of Crouzeix's conjecture*, Linear Algebra Appl. **542** (2018), 225–245, DOI 10.1016/j.laa.2017.04.035. MR3778739

[52] K. E. Gustafson and D. K. M. Rao, *Numerical range*, Universitext, Springer-Verlag, New York, 1997. The field of values of linear operators and matrices, DOI 10.1007/978-1-4613-8498-4. MR1417493

[53] P. R. Halmos, *Invariant subspaces of polynomially compact operators*, Pacific J. Math. **16** (1966), 433–437. MR193505

[54] W. K. Hayman, *Research problems in function theory*, The Athlone Press [University of London], London, 1967. MR217268

[55] D. Khavinson, R. Pereira, M. Putinar, E. B. Saff, and S. Shimorin, *Borcea's variance conjectures on the critical points of polynomials*, Notions of positivity and the geometry of polynomials, Trends Math., Birkhäuser/Springer Basel AG, Basel, 2011, pp. 283–309, DOI 10.1007/978-3-0348-0142-3_16. MR3051172

[56] D. S. Keeler, L. Rodman, and I. M. Spitkovsky, *The numerical range of* 3×3 *matrices*, Linear Algebra Appl. **252** (1997), 115–139, DOI 10.1016/0024-3795(95)00674-5. MR1428632

[57] V. I. Lomonosov, *Invariant subspaces of the family of operators that commute with a completely continuous operator* (Russian), Funkcional. Anal. i Priložen. **7** (1973), no. 3, 55–56. MR420305

[58] F. Lucas, *Propriétés géométriques des fractions rationnelles*, C. R. Acad. Sci. Paris **78** (1874), 271–275.

[59] M. Marden, *Geometry of polynomials*, 2nd ed., Mathematical Surveys, No. 3, American Mathematical Society, Providence, RI, 1966. MR225972

[60] A. Meir and A. Sharma, *On Ilyeff's conjecture*, Pacific J. Math. **31** (1969), 459–467. MR249587

[61] B. Mirman, *Numerical ranges and Poncelet curves*, Linear Algebra Appl. **281** (1998), no. 1-3, 59–85, DOI 10.1016/S0024-3795(98)10037-X. MR1645335

[62] V. Müller, *Spectral theory of linear operators and spectral systems in Banach algebras*, Operator Theory: Advances and Applications, vol. 139, Birkhäuser Verlag, Basel, 2003, DOI 10.1007/978-3-0348-7788-6. MR1975356

[63] A. Nicolau and A. Reijonen, *A characterization of one-component inner functions*, Bull. Lond. Math. Soc. **53** (2021), no. 1, 42–52, DOI 10.1112/blms.12395. MR4224508

[64] E. A. Nordgren, *Composition operators*, Canadian J. Math. **20** (1968), 442–449, DOI 10.4153/CJM-1968-040-4. MR223914

[65] E. A. Nordgren, P. Rosenthal, and F. S. Wintrobe, *Composition operators and the invariant subspace problem*, C. R. Math. Rep. Acad. Sci. Canada **6** (1984), no. 5, 279–283. MR764103

[66] J. R. Partington and E. Pozzi, *Universal shifts and composition operators*, Oper. Matrices **5** (2011), no. 3, 455–467, DOI 10.7153/oam-05-33. MR2858500

[67] C. Pearcy and A. L. Shields, *A survey of the Lomonosov technique in the theory of invariant subspaces*, Topics in operator theory, Math. Surveys, No. 13, Amer. Math. Soc., Providence, RI, 1974, pp. 219–229. MR355639

[68] E. Pozzi, *Universality of weighted composition operators on* $L^2([0,1])$ *and Sobolev spaces*, Acta Sci. Math. (Szeged) **78** (2012), no. 3-4, 609–642. MR3057690

[69] E. Pozzi, *Propriétés spectrales et universalité des opérateurs de composition pondérés*, PhD thesis, Université de Lyon, 2011.

[70] Q. I. Rahman and G. Schmeisser, *Analytic theory of polynomials*, London Mathematical Society Monographs. New Series, vol. 26, The Clarendon Press, Oxford University Press, Oxford, 2002. MR1954841

[71] T. Ransford and F. L. Schwenninger, *Remarks on the Crouzeix-Palencia proof that the numerical range is a* $(1+\sqrt{2})$*-spectral set*, SIAM J. Matrix Anal. Appl. **39** (2018), no. 1, 342–345, DOI 10.1137/17M1143757. MR3769702

[72] P. X. Rault, T. Sendova, and I. M. Spitkovsky, *3-by-3 matrices with elliptical numerical range revisited*, Electron. J. Linear Algebra **26** (2013), 158–167, DOI 10.13001/1081-3810.1646. MR3065855

[73] C. J. Read, *A solution to the invariant subspace problem*, Bull. London Math. Soc. **16** (1984), no. 4, 337–401, DOI 10.1112/blms/16.4.337. MR749447
[74] C. J. Read, *A solution to the invariant subspace problem on the space l_1*, Bull. London Math. Soc. **17** (1985), no. 4, 305–317, DOI 10.1112/blms/17.4.305. MR806634
[75] G.-C. Rota, *On models for linear operators*, Comm. Pure Appl. Math. **13** (1960), 469–472, DOI 10.1002/cpa.3160130309. MR112040
[76] E. B. Saff and J. B. Twomey, *A note on the location of critical points of polynomials*, Proc. Amer. Math. Soc. **27** (1971), 303–308, DOI 10.2307/2036310. MR271312
[77] D. Sarason, *Generalized interpolation in H^∞*, Trans. Amer. Math. Soc. **127** (1967), 179–203, DOI 10.2307/1994641. MR208383
[78] G. Schmeisser, *On Ilieff's conjecture*, Math. Z. **156** (1977), no. 2, 165–173, DOI 10.1007/BF01178761. MR486436
[79] R. Schroderus and H.-O. Tylli, *On universal operators and universal pairs*, Proc. Edinb. Math. Soc. (2) **61** (2018), no. 3, 891–908, DOI 10.1017/s0013091518000068. MR3834736
[80] B. Sendov, *New conjectures in the geometry of polynomials*, C. R. Acad. Bulgare Sci. **63** (2010), no. 5, 659–664. MR2681436
[81] T. Sheil-Small, *Complex polynomials*, Cambridge Studies in Advanced Mathematics, vol. 75, Cambridge University Press, Cambridge, 2002, DOI 10.1017/CBO9780511543074. MR1962935
[82] P. Siebeck, *Ueber eine neue analytische Behandlungsweise der Brennpunkte* (German), J. Reine Angew. Math. **64** (1865), 175–182, DOI 10.1515/crll.1865.64.175. MR1579291
[83] B. Simon, *The strong Gauss-Lucas theorem and analyticity of correlation functions via the Lee-Yang theorem*, J. Math. Phys. **63** (2022), no. 3, Paper No. 033302, 11, DOI 10.1063/5.0077229. MR4387141
[84] D. A. Singer, *The location of critical points of finite Blaschke products*, Conform. Geom. Dyn. **10** (2006), 117–124, DOI 10.1090/S1088-4173-06-00145-7. MR2223044
[85] K. Stephenson and C. Sundberg, *Level curves of inner functions*, Proc. London Math. Soc. (3) **51** (1985), no. 1, 77–94, DOI 10.1112/plms/s3-51.1.77. MR788850
[86] B. Sz.-Nagy, C. Foias, H. Bercovici, and L. Kérchy, *Harmonic analysis of operators on Hilbert space*, Revised and enlarged edition, Universitext, Springer, New York, 2010, DOI 10.1007/978-1-4419-6094-8. MR2760647
[87] T. Tao, *Sendov's conjecture for sufficiently-high-degree polynomials*, Acta Mathematica **229** (2022), no. 2, 347–392.
[88] J. L. Walsh, *On the critical points of functions possessing central symmetry on the sphere*, Amer. J. Math. **70** (1948), 11–21, DOI 10.2307/2371928. MR23949

Department of Mathematics, Bucknell University, 360 Olin Science Building, Lewisburg, Pennsylvania 17837
Email address: pgorkin@bucknell.edu

Department of Mathematics and Statistics, Saint Louis University, 220 North Grand Boulevard, Saint Louis, Missouri 63103
Email address: elodie.pozzi@slu.edu

Contemporary Mathematics
Volume **799**, 2024
https://doi.org/10.1090/conm/799/16018

On the valence of logharmonic polynomials

Dmitry Khavinson, Erik Lundberg, and Sean Perry

ABSTRACT. Investigating a problem posed by W. Hengartner (2000), we study the maximal valence (number of preimages of a prescribed point in the complex plane) of logharmonic polynomials, i.e., complex functions that take the form $f(z) = p(z)\overline{q(z)}$ of a product of an analytic polynomial $p(z)$ of degree n and the complex conjugate of another analytic polynomial $q(z)$ of degree m. In the case $m = 1$, we adapt an indirect technique utilizing anti-holomorphic dynamics to show that the valence is at most $3n - 1$. This confirms a conjecture of Bshouty and Hengartner (2000). Using a purely algebraic method based on Sylvester resultants, we also prove a general upper bound for the valence showing that for each $n, m \geq 1$ the valence is at most $n^2 + m^2$. This improves, for every choice of $n, m \geq 1$, the previously established upper bound $(n + m)^2$ based on Bezout's theorem. We also consider the more general setting of polyanalytic polynomials where we show that this latter result can be extended under a nondegeneracy assumption.

1. Introduction

The following problem was posed by W. Hengartner at the second international workshop on planar harmonic mappings at the Technion, Haifa, January 7–13, 2000. The problem has been restated several times (see [**8**], [**7**], and [**1**]).

PROBLEM 1.1 (Hengartner's valence problem). Let $f(z) = p(z)\overline{q(z)}$ be a logharmonic polynomial of degree $n = \deg p$ and $m = \deg q$ with p not a constant multiple of q. Find a sharp upper bound for the valence (number of preimages of a prescribed $w \in \mathbb{C}$) of f, i.e., determine the maximal valence in terms of n and m.

The condition that p is not a constant multiple of q ensures that the valence is finite, see [**1**, Sec. 3]. The term *logharmonic* refers to the fact, which is apparent, that the log of such an f is locally harmonic on the set excluding its zeros. Concerning the case $m = 1$, Bshouty and Hengartner made the following conjecture in [**7**].

CONJECTURE 1.2 (Bshouty and Hengartner, 2000). *In the case $m = 1$, the maximal valence is $3n - 1$.*

In referring to the maximal valence, this conjecture implicitly contains two assertions: (1) in the case $m = 1$, for each $n \geq 1$ the valence is at most $3n - 1$ and

2020 *Mathematics Subject Classification.* Primary 30C55, 37F10, 13P15.
The first-named author acknowledges support from the Simons Foundation (grant 513381).
The second-named author acknowledges support from the Simons Foundation (grant 712397).

(2) for each $n \geq 1$ there exists a polynomial $p(z)$ of degree n, a linear $q(z)$, and a $w \in \mathbb{C}$ such that there are $3n-1$ preimages of w under f. In this paper, as our first main result, we prove Theorem 1.3 confirming the first assertion (1), i.e., the upper bound that is entailed in Bshouty and Hengartner's conjecture.

THEOREM 1.3. *Let p be a polynomial of degree $n > 1$ and let $q(z)$ be linear. For each $w \in \mathbb{C}$, the number of solutions of the equation $p(z)\overline{q(z)} = w$ is at most $3n-1$.*

Regarding the assertion (2), i.e., the sharpness part of the conjecture, examples are provided in [**7**] having valence $3n-3$, which shows that Theorem 1.3 is asymptotically sharp. There is additionally an example presented in [**7**] for $n=2$ with valence $3n-1=5$ showing sharpness in the case $n=2$. In Section 3, we present additional examples that, according to our numerical simulations, attain the upper bound $3n-1$ in the case $n=3,4$. It is an enticing open problem to confirm the second statement (2) implicit in Bshouty and Hengartner's conjecture by producing examples showing that Theorem 1.3 is sharp for each n.

Returning to the general setting of Hengartner's Problem, it was observed in [**1**] that Bezout's Theorem gives an upper bound $(n+m)^2$ on the valence, and the authors conjectured the following.

CONJECTURE 1.4 (Abdulhadi and Hengartner, 2001). *The Bezout bound fails to be sharp for every $m, n \geq 1$, i.e., the maximal valence is strictly less than $(m+n)^2$.*

Since $n^2 + m^2 < (m+n)^2$ when $m, n \geq 1$, Theorem 1.5, our second main result of this paper, improves Bezout's bound for every $m, n \geq 1$, hence confirming Conjecture 1.4 of Abdulhadi and Hengartner.

THEOREM 1.5. *Let $f(z) = p(z)\overline{q(z)}$ be a logharmonic polynomial with $n := \deg p \geq 1$, $m := \deg q \geq 1$, and p not a constant multiple of q. Then the valence of f is at most $n^2 + m^2$.*

This result admits further improvement, at least in the case $m=1$, in light of Theorem 1.3. The question was raised in [**7**] whether the maximal valence is actually linear in m, n. We suspect that the maximal valence grows linearly in n for each fixed m but grows quadratically when $n = m$ or when n is asymptotically proportional to m.

Our proof of Theorem 1.5 uses a purely algebraic technique based on polynomial resultants applied to the system of equations obtained by considering $p(z)\overline{q(z)} - 1 = 0$ and its complex conjugate $\overline{p(z)}q(z) - 1 = 0$. A technical obstacle in using this method is establishing the following result (stated as Lemma 4.7 in Section 4) showing that the left hand sides of these equations, viewed as bivariate polynomials in $z, \overline{z}$, are coprime.

LEMMA 1.6. *Given two complex analytic univariate polynomials p and q with p not a constant multiple of q, the polynomials $P(z,\overline{z}) = p(z)\overline{q(z)} - 1$ and $Q(z,\overline{z}) = \overline{p(z)}q(z) - 1$, viewed as bivariate polynomials in $z, \overline{z}$ are coprime.*

As a corollary of Lemma 1.6 we recover the following aforementioned result that was proved in [**1**, Sec. 3] using several complex analytic tools (including the Riemann mapping theorem, Schwarz reflection, analytic continuation, and properties of Blaschke products), whereas our proof of Lemma 1.6 is purely algebraic and essentially relies on a single property of resultants (namely, that coprimacy is equivalent to the resultant not vanishing identically).

COROLLARY 1.7. *If p is not a constant multiple of q, then the logharmonic polynomial $p(z)\overline{q(z)}$ has finite valence.*

Concerning the *minimal* valence we note, as was observed in [**1**, Thm. 3.6], that for $n > m$ it follows from a generalization of the argument principle that the valence is bounded below by $n - m$. This lower bound is sharp (and hence gives the minimal valence) as it is easy to check (especially in polar coordinates $z = re^{i\theta}$) that the equation $z^n\overline{z}^m = 1$ has exactly $n - m$ solutions (located at roots of unity).

1.1. Sheil-Small's valence problem for harmonic polynomials. Hengartner's valence problem for logharmonic polynomials is reminiscent of another valence problem posed by T. Sheil-Small, where instead of a product of $p(z)$ and $\overline{q(z)}$ a sum $p(z) + \overline{q(z)}$ is considered.

PROBLEM 1.8 (Sheil-Small's valence problem). Determine the maximal valence of harmonic polynomials $p(z) + \overline{q(z)}$ with $n = \deg p > m = \deg q$.

In his thesis [**42**], A. S. Wilmshurst used the maximum principle for harmonic functions together with Bezout's theorem to show that the valence is at most n^2, and he constructed n^2-valent examples with $m = n-1$. This confirmed a conjecture of Sheil-Small that n^2 is the maximum valence. Wilmshurst conjectured [**42**] (cf. [**41**, Remark 2]) that for each pair of integers $n > m \geq 1$, the maximum valence is $3n - 2 + m(m-1)$. This conjecture was confirmed for $n = m - 1$ by the previously mentioned results of Wilmshurst and for $m = 1$ by a result of the first-named author and G. Swiatek [**22**] together with L. Geyer's proof [**14**] of the Crofoot-Sarason conjecture. However, counterexamples to Wilmshurst's conjecture for the case $m = n - 3$ were provided in [**23**]. Further counterexamples were provided in [**25**], [**17**], and [**31**]. In spite of these counterexamples, it is still seems likely that, in the spirit of Wilmshurst's conjecture, the maximal valence increases linearly in n for each fixed m [**23**], [**20**], [**31**]. The latter result [**31**] used a nonconstructive probabilistic method, thus connecting the study of the extremal problem with a parallel line of research on the average number of zeros of random harmonic polynomials that had been investigated (for a variety of models) in [**27**], [**26**], [**39**], [**40**].

Harmonic polynomials and logharmonic polynomials are just two species of so-called polyanalytic polynomials, that is polynomials $P(z,\overline{z})$ in z and $\overline{z}$ that satisfy $\left(\frac{\partial}{\partial \overline{z}}\right)^{m+1} P(z,\overline{z}) = 0$ with $\frac{\partial}{\partial \overline{z}} := \frac{1}{2}\left(\frac{\partial}{\partial x} + i\frac{\partial}{\partial y}\right)$ and hence take the form

$$P(z,\overline{z}) = \sum_{k=0}^{m} p_k(z)\overline{z}^k,$$

where $p_k(z)$ are complex-analytic polynomials. Problems 1.1 and 1.8 thus fit into a broader setting concerning the valence of polyanalytic polynomials. Polyanalytic polynomials have been studied classically [**3**] and have also arisen in more recent studies in gravitational lensing [**34**], [**32**], approximation theory [**12**], [**9**], [**19**] numerical linear algebra [**16**], random matrix theory [**18**], and determinantal point processes [**2**]. Motivated by the valence problems for harmonic and logharmonic polynomials as well as the prominent role played by zero sets in several of the aforementioned studies on polyanalytic polynomials, it seems natural to pose the following general valence problem for polyanalytic polynomials.

PROBLEM 1.9. Determine conditions on the coefficient polynomials $p_k(z)$ which guarantee that $P(z,\overline{z})$ has finite valence, and estimate the maximal valence in terms of m along with the degrees of $p_k(z)$.

The proof of Theorem 1.5 in Section 4 leads to the following result (see Lemma 4.6 in Section 4) estimating the valence of polyanalytic polynomials for which $\deg p_k \leq n$ for each $k = 0, 1, \ldots, m$.

PROPOSITION 1.10. *Let $P(z,\overline{z}) = \sum_{k=0}^{m} p_k(z)\overline{z}^k$ be a polyanalytic polynomial with $\deg p_k \leq n$ for each $k = 0, 1, \ldots, m$. Assume $P(z,\overline{z})$ and $Q(z,\overline{z}) = \overline{P(z,\overline{z})}$ are coprime. Then the valence is at most $n^2 + m^2$.*

1.2. Proof techniques. The proof of Theorem 1.3 begins with a reformulation of the valence problem in terms of rational harmonic functions and then specializes a technique previously used in that setting by the first named author and G. Neumann in [**21**], where the key idea is to use a theorem of Fatou from holomorphic dynamics. This indirect method was introduced by the first named author and G. Swiatek in [**22**] in their previously mentioned proof of the $m = 1$ case of Wilmshurst's conjecture. Subsequent adaptations of this method have led to solutions to additional problems including sharp estimates for the topology of quadrature domains [**24**], classification of the number of critical points of Green's function on a torus [**4**], and sharp estimates for the number of solutions of certain transcendental equations [**5**]. The results of [**21**], which are directly relevant to the current paper, confirmed astronomer S. Rhie's conjecture in gravitational lensing (motivated by the connection to gravitational lensing the zeros of rational harmonic functions were investigated further in [**6**], [**30**], [**35**], [**36**], [**29**], [**28**]).

The proof of Theorem 1.5 utilizes another technique that has been applied in studies of gravitational lensing, in this instance, going in a separate, purely algebraic direction utilizing Sylvester resultants. The application of Sylvester resultants to bound the valence of certain polynomials in z and its complex conjugate $\overline{z}$ was introduced by A.O. Petters in the study of gravitational lensing in [**34**]. Our proof of Theorem 1.5 particularly resembles the more recent application of resultants in the study of gravitational lensing by multi-plane point mass ensembles carried out by the third-named author in [**32**].

1.3. Outline of the paper. We present the proof of Theorem 1.3 in Section 2 where we begin by reviewing some preliminary results in Section 2.1 before proceeding to the proof in Section 2.2. Our proof of Theorem 1.3 adapts the method from [**21**] with an important modification, see Remark 2.7. We discuss Bshouty and Hengartner's $(3n-3)$-valent examples from the perspective of holomorphic dynamics in Section 3, where we also present the results of some numerical experiments. We present the proof of Theorem 1.5 in Section 4, where we include review of some essential algebraic preliminaries related to Sylvester resultants.

2. Anti-holomorphic dynamics and the case $m = 1$

2.1. Preliminaries. In this section, we collect some notation and terminology along with several lemmas that will be used in the proof of Theorem 1.3.

2.1.1. *Anti-holomorphic dynamics.* We will need the following result that adapts a classical theorem of Fatou from the setting of holomorphic dynamics to the setting of anti-holomorphic dynamics.

A point $z_0 \in \hat{\mathbb{C}}$ is a *critical point* of r if the spherical derivative of r vanishes at z_0. The complex conjugate of a rational function is referred to as an *anti-rational map*. A fixed point z_0 of the anti-rational map $z \mapsto \overline{r(z)}$ is referred to as an *attracting* fixed point if $|r'(z_0)| < 1$. A fixed point z_0 is said to attract some point $w \in \hat{\mathbb{C}}$ if the iterates of $z \mapsto \overline{r(z)}$ starting at w converge to z_0.

The following result is taken from [**15**, Thm. 2.6].

LEMMA 2.1 (Fatou's Theorem for anti-rational maps). *Let r be a rational function with degree greater than 1. If z_0 is an attracting fixed point of the anti-rational map $z \mapsto \overline{r(z)}$, then z_0 attracts at least one critical point of r.*

2.1.2. *The argument principle for harmonic maps.* We will also need the following generalization of the argument principle for harmonic functions. We recall that a zero z_0 of a harmonic function is referred to as a *singular zero* of $F(z) = h(z) + \overline{g(z)}$ if the Jacobian of F, that is easily computed to be $|h'(z)|^2 - |g'(z)|^2$, vanishes at z_0. Moreover, F is said to be sense-preserving (respectively sense-reversing) at z_0 if the Jacobian is positive (respectively negative) at z_0. The *order* of a sense-preserving zero is defined to be the smallest positive integer k such that $h^{(k)}(z_0) \neq 0$. The order of a sense-reversing zero z_0 is defined to be $-k$ where k is the order of z_0 as a sense-preserving zero of $\overline{F(z)}$. If F is harmonic in a punctured neighborhood of z_0 and $F \to \infty$ as $z \to z_0$ then z_0 is referred to as a *pole*. As defined in [**38**], the *order of a pole* is given by $\frac{1}{2\pi}\Delta_T \arg F$, where T is a sufficiently small circle centered at the pole, and $\Delta_T \arg F$ denotes the increment of the argument of $F(z)$ along T. It follows that the order of a pole z_0 is negative when F is sense-reversing in a punctured neighborhood of z_0, and elaborating on this point we set aside Remark 2.2 that will be used later.

REMARK 2.2. In the sequel, we will consider the harmonic rational function $\frac{1}{\overline{p(z)}} - b - z$. We notice that this function has a pole at each zero of p and is sense-reversing in a neighborhood of each pole. Hence, the orders of the poles are all negative, and the sum of the orders is $-n$.

The following version of the argument principle is taken from [**38**, Thm. 2.2]. We note that there are topological versions of the argument principle (having classical roots going back to the Poincaré-Hopf Theorem for vector fields) that hold in more general settings [**37**], [**11**].

LEMMA 2.3 (The Generalized Argument Principle for harmonic maps). *Let F be harmonic, except for finitely many poles, in a region D, and let C be a Jordan curve that does not pass through any zeros or poles of F. Let Ω denote the interior of C and assume F has no singular zeros in Ω. Let $\Delta_C \arg F$ denote the increment of the argument of $F(z)$ as z traverses C. Let n_+, $-n_-$ be the sum of the orders of the sense-preserving and sense-reversing (respectively) zeros of F in Ω and let $p_+, -p_-$ be the sum of the orders of the sense preserving and sense reversing (respectively) poles of F in Ω. Then*

$$\frac{1}{2\pi}\Delta_C \arg F = n_+ - n_- - (p_+ - p_-).$$

2.1.3. *Tools for reducing the general case to the generic case.* The generalized argument principle stated in Lemma 2.3 can only be applied in the nondegenerate case where the harmonic mapping is free of singular zeros. The next two lemmas will be used to reduce the problem of establishing the upper bound in the general case (where singular zeros are allowed) to establishing it for the nondegenerate case. Harmonicity comes into play in this reduction, and there is no analogous strategy in general smooth settings. To elaborate, the first of these lemmas establishes a genericity result for nondegenerate examples. It is an important principle of differential topology that results of this type hold more generally for smooth maps. On the other hand, the second lemma establishes a stability result for extremal examples, and as pointed out in [**4**] this type of result does not hold for arbitrary smooth maps.

The following result is taken from [**21**, Lem. 1].

LEMMA 2.4 (Nondegenerate examples are generic). *If $r(z)$ is a rational function of degree greater than 1, then the set of complex numbers c for which $\overline{r(z)} - z - c$ does not have singular zeros is open and dense in $\mathbb{C}$.*

The following result is taken from [**4**, Prop. 3].

LEMMA 2.5 (Extremal examples are stable). *Let $f : D \to \mathbb{C}$ be a harmonic function defined in a region D in $\mathbb{C}$. Suppose that every $w \in \mathbb{C}$ has at most μ preimages, where $\mu < \infty$. Then the set of points which have μ preimages is open.*

2.2. Proof of Theorem 1.3. The problem is to bound the number of solutions of

$$p(z)\overline{q(z)} = w \tag{2.1}$$

when $q(z)$ is of degree $m = 1$ and p is a polynomial of degree $n > 1$. By factoring out the leading coefficient of q and assimilating it into p, we may assume without loss of generality that $q(z) = z + b$, $b \in \mathbb{C}$.

Since we easily get an upper bound of $n + 1$ in the case $w = 0$ (by setting each factor $p(z)$ and $\overline{z + b}$ to zero), we may assume $w \neq 0$. By multiplying p by $1/w$ we may further assume without loss of generality $w = 1$. Applying these reductions and rearranging (2.1), the problem is then to bound the number of solutions of the equation

$$\frac{1}{\overline{p(z)}} - b - z = 0, \tag{2.2}$$

i.e., zeros of the harmonic rational function $z \mapsto \frac{1}{\overline{p(z)}} - b - z$. We first estimate the number of sense-reversing zeros in Proposition 2.6 that is a specialized version of [**21**, Prop. 1]. The proof of the proposition essentially follows [**21**] with one important modification (see Remark 2.7). We simplify the presentation of the proof (in comparison with the proof of [**21**, Prop. 1]) by making use of Lemma 2.1 in place of the classical theorem of Fatou and by settling for a slightly weaker (yet sufficient for our purposes) result that avoids consideration of so-called "neutral" fixed points.

PROPOSITION 2.6. *Let p be a polynomial of degree $n \geq 2$. Then the number of sense-preserving zeros of $\dfrac{1}{\overline{p(z)}} - b - z$ is at most n.*

PROOF. Let $r(z) = \frac{1}{p(z)} - \bar{b}$. The sense-preserving zeros of $\dfrac{1}{\overline{p(z)}} - b - z$ are attracting fixed points of the anti-rational map $z \mapsto \overline{r(z)}$. By Lemma 2.1, the number of such attracting fixed points is at most the number of critical points of r, which is a rational function of degree n. By the Riemann-Hurwitz relation [**13**, Sec. 17.14], a rational function of degree n has exactly $2n-2$ critical points (counted with multiplicity), but the special case $r(z) = \dfrac{1}{p(z)} - \bar{b}$ of the shifted reciprocal of a polynomial of degree n has a critical point of order $n-1$ at infinity along with $n-1$ finite critical points (located at the zeros of p). Consequently, the anti-rational map $z \mapsto \overline{r(z)}$ has at most $n-1+1=n$ attracting fixed points, i.e., the harmonic rational function $\overline{r(z)} - z = \dfrac{1}{\overline{p(z)}} - b - z$ has at most n sense-preserving zeros, as desired. □

REMARK 2.7. To reiterate an important point in the proof of Proposition 2.6, we note that while being of high multiplicity (counted $n-1$ times in the Riemann-Hurwitz relation), the critical point at infinity only counts once in the dynamics argument, since it can be attracted to at most one attracting fixed point. In comparison with general anti-rational maps, this diminishes the possible number of attracting fixed points by $n-2$ which is responsible for the ultimate improvement in the upper bound $3n-1$ appearing in Theorem 1.3 in comparison with the (sharp) upper bound $5n-5$ [**21**] for the valence of harmonic rational functions of the form $\overline{r(z)} - z$ when r is a general analytic rational function (the improvement $n-2$ gets doubled to $2n-4$ in the proof of Theorem 2.8 at the step where the argument principle is applied).

From the discussion preceding Proposition 2.6, it follows that the problem of establishing Theorem 1.3 reduces to proving the following result.

THEOREM 2.8. *The number of solutions of* (2.2) *is at most* $3n-1$, *where as before* $n := \deg p$.

PROOF. As in the proof of Proposition 2.6 let us use the notation

$$r(z) = \frac{1}{p(z)} - \bar{b}. \tag{2.3}$$

Step 1. We first consider the case that $\overline{r(z)} - z$ does not have singular zeros. Then we apply Lemma 2.3 on a disk D_R of radius R large enough that D_R contains all zeros and poles of $\overline{r(z)} - z$. Since $r(z) = \mathcal{O}(1)$ as $z \to \infty$, the increment of the argument of $\overline{r(z)} - z$ along the boundary of D_R is 2π for any R sufficiently large. By Lemma 2.3,

$$1 = n_+ - n_- - (p_+ - p_-).$$

As noted in Remark 2.2, we have $p_+ = 0$ and $p_- = n$, so that the above equation becomes

$$n_- = n_+ + n - 1.$$

By Proposition 2.6 we have $n_+ \leq n$ which together with the above equation gives the following estimate on n_-.

$$n_- \leq 2n - 1,$$

Therefore, the total number of zeros satisfies $n_+ + n_- \leq 3n - 1$. This establishes the theorem in the case when $\overline{r(z)} - z$ has no singular zeros.

Step 2. We extend the result from Step 1 to the general case (where singular zeros are allowed) by employing Lemmas 2.4 and 2.5 as follows. Let $\mu(n)$ denote the maximal number of zeros, and suppose $\hat{r}$ is a rational function of the specified form (2.3) such that $\overline{\hat{r}(z)} - z$ has $\mu(n)$ zeros. By Lemma 2.4, there is a sequence of numbers $c_k \in \mathbb{C}$ with $c_k \to 0$ such that $\overline{\hat{r}(z)} - c_k - z$ is free of singular zeros for each k. By Lemma 2.5, the set of c such that $\overline{\hat{r}(z)} - c - z$ attains $\mu(n)$ zeros is open. Hence, for k large enough $\overline{\hat{r}(z)} - c_k - z$ is both free of singular zeros and attains $\mu(n)$ zeros. It then follows from Step 1 of the proof that $\mu(n) \leq 3n - 1$, and this completes the proof of the theorem.

□

3. Numerical results

The example from [7] with valence at least $3n-3$ is the following. Let

$$f(z) = \left(\frac{z^n}{n} - z\right)\overline{z} = |z|^2\left(\frac{z^{n-1}}{n} - 1\right).$$

As pointed out in [7], for $w = -\frac{n-1}{n}$ the roots of unity $z_k = e^{2\pi i k/(n-1)}$, $(k = 1, 2, \ldots, n-1)$ are solutions of $f(z) = w$.

Following the reformulation we used in the proof of Theorem 2.8, let us rewrite $f(z) = w$ as

$$\overline{r(z)} = z,$$

where

$$r(z) := \frac{nw}{z^n - nz} = -\frac{n-1}{z^n - nz}.$$

Then the roots of unity z_k are attracting fixed points of $\overline{r(z)}$. Indeed, we have

$$\begin{aligned} r(z_k) &= -\frac{n-1}{z_k(z_k^{n-1} - n)} \\ &= -\frac{n-1}{z_k(1-n)} = \frac{1}{z_k} = \overline{z_k}, \end{aligned}$$

where we used $|z_k| = 1$ in the last equality. This shows that each root of unity z_k is a fixed point. To see that each such z_k is attracting, we compute

$$r'(z) = (nz^{n-1} - n)\frac{n-1}{(z^n - nz)^2},$$

and notice that this vanishes when evaluated at a root of unity, i.e., we have $r'(z_k) = 0$. Thus, they are actually super-attracting fixed points.

Attracting fixed points of $\overline{r(z)}$ are orientation-preserving solutions of $\overline{r(z)} - z = 0$. By the generalized argument principle, this leads to at least $3n-3$ total solutions.

In fact, we can see from inspecting the dynamics of iteration of the map $z \mapsto \overline{r(z)}$ that there are exactly $3n-3$ for this example, since the critical point at infinity is not attracted to a fixed point. On the contrary, it is in a period-two orbit $\infty \to 0 \to \infty \to 0 \cdots$. This diminishes the possible number of attracting fixed points by one, i.e., inspecting the proof of Proposition 2.6 the upper bound stated in the proposition becomes $n-1$. This then diminishes the total count by two. Indeed, following the argument in the proof of Theorem 2.8 using the generalized argument principle, we have $n_+ \leq n-1$, $n_- \leq 2n-2$, and $N = n_+ + n_- \leq 3n-3$, so that the valence of this example is at most $3n-3$ (and hence exactly $3n-3$).

In order to show sharpness of the upper bound in Theorem 1.3, we would like to modify this example to have an additional attracting fixed point that attracts the critical point at infinity, leading to a total of $3n-1$ solutions of $\overline{r(z)} = z$. According to numerics presented in Section 3.1, this strategy succeeds at least for $n = 3, 4$, but it remains an open problem to verify this analytically and extend it to each n.

3.1. Numerical examples showing sharpness for $n = 3, 4$. Let us consider a modified version of Bshouty and Hengartner's example with the following slightly different choice of r that includes two real parameters $A > 0, C \geq 0$ (adding C prevents infinity from going to the origin under iteration, and including A gives some extra flexibility while trying to tune parameters to produce an additional attracting fixed point)

$$r(z) := \frac{A}{z\left[z^{n-1} + n\left(\frac{A}{n-1}\right)^{(n-1)/(n+1)}\right]} + C.$$

When $C = 0$ and $A = 1$, this reduces to (a rotation of) the Bshouty-Hengartner example, with attracting fixed points at roots of (-1) instead of roots of unity. When $C = 0$ and A varies, there are still $n-1$ attracting fixed points on a common circle,

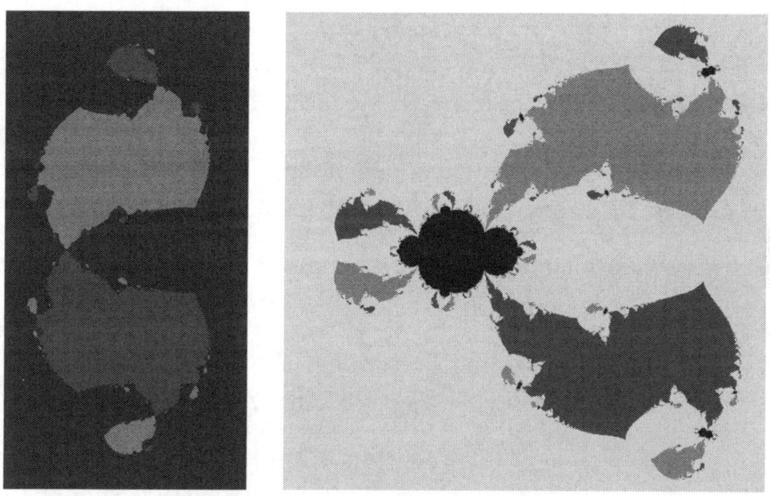

FIGURE 1. Left: The basins of attraction for each of the three attracting fixed points of the anti-rational map $z \mapsto \dfrac{100}{\overline{z}\left[\overline{z}^2 + 3\sqrt{50}\right]} +$ 0.68 which is an extremal example for the case $n = 3$. Right: The basins of attraction for each of the four attracting fixed points of the anti-rational map $z \mapsto \dfrac{250}{\overline{z}\left[\overline{z}^3 + 4\left(\frac{250}{3}\right)^{3/5}\right]} + 0.86$ which is an extremal example for the case $n = 4$. In both pictures, regions with the same color/shade are in a common basin of attraction, i.e., are attracted to a common fixed point. Image Credit: Both pictures were created by Walter Bergweiler.

namely, they are at the $(n-1)$th roots of (-1) scaled by $R(A,n) := \left(\frac{A}{n-1}\right)^{\frac{1}{n+1}}$, i.e., for each $k = 1, \ldots, n-1$, the point $z_k = R(A,n)e^{i\pi\frac{1+2k}{n-1}}$ is a critical point for $r(z)$ that is also a fixed point of $\overline{r(z)}$.

With specific choices of parameters, we used numerical evaluation of the built-in Mathematica command NestList in order to iterate the anti-rational map $z \mapsto \overline{r(z)}$ while tracking the trajectory of each critical point. According to these numerical experiments, when $n = 3$, $A = 100$, $C = 0.68$, this example has three attracting fixed points and hence $8 = 3n - 1$ solutions. The attracting fixed points are located at $z \approx 1.257 \pm 2.069i$ and $z \approx 2.31$. The critical point at infinity is attracted to the latter fixed point, and the symmetric pair of critical points on the imaginary axis are each attracted to the nearest of the other two fixed points. Similarly, we were able to locate parameter values that, according to numerical simulation, produce an example showing sharpness of Theorem 1.3 in the case $n = 4$. Namely, with $A = 250$ and $C = .86$ we find four attracting fixed points and hence $11 = 3n - 1$ solutions. The basins of attraction for each of these extremal examples are depicted in Figure 1.

4. An improvement on the Bezout bound

Recall that Hengartner's valence problem asks for a bound on the number of solutions to

$$p(z)\overline{q(z)} = w,$$

where w is an arbitrary complex number, and p and q are (analytic) complex polynomials, of degree n and m respectively, such that p is not a constant multiple of q. The case $w = 0$ admits a bound of $m + n$ on the valence (by separately setting $p(z) = 0$, $\overline{q(z)} = 0$), so we may assume $w \neq 0$. Furthermore, without loss of generality (by multiplying by $1/w$ and renaming the coefficients of, say, p) we assume $w = 1$.

Let us further reformulate the problem as one of bounding the number of zeros of the polynomial $P(z, \overline{z}) = p(z)\overline{q(z)} - 1$. Writing $Q(z, \overline{z}) = \overline{P(z, \overline{z})}$, we note that $P(z, \overline{z})$ and $Q(z, \overline{z})$ have the same zeros. Hence, the number of zeros $z \in \mathbb{C}$ of $P(z, \overline{z})$ is bounded from above by the number of common zeros $(z_1, z_2) \in \mathbb{C}^2$ of $P(z_1, z_2)$ and $Q(z_1, z_2)$. An application of Bezout's theorem (which can be justified when p is not a constant multiple of q [**1**]) then gives an upper bound of $(n + m)^2$ on the number of zeros, an estimate that we will refer to as the *Bezout bound*.

The main goal of this section is to improve the Bezout bound by proving Theorem 1.5 which in light of the above discussion reduces to establishing the following result.

THEOREM 4.1. *Let p and q be complex polynomials of degree n and m respectively. If p is not a constant multiple of q, then the number of zeros of $P(z, \overline{z}) = p(z)\overline{q(z)} - 1$ is bounded above by $n^2 + m^2$.*

Before presenting the proof of this result, we review some preliminaries concerning Sylvester resultants.

REMARK 4.2. It might seem surprising that an application of resultants leads to an improvement on the Bezout bound considering that Bezout's theorem itself is often proved using resultants [**37**]; the proof of Theorem 4.1 indeed uses resultants in an essentially different manner, as in [**33**, Sec. 11.5.1], that combines fruitfully with the constraint that the second variable is taken to be the complex conjugate of the first variable, see Remark 4.3.

While our proof of Theorem 4.1 uses essentially the same ideas as in [**34**], [**32**], in the current setting we are able to establish coprimacy of the polynomials P, Q without imposing an additional nondegeneracy condition. It is an open problem in the theory of gravitational lensing to determine whether the nondegeneracy assumption can be removed from the results of [**34**], [**32**], bounding the number of apparent images of a single background source lensed by a multiplane gravitational lens comprised of point masses.

4.1. Sylvester resultants. Let $P \in \mathbb{C}[z_1, z_2]$ be a bivariate polynomial with complex coefficients and variables. Written variously

$$P(z_1, z_2) = \sum_{i=0}^{n} a_i(z_2) z_1^i = \sum_{j=0}^{m} b_j(z_1) z_2^j = \sum_{i,j} c_{i,j} z_1^i z_2^j \tag{4.1}$$

we denote by $\deg_{z_1} P = n$, $\deg_{z_2} P = m$, and $\deg P = N = \max\{i + j \mid c_{i,j} \neq 0\}$ the *degree in* z_1, the *degree in* z_2, and the *total degree* of P respectively. Of course, $\deg P \leq \deg_{z_1} P + \deg_{z_2} P$. Let $Q \in \mathbb{C}[z_1, z_2]$ be another such polynomial written

$$Q(z_1, z_2) = \sum_{i=0}^{s} d_i(z_2) z_1^i = \sum_{j=0}^{t} e_j(z_1) z_2^j = \sum_{i,j} f_{i,j} z_1^i z_2^j. \tag{4.2}$$

Denote $\deg_{z_1} Q = s$, $\deg_{z_2} Q = t$, and $\deg Q = M$. Writing P and Q as polynomials in z_1, their coefficients are polynomials in z_2, namely $a_0, \ldots, a_n$ and $d_0, \ldots, d_s$ respectively. The z_1-*Sylvester Matrix* $\mathcal{S}_{z_1}(P, Q)$ of P and Q is an $(n + s) \times (n + s)$ matrix which is constructed by diagonally stacking s copies of the polynomials a_i and n copies of the d_i with 0 used for the other entries as follows.

$$\mathcal{S}_{z_1}(P, Q) = \left[\begin{array}{ccccc|cccc}
a_n & a_{n-1} & a_{n-2} & \cdots & a_1 & a_0 & 0 & \cdots & 0 \\
0 & a_n & a_{n-1} & a_{n-2} & \cdots & a_1 & a_0 & \cdots & 0 \\
\vdots & & \ddots & & & & & \ddots & \\
0 & \cdots & 0 & a_n & a_{n-1} & a_{n-2} & \cdots & a_1 & a_0 \\
\hline
d_s & d_{s-1} & \cdots & d_0 & 0 & 0 & \cdots & & 0 \\
0 & d_s & d_{s-1} & \cdots & d_0 & 0 & \cdots & & 0 \\
0 & 0 & d_s & d_{s-1} & \cdots & d_0 & \cdots & & 0 \\
\vdots & & & \ddots & & & \ddots & & \\
0 & \cdots & & 0 & d_s & d_{s-1} & \cdots & & d_0
\end{array}\right]. \tag{4.3}$$

Formally, for $1 \leq i \leq s$, entry $\mathcal{S}_{z_1}(P, Q)_{i,j} = a_{n-j+i}$ for $i \leq j \leq i + n$ and 0 otherwise. Likewise for $s + 1 \leq i \leq s + n$ entry $\mathcal{S}_{z_1}(P, Q)_{i,j} = d_{i-j}$ for $i - s \leq j \leq i$ and 0 otherwise. Note that the exact location of columns in the top and bottom halves of (4.3) depends on the values of n and s, but the matrix always has the indicated block structure where the top right $s \times s$ block is lower triangular and the bottom left $n \times n$ block is upper triangular.

REMARK 4.3. Note that there is an important, yet subtle, difference between the Sylvester matrix formulation in [**37**], which uses total degree, and [**33**] which uses the degree in z_1. This difference (along with the fact that we will be interested in solutions satisfying the constraint $z_2 = \overline{z_1}$) is the source of the relative advantage of one method over the other, as will be seen below.

Here, the z_1-*Sylvester resultant* of P and Q is the determinant of the z_1-Sylvester matrix of P and Q and will be denoted

$$\mathcal{R}_{z_1}(P,Q) = \det \mathcal{S}_{z_1}(P,Q)$$

or as $\mathcal{R}$ where the context is unambiguous. It is itself a polynomial in z_2 whose degree is bounded by $nt + sm$. [**33**, pp. 437–438].

We say that P and Q are *coprime* if their greatest common factor has degree zero, i.e. $P = CP_1$ and $Q = CQ_2 \implies \deg C = 0$. The following lemma is shown in [**10**, Prop. 1, p. 163].

LEMMA 4.4. *Let P and Q be bivariate polynomials in z_1 and z_2. Then $\mathcal{R}$ is identically zero if and only if P and Q are not coprime.*

The next result follows from [**33**, Thm. 11.10] and Lemma 4.4.

THEOREM 4.5 (The Resultant Theorem). *Let P and Q be complex polynomials as in* (4.1) *and* (4.2). *Then z_2 satisfies $\mathcal{R}(z_2) = 0$ if and only if either $a_n(z_2) = d_s(z_2) = 0$ or there exists some z_1 such that $P(z_1, z_2) = Q(z_1, z_2) = 0$. Hence, if P and Q are coprime then the number of values of z_2 that appear in the set of common zeros (z_1, z_2) of both P and Q is bounded from above by $nt + ms$.*

The Resultant Theorem is useful when paired with the additional constraint that the second variable is the conjugate of the first (as will be considered in Section 4.2). We note in passing that in the absence of such a constraint, Theorem 4.5 provides no advantage over the Bezout bound. Indeed, the Bezout bound [**37**] on the number of common zeros of P and Q as previously defined is

$$\#\{(z_1, z_2) \mid P(z_1, z_2) = 0 = Q(z_1, z_2)\} \le MN.$$

The Resultant Theorem, on the other hand, gives

$$\#\{z_2 \mid P(z_1, z_2) = 0 = Q(z_1, z_2) \text{ for some } z_1\} \le nt + ms$$

or symmetrically:

$$\#\{z_1 \mid P(z_1, z_2) = 0 = Q(z_1, z_2) \text{ for some } z_2\} \le sm + tn.$$

But since each z_1 solution could, a priori, have the full set of z_2 solutions as partners, the above two estimates together produce the rather poor estimate

$$\#\{(z_1, z_2) \mid P(z_1, z_2) = 0 = Q(z_1, z_2)\} \le (nt + ms)^2,$$

which is quite a bit larger than the Bezout bound.

4.2. Polynomials in z and $\overline{z}$. As in the statement of Theorem 4.1, we now consider the additional restriction that $z_2 = \overline{z_1}$, so that $P(z_1, z_2) = P(z, \overline{z})$ is a polyanalytic polynomial, and we also restrict to the case that $Q(z, \overline{z}) = \overline{P(z, \overline{z})}$.

Formally treating z and $\overline{z}$ as separate variables we have $\deg_z P = n = \deg_{\overline{z}} Q$ and $\deg_{\overline{z}} P = m = \deg_z Q$. Let $N = \deg P = \deg Q$. Following [**37**, Sec. 1.2.4], the Bezout bound is then

$$\#\{(z, \overline{z}) \mid P(z, \overline{z}) = 0\} \le N^2.$$

Note that $N \le n+m$ and hence $N^2 \le (n+m)^2$. The corresponding bound provided by Theorem 4.5 is

$$\#\{z_1 \mid P(z_1, z_2) = 0 \text{ for some } z_2\} \le m^2 + n^2.$$

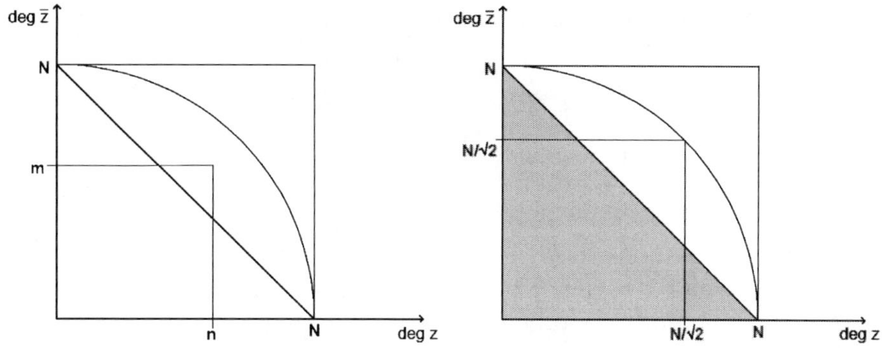

FIGURE 2. Left: A depiction of the condition $n^2 + m^2 < N^2$. The monomials of a polyanalytic polynomial $P(z, \overline{z})$ of total degree N correspond to integer lattice points that lie on and within the right triangle. Note that the lattice point (n, m), which is constructed from the maximum degrees in z and $\overline{z}$, need not correspond to an individual monomial. Right: A depiction of a simple sufficient condition $n, m < N/\sqrt{2}$. If the monomials are limited to the pentagonal region, the Resultant Bound is less than the Bezout bound.

But the restriction $z_2 = \overline{z_1}$ means in this case each solution matches with only one partner. In other words:

$$\#\{(z, \overline{z}) \mid P(z, \overline{z}) = 0\} \leq m^2 + n^2.$$

This proves Lemma 4.6.

LEMMA 4.6 (The Polyanalytic Resultant Lemma). *Let P be a polyanalytic polynomial of the form as in equation* (4.1). *Assume $P(z, \overline{z})$ and $Q(z, \overline{z}) = \overline{P(z, \overline{z})}$, viewed as bivariate polynomials in $z, \overline{z}$, are coprime. Then the number of zeros of $P(z, \overline{z})$ is bounded above by $n^2 + m^2$.*

Let us refer to the bound $n^2 + m^2$ provided by this lemma as the *Resultant Bound*. Trivially, the resultant bound improves the Bezout bound N^2 if and only if

$$n^2 + m^2 < N^2. \tag{4.4}$$

This necessary and sufficient condition is depicted in Figure 2. We note that for the case $P(z, \overline{z}) = p(z)\overline{q(z)} - w$ related to logharmonic polynomials we have $N = n+m$, so that the condition (4.4) holds for all $m, n \geq 1$. On the other hand, the condition (4.4) does not hold for any nontrivial harmonic polynomials, i.e., for polynomials of the form $f(z) + \overline{g(z)}$ having $\deg f, \deg g > 0$.

4.3. Coprimacy of P and $\overline{P}$ and proof of Theorem 4.1. Let p and q be complex univariate polynomials of degree n and m respectively. We write these as

$$p(z) = \sum_{k=0}^{n} p_k z^k \quad \text{and} \quad q(z) = \sum_{k=0}^{m} q_k z^k$$

where $p_k, q_k \in \mathbb{C}$ and $p_n, q_m \neq 0$.

As in the statement of Theorem 4.1, let us denote

$$P(z,\overline{z}) = p(z)\overline{q(z)} - 1,$$

and as in the statement of Lemma 4.6 we denote

$$Q(z,\overline{z}) = \overline{P(z,\overline{z})} = \overline{p(z)}q(z) - 1.$$

In view of Lemma 4.6, all that remains to prove Theorem 4.1 is to establish Lemma 4.7 ensuring that P and Q are coprime if p is not a constant multiple of q.

LEMMA 4.7. *Given two complex polynomials p and q, let $P(z,\overline{z}) = p(z)\overline{q(z)} - 1$ and $Q(z,\overline{z}) = \overline{p(z)}q(z) - 1$. Then either $p(z) = cq(z)$ for some complex constant c, or P and Q are coprime.*

The proof of the lemma will use the characterization of coprimacy in terms of resultants provided by Lemma 4.4. Note that P and Q may be written

$$P(z,\overline{z}) = -1 + \sum_{k=1}^{n} p_k\overline{q(z)}z^k, \quad Q(z,\overline{z}) = q_0\overline{p(z)} - 1 + \sum_{k=1}^{m} q_k\overline{p(z)}z^k,$$

where we have assumed, (by precomposing with a translation), $p_0 = 0$.

Having written P, Q as polynomials in z with coefficients depending on $\overline{z}$, recall that the resultant $\mathcal{R}_z(P,Q)$ of P, Q is expressed in terms of those coefficients as the determinant of the Sylvester matrix (4.3):

$$\left[\begin{array}{cccc|cccc}
p_n\overline{q(z)} & p_{n-1}\overline{q(z)} & \cdots & p_1\overline{q(z)} & -1 & 0 & \cdots & 0 \\
0 & p_n\overline{q(z)} & p_{n-1}\overline{q(z)} & \cdots & p_1\overline{q(z)} & -1 & \cdots & 0 \\
\vdots & & \ddots & & & & \ddots & \\
0 & \cdots & 0 & p_n\overline{q(z)} & & \cdots & & -1 \\
\hline
q_m\overline{p(z)} & \cdots & q_1\overline{p(z)} & q_0\overline{p(z)} - 1 & 0 & 0 & \cdots & 0 \\
0 & q_m\overline{p(z)} & \cdots & q_1\overline{p(z)} & q_0\overline{p(z)} - 1 & 0 & \cdots & 0 \\
\vdots & & \ddots & & & & & \\
0 & \cdots & 0 & q_m\overline{p(z)} & q_{m-1}\overline{p(z)} & & \cdots & q_0\overline{p(z)} - 1
\end{array}\right]. \tag{4.5}$$

To prove Lemma 4.7, we first establish the following intermediate step.

LEMMA 4.8. *With P, Q as in the statement of Lemma 4.7, if P and Q are not coprime then*

$$\{z \in \mathbb{C} : p(z) = 0\} = \{z \in \mathbb{C} : q(z) = 0\}.$$

PROOF OF LEMMA 4.8. Proving the contrapositive, we assume without loss of generality that there exists $z_0 \in \mathbb{C}$ such that $p(z_0) \neq 0$ while $q(z_0) = 0$. We then show that the resultant $\mathcal{R}_z(P,Q)$ is not identically zero and hence P and Q are coprime by Lemma 4.4. Substituting z_0 into (4.5) and evaluating, we have a matrix of the form

$$\left[\begin{array}{cccc|cccc} 0 & 0 & \cdots & 0 & -1 & 0 & \cdots & 0 \\ 0 & 0 & \cdots & 0 & 0 & -1 & \cdots & 0 \\ & \vdots & & & & & \ddots & \\ 0 & 0 & \cdots & 0 & 0 & 0 & \cdots & -1 \\ \hline q_m\overline{p(z_0)} & q_{m-1}\overline{p(z_0)} & \cdots & & & & \cdots & \\ 0 & q_m\overline{p(z_0)} & \cdots & & & & \cdots & \\ \vdots & & \ddots & & & & \cdots & \\ 0 & \cdots & 0 & q_m\overline{p(z_0)} & & & \cdots & \end{array}\right]$$

whose determinant is $(-1)^m(q_m\overline{p(z_0)})^n \neq 0$. Since $\mathcal{R}_z(P,Q)$ is not identically zero, it follows from Lemma 4.4 that P and Q are coprime. □

We now prove Lemma 4.7 (and hence conclude Theorem 4.1 as explained at the beginning of the current subsection).

Proof of Lemma 4.7. Suppose that p is not a constant multiple of q and, working toward a contradiction, suppose that P and Q are not coprime. From Lemma 4.8 we have that $\{p=0\}=\{q=0\}$, but since p is not a constant multiple of q there exists $z_0 \in \{z : p(z)=q(z)=0\}$ where p and q have distinct order of vanishing, i.e., as $z \to z_0$

$$\begin{aligned} p(z) &= \alpha(z-z_0)^k + \mathcal{O}((z-z_0)^{(k+1)}), \quad \text{and} \\ q(z) &= \beta(z-z_0)^\ell + \mathcal{O}((z-z_0)^{(\ell+1)}), \end{aligned}$$

with $k \neq \ell$. Assume $k > \ell$ without loss of generality. We then have

$$\frac{\mathcal{R}_z(P,Q)}{(\overline{z}-\overline{z_0})^{kn}} = \det \left[\begin{array}{cccc|cccc} & & & & -1 & 0 & \cdots & 0 \\ & \mathcal{A} & & & & -1 & \cdots & 0 \\ & & & & & & \ddots & \\ & & & & \mathcal{B} & & & -1 \\ \hline \frac{q_m\overline{p(z)}}{(\overline{z}-\overline{z_0})^k} & & \mathcal{C} & & & & & \\ 0 & \frac{q_m\overline{p(z)}}{(\overline{z}-\overline{z_0})^k} & & & & \mathcal{D} & & \\ 0 & 0 & \cdots & \frac{q_m\overline{p(z)}}{(\overline{z}-\overline{z_0})^k} & & & & \end{array}\right]$$

where we have multiplied each of the first n columns by $1/(\overline{z}-\overline{z_0})^k$. Note that in submatrix $\mathcal{A}$ the term $\overline{q(z)}/(\overline{z}-\overline{z_0})^k$ appears as a factor in each entry. Using row operations, each entry in triangular component $\mathcal{C}$ may be eliminated (utilizing the nonzero diagonal entries below component $\mathcal{C}$) while only disturbing entries in submatrix $\mathcal{D}$. Letting $z \to z_0$ and using

$$\lim_{z \to z_0} \frac{p(z)}{(z - z_0)^k} = \alpha \quad \text{and} \quad \lim_{z \to z_0} \frac{q(z)}{(z - z_0)^k} = 0$$

we obtain

$$\lim_{z \to z_0} \frac{\mathcal{R}_z(P,Q)}{(\overline{z} - \overline{z_0})^{kn}} = \det \left[\begin{array}{cccc|cccc} 0 & 0 & \cdots & 0 & -1 & 0 & \cdots & 0 \\ 0 & 0 & \cdots & 0 & & -1 & \cdots & 0 \\ \vdots & & & \vdots & & & \ddots & \\ 0 & 0 & \cdots & 0 & \mathcal{B} & & & -1 \\ \hline q_m\overline{\alpha} & 0 & \cdots & 0 & & & & \\ 0 & q_m\overline{\alpha} & \cdots & 0 & & & \mathcal{D}^* & \\ & & \ddots & & & & & \\ 0 & 0 & \cdots & q_m\overline{\alpha} & & & & \end{array} \right]$$
$$= (-1)^m (q_m\overline{\alpha})^n \neq 0,$$

which contradicts the assumption that P and Q are not coprime and concludes the proof of the lemma. □

Acknowledgments

The authors are grateful to Walter Bergweiler for producing the images of the basins of attraction shown in Figure 1. We also thank Alexandre Eremenko for helpful comments.

References

[1] Z. Abdulhadi and W. Hengartner, *Polynomials in $H\overline{H}$*, Complex Variables Theory Appl. **46** (2001), no. 2, 89–107, DOI 10.1080/17476930108815400. MR1867260

[2] L. D. Abreu, K. Gröchenig, and J. L. Romero, *Harmonic analysis in phase space and finite Weyl-Heisenberg ensembles*, J. Stat. Phys. **174** (2019), no. 5, 1104–1136, DOI 10.1007/s10955-019-02226-2. MR3921391

[3] M. B. Balk, *The fundamental theorem of algebra for polyanalytic polynomials* (Russian, with English and Lithuanian summaries), Litovsk. Mat. Sb. **8** (1968), 401–404. MR245813

[4] W. Bergweiler and A. Eremenko, *Green's function and anti-holomorphic dynamics on a torus*, Proc. Amer. Math. Soc. **144** (2016), no. 7, 2911–2922, DOI 10.1090/proc/13044. MR3487224

[5] W. Bergweiler and A. Eremenko, *On the number of solutions of some transcendental equations*, Anal. Math. Phys. **8** (2018), no. 2, 185–196, DOI 10.1007/s13324-017-0204-6. MR3817849

[6] P. M. Bleher, Y. Homma, L. L. Ji, and R. K. W. Roeder, *Counting zeros of harmonic rational functions and its application to gravitational lensing*, Int. Math. Res. Not. IMRN **8** (2014), 2245–2264, DOI 10.1093/imrn/rns284. MR3194017

[7] D. Bshouty and W. Hengartner, *On the valency of a polynomial in $H\overline{H}$*, Ann. Univ. Mariae Curie-Skłodowska Sect. A **54** (2000), 21–25. Dedicated to Professor Zdzisław Lewandowski on the occasion of his seventieth birthday. MR1825300

[8] D. Bshouty and A. Lyzzaik, *Problems and conjectures in planar harmonic mappings*, J. Anal. **18** (2010), 69–81. MR2850236

[9] J. J. Carmona and K. Yu. Fedorovskiy, *New conditions for uniform approximation by polyanalytic polynomials*, Tr. Mat. Inst. Steklova **279** (2012), no. Analiticheskie i Geometricheskie Voprosy Kompleksnogo Analiza, 227–241, DOI 10.1134/s0081543812080159; English transl., Proc. Steklov Inst. Math. **279** (2012), no. 1, 215–229. MR3086767

[10] D. Cox, J. Little, and D. O'Shea, *Ideals, varieties, and algorithms*, 3rd ed., Undergraduate Texts in Mathematics, Springer, New York, 2007. An introduction to computational algebraic geometry and commutative algebra, DOI 10.1007/978-0-387-35651-8. MR2290010

[11] M. Cristea, *A generalization of the argument principle*, Complex Variables Theory Appl. **42** (2000), no. 4, 333–345, DOI 10.1080/17476930008815293. MR1789408

[12] K. Yu. Fedorovskiy, *C^m-approximation by polyanalytic polynomials on compact subsets of the complex plane*, Complex Anal. Oper. Theory **5** (2011), no. 3, 671–681, DOI 10.1007/s11785-010-0099-9. MR2836315

[13] O. Forster, *Lectures on Riemann surfaces*, Graduate Texts in Mathematics, vol. 81, Springer-Verlag, New York, 1991. Translated from the 1977 German original by Bruce Gilligan; Reprint of the 1981 English translation. MR1185074

[14] L. Geyer, *Sharp bounds for the valence of certain harmonic polynomials*, Proc. Amer. Math. Soc. **136** (2008), no. 2, 549–555, DOI 10.1090/S0002-9939-07-08946-0. MR2358495

[15] L. Geyer, *Classification of crtitically fixed anti-rational maps*, Preprint, `arXiv:2006.10788v3`, 2022.

[16] A. Haimi and H. Hedenmalm, *The polyanalytic Ginibre ensembles*, J. Stat. Phys. **153** (2013), no. 1, 10–47, DOI 10.1007/s10955-013-0813-x. MR3100813

[17] J. D. Hauenstein, A. Lerario, E. Lundberg, and D. Mehta, *Experiments on the zeros of harmonic polynomials using certified counting*, Exp. Math. **24** (2015), no. 2, 133–141, DOI 10.1080/10586458.2014.966180. MR3350520

[18] M. Huhtanen, *Orthogonal polyanalytic polynomials and normal matrices*, Math. Comp. **72** (2003), no. 241, 355–373, DOI 10.1090/S0025-5718-02-01417-5. MR1933825

[19] D. Khavinson, *A thought on approximation by bi-analytic functions*, New trends in approximation theory, Fields Inst. Commun., vol. 81, Springer, New York, 2018, pp. 131–136. MR3586185

[20] D. Khavinson, S.-Y. Lee, and A. Saez, *Zeros of harmonic polynomials, critical lemniscates, and caustics*, Complex Anal. Synerg. **4** (2018), no. 1, Paper No. 2, 20, DOI 10.1186/s40627-018-0012-2. MR3772098

[21] D. Khavinson and G. Neumann, *On the number of zeros of certain rational harmonic functions*, Proc. Amer. Math. Soc. **134** (2006), no. 4, 1077–1085, DOI 10.1090/S0002-9939-05-08058-5. MR2196041

[22] D. Khavinson and G. Świątek, *On the number of zeros of certain harmonic polynomials*, Proc. Amer. Math. Soc. **131** (2003), no. 2, 409–414, DOI 10.1090/S0002-9939-02-06476-6. MR1933331

[23] S.-Y. Lee, A. Lerario, and E. Lundberg, *Remarks on Wilmshurst's theorem*, Indiana Univ. Math. J. **64** (2015), no. 4, 1153–1167, DOI 10.1512/iumj.2015.64.5526. MR3385788

[24] S.-Y. Lee and N. G. Makarov, *Topology of quadrature domains*, J. Amer. Math. Soc. **29** (2016), no. 2, 333–369, DOI 10.1090/jams828. MR3454377

[25] S.-Y. Lee and A. Saez, *A new lower bound for the maximal valence of harmonic polynomials*, Comput. Methods Funct. Theory **17** (2017), no. 1, 139–149, DOI 10.1007/s40315-016-0175-x. MR3620883

[26] A. Lerario and E. Lundberg, *On the zeros of random harmonic polynomials: the truncated model*, J. Math. Anal. Appl. **438** (2016), no. 2, 1041–1054, DOI 10.1016/j.jmaa.2016.02.039. MR3466078

[27] W. V. Li and A. Wei, *On the expected number of zeros of a random harmonic polynomial*, Proc. Amer. Math. Soc. **137** (2009), no. 1, 195–204, DOI 10.1090/S0002-9939-08-09555-5. MR2439441

[28] J. Liesen and J. Zur, *How constant shifts affect the zeros of certain rational harmonic functions*, Comput. Methods Funct. Theory **18** (2018), no. 4, 583–607, DOI 10.1007/s40315-018-0240-8. MR3874883

[29] J. Liesen and J. Zur, *The maximum number of zeros of $r(z) - \overline{z}$ revisited*, Comput. Methods Funct. Theory **18** (2018), no. 3, 463–472, DOI 10.1007/s40315-017-0231-1. MR3844663

[30] R. Luce, O. Sète, and J. Liesen, *A note on the maximum number of zeros of $r(z) - \overline{z}$*, Comput. Methods Funct. Theory **15** (2015), no. 3, 439–448, DOI 10.1007/s40315-015-0110-6. MR3393817

[31] E. Lundberg, *The valence of harmonic polynomials viewed through the probabilistic lens*, Proc. Amer. Math. Soc. **151** (2023), no. 7, 2963–2973, DOI 10.1090/proc/16152. MR4579370

[32] S. Perry, *An upper bound for the number of gravitationally lensed images in a multiplane point-mass ensemble*, Anal. Math. Phys. **11** (2021), no. 2, Paper No. 52, 13, DOI 10.1007/s13324-021-00478-4. MR4213761

[33] A. O. Petters, H. Levine, and J. Wambsganss, *Singularity theory and gravitational lensing*, Progress in Mathematical Physics, vol. 21, Birkhäuser Boston, Inc., Boston, MA, 2001. With a foreword by David Spergel, DOI 10.1007/978-1-4612-0145-8. MR1836154

[34] A. O. Petters, *Multiplane gravitational lensing. III. Upper bound on number of images*, J. Math. Phys. **38** (1997), no. 3, 1605–1613, DOI 10.1063/1.531818. MR1435685

[35] O. Sète, R. Luce, and J. Liesen, *Creating images by adding masses to gravitational point lenses*, Gen. Relativity Gravitation **47** (2015), no. 4, Art. 42, 8, DOI 10.1007/s10714-015-1876-6. MR3322444

[36] O. Sète, R. Luce, and J. Liesen, *Perturbing rational harmonic functions by poles*, Comput. Methods Funct. Theory **15** (2015), no. 1, 9–35, DOI 10.1007/s40315-014-0083-x. MR3318304

[37] T. Sheil-Small, *Complex polynomials*, Cambridge Studies in Advanced Mathematics, vol. 75, Cambridge University Press, Cambridge, 2002, DOI 10.1017/CBO9780511543074. MR1962935

[38] T. J. Suffridge and J. W. Thompson, *Local behavior of harmonic mappings*, Complex Variables Theory Appl. **41** (2000), no. 1, 63–80, DOI 10.1080/17476930008815237. MR1758598

[39] A. Thomack, *On the zeros of random harmonic polynomials: the naive model*, Preprint, arXiv:1610.02611, 2016.

[40] A. Thomack and Z. Tyree, *On the zeros of random harmonic polynomials: the Weyl model*, Anal. Math. Phys. **8** (2018), no. 2, 237–253, DOI 10.1007/s13324-018-0220-1. MR3817853

[41] A. S. Wilmshurst, *The valence of harmonic polynomials*, Proc. Amer. Math. Soc. **126** (1998), no. 7, 2077–2081, DOI 10.1090/S0002-9939-98-04315-9. MR1443416

[42] A. S. Wilmshurst, *Complex harmonic mappings and the valence of harmonic polynomials*, D. Phil. thesis, University of York, England, 1994.

Department of Mathematics and Statistics, University of South Florida, Tampa, Florida 33620

Email address: dkhavins@usf.edu

Department of Mathematical Sciences, Florida Atlantic University, Boca Raton, Florida 33431

Email address: elundber@fau.edu

Department of Mathematics and Statistics, University of South Florida, Tampa, Florida 33620

Email address: perry2@usf.edu

Contemporary Mathematics
Volume **799**, 2024
https://doi.org/10.1090/conm/799/16019

A generalized cross ratio

Michael R. Pilla

ABSTRACT. In one complex variable, the cross ratio is a well-known quantity associated with four given points in the complex plane that remains invariant under linear fractional maps. In particular, if one knows where three points in the complex plane are mapped under a linear fractional map, one can use this invariance to explicitly determine the map and to show that linear fractional maps are 3-transitive. In this paper, we define a generalized cross ratio and determine some of its basic properties. In particular, we determine which hypotheses must be made to guarantee that our generalized cross ratio is well defined. We thus obtain a class of maps that obey similar transitivity properties as in one complex dimension, under some more restrictive conditions.

1. Background

In 1872, Felix Klein famously introduced a point of view regarding what geometry should be about [**3**]. Known as the Erlangen program, Klein viewed geometry as a study of invariants under group transformations. An important and robust example of such an invariant quantity is given by the cross ratio.

Given four finite distinct points z_1, z_2, z_3, and z_4 in the complex plane, the cross ratio is defined as

$$(z_1, z_2, z_3, z_4) = \frac{(z_1 - z_3)(z_2 - z_4)}{(z_1 - z_4)(z_2 - z_3)}. \tag{1.1}$$

If $z_i = \infty$ for some $i = 1, 2, 3, 4$ then we cancel the terms with z_i. For example, if $z_1 = \infty$, then

$$(z_1, z_2, z_3, z_4) = \frac{(z_2 - z_4)}{(z_2 - z_3)}.$$

With this definition, we may view the cross ratio as a function of z given by (z, z_2, z_3, z_4). Recall that a linear fractional map, also known as a Mobius transformation, is defined as

$$f(z) = \frac{az + b}{cz + d},$$

where the coefficients a, b, c, and d are complex numbers such that $ad - bc \neq 0$ (otherwise $f(z)$ is a constant function).

2020 *Mathematics Subject Classification.* Primary 32A08.
Key words and phrases. Linear fractional maps, cross ratio, transitivity.

It is well known that the cross ratio is preserved under linear fractional maps. Thus if ϕ is a linear fractional map in one complex variable and $\phi(z_i) = w_i$ for $i = 1, 2, 3$, then

$$\frac{(w - w_2)(w_1 - w_3)}{(w - w_3)(w_1 - w_2)} = \frac{(z - z_2)(z_1 - z_3)}{(z - z_3)(z_1 - z_2)},$$

where $w = \phi(z)$. In such a case one may solve for w to determine the map explicitly.

2. Linear fractional maps in several complex variables

To define the cross ratio in several complex variables, we must first say what it means to be a linear fractional map in $\mathbb{C}^N$ for $N > 1$. Our approach to generalization will utilize homogeneous coordinates. We associate the point $z = (z_1', z_2)$ where $z_1' \in \mathbb{C}^N$ and $z_2 \in \mathbb{C}$, $z \neq 0$, with the point $\frac{z_1'}{z_2} \in \mathbb{C}^N$. This associated space is known as the complex projective space $\mathbf{CP}^N$. We will write $z \sim u$ if $z \in \mathbb{C}^N$ is the point associated with $u \in \mathbf{CP}^N$. In $\mathbb{C}$, one may show that linear fractional maps are precisely the linear transformations acting on homogeneous coordinates in $\mathbb{C}^2$. We take the perspective that linear fractional maps in $\mathbb{C}^N$ should be defined likewise.

For vectors $\begin{pmatrix} z_1' \\ z_2 \end{pmatrix}$ and $\begin{pmatrix} w_1' \\ w_2 \end{pmatrix}$ in $\mathbb{C}^{N+1}$ with $z_2, w_2 \in \mathbb{C}$, we consider a linear transformation in $\mathbb{C}^{N+1}$ represented by a complex $(N+1) \times (N+1)$ matrix as

$$\begin{pmatrix} w_1' \\ w_2 \end{pmatrix} = \begin{pmatrix} A & B \\ C^* & D \end{pmatrix} \begin{pmatrix} z_1' \\ z_2 \end{pmatrix} = \begin{pmatrix} \langle a_1, \overline{z_1'} \rangle + b_1 z_2 \\ \vdots \\ \langle a_N, \overline{z_1'} \rangle + b_N z_2 \\ \langle z_1', C \rangle + D z_2 \end{pmatrix},$$

where A is an $N \times N$ matrix with rows denoted by a_i for $i = 1, \dots, N$, $B = \begin{pmatrix} b_1 & \cdots & b_N \end{pmatrix}^T$ and C are column vectors in $\mathbb{C}^N$, $D \in \mathbb{C}$, C^* represents the conjugate transpose of C, and $\langle \cdot, \cdot \rangle$ is the standard inner product.

If we now think of our vectors $\begin{pmatrix} z_1' \\ z_2 \end{pmatrix}$ and $\begin{pmatrix} w_1' \\ w_2 \end{pmatrix}$ as homogeneous coordinates of z and w so that $z \sim \begin{pmatrix} z_1' \\ z_2 \end{pmatrix}$ and $w \sim \begin{pmatrix} w_1' \\ w_2 \end{pmatrix}$, then we can associate the previous linear transformation in $\mathbb{C}^{N+1}$ with the not necessarily linear transformation in $\mathbb{C}^N$ given by

$$\begin{aligned} w = \frac{w_1'}{w_2} &= \left(\frac{\langle a_1, \overline{z_1'} \rangle + b_1 z_2}{\langle z_1', C \rangle + D z_2}, \dots, \frac{\langle a_N, \overline{z_1'} \rangle + b_N z_2}{\langle z_1', C \rangle + D z_2} \right) \\ &= \left(\frac{\langle a_1, \overline{\frac{z_1'}{z_2}} \rangle + b_1}{\langle \frac{z_1'}{z_2}, C \rangle + D}, \dots, \frac{\langle a_N, \overline{\frac{z_1'}{z_2}} \rangle + b_N}{\langle \frac{z_1'}{z_2}, C \rangle + D} \right) \\ &= \left(\frac{\langle a_1, \overline{z} \rangle + b_1}{\langle z, C \rangle + D}, \dots, \frac{\langle a_N, \overline{z} \rangle + b_N}{\langle z, C \rangle + D} \right) \\ &= \frac{Az + B}{\langle z, C \rangle + D}. \end{aligned}$$

This is precisely the definition given by Cowen and MacCluer [**1**] and is the one we will adopt.

DEFINITION 2.1. We say ϕ is a linear fractional map in $\mathbb{C}^N$ if

$$\phi(z) = \frac{Az + B}{\langle z, C\rangle + D}, \tag{2.1}$$

where A is an $N \times N$ matrix, B and C are column vectors in $\mathbb{C}^N$, $D \in \mathbb{C}$, and $\langle \cdot, \cdot \rangle$ is the standard inner product.

We define the associated matrix m_ϕ of the linear fractional map $\phi(z) = \frac{Az+B}{\langle z,C\rangle+D}$ to be given by

$$m_\phi = \begin{pmatrix} A & B \\ C^* & D \end{pmatrix}.$$

If $\phi(z) = w$ and the point $u \in \mathbb{C}^{N+1}$ is associated with z, then $v = m_\phi u$ is associated with the point w and vice versa. A routine calculation also shows that $m_{\phi_1 \circ \phi_2} = m_{\phi_1} m_{\phi_2}$ and $m_{\phi^{-1}} = (m_\phi)^{-1}$. That is, if $z \sim u$, then $\phi(z) \sim m_\phi u$. Thus, we can toggle back and forth between the complex space $\mathbb{C}^N$ and $\mathbf{CP}^N$ at our convenience.

3. The cross ratio in several complex variables

While generalized cross ratios have been defined previously in terms of the Schwarzian derivative [**2**] and from the perspective of symplectic geometry [**5**], this paper takes a different perspective that is more tractable and attempts to be in the elementary spirit of the traditional cross ratio. For one complex variable, one may show that the cross ratio is invariant under the maps ϕ and m_ϕ. We would like the same to be true for the cross ratio in $\mathbb{C}^N$. The cross ratio can also be utilized to show that any three pairwise distinct points in $\mathbb{C}$ that are sent to another set of three pairwise distinct points uniquely determine a linear fractional map in $\mathbb{C}$. The proof of this uses the fact that one can utilize cross ratios to define a unique map that sends a pairwise distinct triple to the standardized points $(1, 0, \infty)$. Generalizing to higher dimensions, however, does not come without cost. Even for $N = 2$, one may recall that projective transformations map lines to lines. Without further hypotheses to avoid problems of collinearity, the task of mapping a pairwise distinct quadruple to four standardized points is hopeless. Treading carefully, we use these facts to motivate our definition of a cross ratio in $\mathbb{C}^N$ for $N > 1$. In particular we seek a minimal definition that will preserve the properties of the cross ratio just described. Since we have reserved subscripts for the components of a point in several variables, for the remainder of our discussion we choose to designate *superscripts* for distinct points. As we will only consider linear fractional maps, there should be no confusion about notation. Just remember, superscripts will not designate exponents but rather distinct points.

Thus, as subscripts represent coordinates (components of a vector) and superscripts distinguish distinct points in higher dimensions, we define the following quantity. If we let $u^i = (u_1^i, u_2^i, \ldots, u_{N+1}^i)$ for $i = 1, \ldots, N+1$ be elements of $\mathbb{C}^{N+1}$, then we define

$$
(3.1) \qquad [u^1, u^2, \ldots, u^{N+1}] = \det \begin{pmatrix} u_1^1 & u_1^2 & u_1^3 & \cdots & u_1^{N+1} \\ u_2^1 & u_2^2 & u_2^3 & \cdots & u_2^{N+1} \\ \vdots & \vdots & \vdots & \ddots & \vdots \\ u_{N+1}^1 & u_{N+1}^2 & u_{N+1}^3 & \cdots & u_{N+1}^{N+1} \end{pmatrix}.
$$

In order to motivate our general definition, we begin by defining a cross ratio in $\mathbb{C}^2$ and proceed to generalize to $\mathbb{C}^N$. In particular, see Definition 3.1.

DEFINITION 3.1. Given five distinct points u^i for $i = 1, \ldots, 5$ in $\mathbf{CP}^2$, we define the cross ratio as

$$
(u^1, u^2, u^3, u^4, u^5) = \frac{[u^1, u^3, u^5][u^2, u^4, u^5]}{[u^1, u^4, u^5][u^2, u^3, u^5]}.
$$

Note that although our notation for the determinant given by Equation 3.1 is only well-defined for vectors in $\mathbb{C}^{N+1}$, the definition for the cross ratio *is* well-defined for points in $\mathbf{CP}^N$.

One motivation for choosing this particular cross ratio is because it "reduces" to our usual cross ratio using the appropriate coordinates. We will also find it useful when addressing questions of transitivity. Let's look at an example of how, under the right conditions, this definition "reduces" to the standard definition of the cross ratio. For the vectors $u^i = (u_1^i, u_2^i, u_3^i)$ with $i = 1, .., 4$, if we standardize coordinates so that $u_3^i = 1$, $z_1^i = u_1^i/u_3^i$, and $z_2^i = u_2^i/u_3^i$, we get vectors in $\mathbb{C}^2/\{0\}$ which we can interpret as $\mathbf{CP}^1$. We still have one extra vector u^5 though. However, since we let $u_3^i = 1$ for $i = 1, \ldots, 4$, why not choose u^5 to be a "point at infinity". In our context, this means choosing the final coordinate to equal 0. If we choose $u^5 = (0, 1, 0)$, which is *a* point on the line at infinity, we have

$$
\begin{aligned}
(u^1, u^2, u^3, u^4, u^5) &= \frac{[u^1, u^3, u^5][u^2, u^4, u^5]}{[u^1, u^4, u^5][u^2, u^3, u^5]} \\
&= \frac{\det \begin{pmatrix} u_1^1 & u_1^3 & 0 \\ u_2^1 & u_2^3 & 1 \\ u_3^1 & u_3^3 & 0 \end{pmatrix} \det \begin{pmatrix} u_1^2 & u_1^4 & 0 \\ u_2^2 & u_2^4 & 1 \\ u_3^2 & u_3^4 & 0 \end{pmatrix}}{\det \begin{pmatrix} u_1^1 & u_1^4 & 0 \\ u_2^1 & u_2^4 & 1 \\ u_3^1 & u_3^4 & 0 \end{pmatrix} \det \begin{pmatrix} u_1^2 & u_1^3 & 0 \\ u_2^2 & u_2^3 & 1 \\ u_3^2 & u_3^3 & 0 \end{pmatrix}} = \frac{\det \begin{pmatrix} z_1^1 & z_1^3 & 0 \\ z_2^1 & z_2^3 & 1 \\ 1 & 1 & 0 \end{pmatrix} \det \begin{pmatrix} z_1^2 & z_1^4 & 0 \\ z_2^2 & z_2^4 & 1 \\ 1 & 1 & 0 \end{pmatrix}}{\det \begin{pmatrix} z_1^1 & z_1^4 & 0 \\ z_2^1 & z_2^4 & 1 \\ 1 & 1 & 0 \end{pmatrix} \det \begin{pmatrix} z_1^2 & z_1^3 & 0 \\ z_2^2 & z_2^3 & 1 \\ 1 & 1 & 0 \end{pmatrix}} \\
&= \frac{(z^1 - z^3)(z^2 - z^4)}{(z^1 - z^4)(z^2 - z^3)} = (z^1, z^2, z^3, z^4).
\end{aligned}
$$

In $\mathbb{C}$, there are $4! = 24$ permutations of (z^1, z^2, z^3, z^4) of which 6 are distinct. In $\mathbb{C}^2$, we recognize that

$$
(z^i, z^j, z^k, z^l, z^m) = (z^k, z^l, z^i, z^j, z^m) = (z^j, z^i, z^l, z^k, z^m) = (z^l, z^k, z^j, z^i, z^m)
$$

and that no other permutation is equal to $(z^i, z^j, z^k, z^l, z^m)$. Thus, we have five ways to choose m and once m is chosen, we start with i (by the previous equation we may, equivalently, start with any of i, j, k, l and have $3! = 6$ ways to choose the remaining three values which gives us $5 \times 6 = 30$ distinct permutations.

Let's see, explicitly, what this definition looks in like our usual complex variables when the z^i's for $i = 1, \ldots, 5$ are finite complex numbers where we are once again putting the notation for the distinct points in the superscript, leaving the subscripts for vector components. This means we set the third coordinate to 1 in each case. Thus we have

$$\begin{aligned}
(u^1, u^2, u^3, u^4, u^5) &= \frac{[u^1, u^3, u^5][u^2, u^4, u^5]}{[u^1, u^4, u^5][u^2, u^3, u^5]} \\
&= \frac{\det\begin{pmatrix} u_1^1 & u_1^3 & u_1^5 \\ u_2^1 & u_2^3 & u_2^5 \\ u_3^1 & u_3^3 & u_3^5 \end{pmatrix} \det\begin{pmatrix} u_1^2 & u_1^4 & u_1^5 \\ u_2^2 & u_2^4 & u_2^5 \\ u_3^2 & u_3^4 & u_3^5 \end{pmatrix}}{\det\begin{pmatrix} u_1^1 & u_1^4 & u_1^5 \\ u_2^1 & u_2^4 & u_2^5 \\ u_3^1 & u_3^4 & u_3^5 \end{pmatrix} \det\begin{pmatrix} u_1^2 & u_1^3 & u_1^5 \\ u_2^2 & u_2^3 & u_2^5 \\ u_3^2 & u_3^3 & u_3^5 \end{pmatrix}} = \frac{\det\begin{pmatrix} z_1^1 & z_1^3 & z_1^5 \\ z_2^1 & z_2^3 & z_2^5 \\ 1 & 1 & 1 \end{pmatrix} \det\begin{pmatrix} z_1^2 & z_1^4 & z_1^5 \\ z_2^2 & z_2^4 & z_2^5 \\ 1 & 1 & 1 \end{pmatrix}}{\det\begin{pmatrix} z_1^1 & z_1^4 & z_1^5 \\ z_2^1 & z_2^4 & z_2^5 \\ 1 & 1 & 1 \end{pmatrix} \det\begin{pmatrix} z_1^2 & z_1^3 & z_1^5 \\ z_2^2 & z_2^3 & z_2^5 \\ 1 & 1 & 1 \end{pmatrix}} \\
&= \frac{\left(z_1^1(z_2^3 - z_2^5) - z_1^3(z_2^1 - z_2^5) + z_1^5(z_2^1 - z_2^3)\right)\left(z_1^2(z_2^4 - z_2^5) - z_1^4(z_2^2 - z_2^5) + z_1^5(z_2^2 - z_2^4)\right)}{\left(z_1^1(z_2^4 - z_2^5) - z_1^4(z_2^1 - z_2^5) + z_1^5(z_2^1 - z_2^4)\right)\left(z_1^2(z_2^3 - z_2^5) - z_1^3(z_2^2 - z_2^5) + z_1^5(z_2^2 - z_2^3)\right)} \\
&= (z^1, z^2, z^3, z^4, z^5).
\end{aligned}$$

This expression is definitely cleaner expressed in terms of determinants! As an exercsie for the reader, try writing down the expression when one of the z^i's is a point on the line at infinity.

Next we recall that by letting $z = z^1$ be a complex variable, the cross ratio in one complex variable becomes the linear fractional map given by

$$(z, z^2, z^3, z^4) = \frac{(z - z^3)(z^2 - z^4)}{(z - z^4)(z^2 - z^3)}.^{1}$$

Notice this cross ratio sends the points z^2, z^3, and z^4 to 1, 0, and ∞, respectively. We would like our generalized cross ratio to be able to recover this highlight in some way. In order to achieve this, we first have to extend our definition. After all, if we let $z = z^1 = (z_1, z_2)$, then the previous expression for $(z^1, z^2, z^3, z^4, z^5)$ does not give us a linear fractional map as given in Definition 2.1. Such a map should map $\mathbb{C}^2$ to $\mathbb{C}^2$, not $\mathbb{C}^2$ to $\mathbb{C}$! Thus we define the cross ratio pair.

DEFINITION 3.2. We define the cross ratio pair in $\mathbf{CP}^2$ as

$$(u^1, u^2, u^3, u^4, u^5)_2 = \left(\frac{[u^1, u^3, u^5][u^2, u^4, u^5]}{[u^1, u^4, u^5][u^2, u^3, u^5]}, \frac{[u^1, u^3, u^4][u^2, u^4, u^5]}{[u^1, u^4, u^5][u^2, u^3, u^4]}\right),$$

where we see that this defines a linear fractional map when the point associated

[1] This is just Equation 1.1 with superscript notation instead.

with u^1 is a variable in $\mathbb{C}^2$.

Motivated by the results of Definition 3.2, we define a generalized cross ratio. Although the notation is a bit more intimidating, this definition is nothing more than the natural extension of the two variable case.

DEFINITION 3.3. Given $N+3$ distinct points u^i for $i = 1, \ldots, N+3$ in $\mathbf{CP}^N$, we define the cross ratio as

$$(u^1, u^2, u^3, u^4, u^5, \ldots, u^{N+3}) = \frac{[u^1, u^3, u^5, \ldots, u^{N+3}][u^2, u^4, u^5, \ldots, u^{N+3}]}{[u^1, u^4, u^5, \ldots, u^{N+3}][u^2, u^3, u^5, \ldots, u^{N+3}]},$$

where each ellipsis represents the sequence $\{u_k\}_{k=6}^{N+2}$.

To simplify the definition of the cross ratio N-tuple, we introduce new notation. We define $[u^i, u^j]_N^c$ by

$$[u^i, u^j]_N^c = \left[\{u^k\}_{k=1}^{N+3}\right] \setminus \{u_i, u_j\} = [u^1, \ldots, u^{i-1}, u^{i+1}, \ldots, u^{j-1}, u^{j+1}, \ldots, u^{N+3}].$$

This sets the stage for our general definition. If the notation appears intimidating, just remind yourself that it is the natural generalization of the definition of the cross ratio pair in $\mathbf{CP}^2$.

DEFINITION 3.4. We define the cross ratio N-tuple to be given by

$$(u^1, u^2, \ldots, u^{N+2}, u^{N+3})_N = \left(\left\{ \frac{[u^2, u^i]_N^c [u^1, u^3]_N^c}{[u^2, u^3]_N^c [u^1, u^i]_N^c} \right\}_{i=4,\ldots,N+3} \right),$$

where the curly brackets denote a sequence over $i = 4, \ldots, N+3$.

As in the two variable case, this defines a linear fractional map when the point associated with u^1 is a variable in $\mathbb{C}^N$.

We saw that we could "reduce" our cross ratio in two variables to the one variable definition. The next theorem tells us that this works in any dimension. The reasoning, again, is a natural extension of what was demonstrated in two variables.

THEOREM 3.5. *Identifying coordinates* $z_j^i = \frac{u_j^i}{u_j^{N+1}}$ *for* $j = 1, \ldots, N$ *and* $u^{N+3} = (0, 0, \ldots, 0, 1, 0)^T$, *the cross ratio in* $\mathbb{C}^N$ *reduces to the cross ratio in* $\mathbb{C}^{N-1}$.

PROOF. We use proof by induction. We have seen that our result is true for $N = 2$. Now suppose $N = k+1$ where $k > 1$, making the appropriate identifications, we have

$$
\begin{aligned}
(u^1, u^2, \ldots, u^{(k+1)+3}) &= \frac{[u^1, u^3, u^5, \ldots, u^{k+4}][u^2, u^4, u^5, \ldots, u^{k+4}]}{[u^1, u^4, u^5, \ldots, u^{k+4}][u^2, u^3, u^5, \ldots, u^{k+4}]} \\
&= \frac{\det\begin{pmatrix} u_1^1 & u_1^3 & u_1^5 & \cdots & u_1^{k+4} \\ u_2^1 & u_2^3 & u_2^5 & \cdots & u_2^{k+4} \\ \vdots & \vdots & \vdots & \ddots & \vdots \\ u_{k+2}^1 & u_{k+2}^3 & u_{k+2}^5 & \cdots & u_{k+2}^{k+4} \end{pmatrix} \det\begin{pmatrix} u_1^2 & u_1^4 & u_1^5 & \cdots & u_1^{k+4} \\ u_2^2 & u_2^4 & u_2^5 & \cdots & u_2^{k+4} \\ \vdots & \vdots & \vdots & \ddots & \vdots \\ u_{k+2}^2 & u_{k+2}^4 & u_{k+2}^5 & \cdots & u_{k+2}^{k+4} \end{pmatrix}}{\det\begin{pmatrix} u_1^1 & u_1^4 & u_1^5 & \cdots & u_1^{k+4} \\ u_2^1 & u_2^4 & u_2^5 & \cdots & u_2^{k+4} \\ \vdots & \vdots & \vdots & \ddots & \vdots \\ u_{k+2}^1 & u_{k+2}^4 & u_{k+2}^5 & \cdots & u_{k+2}^{k+4} \end{pmatrix} \det\begin{pmatrix} u_1^2 & u_1^3 & u_1^5 & \cdots & u_1^{k+4} \\ u_2^2 & u_2^3 & u_2^5 & \cdots & u_2^{k+4} \\ \vdots & \vdots & \vdots & \ddots & \vdots \\ u_{k+2}^2 & u_{k+2}^3 & u_{k+2}^5 & \cdots & u_{k+2}^{k+4} \end{pmatrix}} \\
&= \frac{\det\begin{pmatrix} z_1^1 & z_1^3 & z_1^5 & \cdots & 0 \\ z_2^1 & z_2^3 & z_2^5 & \cdots & 0 \\ \vdots & \vdots & \vdots & \ddots & \vdots \\ z_{k+1}^1 & z_{k+1}^3 & z_{k+1}^5 & \cdots & 1 \\ 1 & 1 & 1 & \cdots & 0 \end{pmatrix} \det\begin{pmatrix} z_1^2 & z_1^4 & z_1^5 & \cdots & 0 \\ z_2^2 & z_2^4 & z_2^5 & \cdots & 0 \\ \vdots & \vdots & \vdots & \ddots & \vdots \\ z_{k+1}^2 & z_{k+1}^4 & z_{k+1}^5 & \cdots & 1 \\ 1 & 1 & 1 & \cdots & 0 \end{pmatrix}}{\det\begin{pmatrix} z_1^1 & z_1^4 & z_1^5 & \cdots & 0 \\ z_2^1 & z_2^4 & z_2^5 & \cdots & 0 \\ \vdots & \vdots & \vdots & \ddots & \vdots \\ z_{k+1}^1 & z_{k+1}^4 & z_{k+1}^5 & \cdots & 1 \\ 1 & 1 & 1 & \cdots & 0 \end{pmatrix} \det\begin{pmatrix} z_1^2 & z_1^3 & z_1^5 & \cdots & 0 \\ z_2^2 & z_2^3 & z_2^5 & \cdots & 0 \\ \vdots & \vdots & \vdots & \ddots & \vdots \\ z_{k+1}^2 & z_{k+1}^3 & z_{k+1}^5 & \cdots & 1 \\ 1 & 1 & 1 & \cdots & 0 \end{pmatrix}} \\
&= \frac{\det\begin{pmatrix} z_1^1 & z_1^3 & z_1^5 & \cdots & z_1^{k+3} \\ z_2^1 & z_2^3 & z_2^5 & \cdots & z_2^{k+3} \\ \vdots & \vdots & \vdots & \ddots & \vdots \\ z_k^1 & z_k^3 & z_k^5 & \cdots & z_k^{k+3} \\ 1 & 1 & 1 & \cdots & 1 \end{pmatrix} \det\begin{pmatrix} z_1^2 & z_1^4 & z_1^5 & \cdots & z_1^{k+3} \\ z_2^2 & z_2^4 & z_2^5 & \cdots & z_2^{k+3} \\ \vdots & \vdots & \vdots & \ddots & \vdots \\ z_k^2 & z_k^4 & z_k^5 & \cdots & z_k^{k+3} \\ 1 & 1 & 1 & \cdots & 1 \end{pmatrix}}{\det\begin{pmatrix} z_1^1 & z_1^4 & z_1^5 & \cdots & z_1^{k+3} \\ z_2^1 & z_2^4 & z_2^5 & \cdots & z_2^{k+3} \\ \vdots & \vdots & \vdots & \ddots & \vdots \\ z_k^1 & z_k^4 & z_k^5 & \cdots & z_k^{k+3} \\ 1 & 1 & 1 & \cdots & 1 \end{pmatrix} \det\begin{pmatrix} z_1^2 & z_1^3 & z_1^5 & \cdots & z_1^{k+3} \\ z_2^2 & z_2^3 & z_2^5 & \cdots & z_2^{k+3} \\ \vdots & \vdots & \vdots & \ddots & \vdots \\ z_k^2 & z_k^3 & z_k^5 & \cdots & z_k^{k+3} \\ 1 & 1 & 1 & \cdots & 1 \end{pmatrix}} \\
&= (z^1, z^2, \ldots, z^{k+3}).
\end{aligned}
$$

□

We next show that this definition satisfies the invariant properties we would expect a cross ratio to have. Recall that for a nonzero parameter λ, the points $(\lambda u, \lambda v)$ and (u, v) represent the same point in $\mathbf{CP}^1$ and are associated with the same point in the complex plane. Which representative we choose, however, doesn't matter.

THEOREM 3.6. *The cross ratio is independent of the representative chosen.*

PROOF. Given $N+3$ nonzero parameters $\{\lambda_i\}_{i=1,\ldots,N+3}$, we have

$$\begin{aligned}(\lambda_1 u^1, \lambda_2 u^2, \ldots, \lambda_{N+2} u^{N+2}, \lambda_{N+3} u^{N+3})\\ &= \frac{[\lambda_1 u^1, \lambda_3 u^3, \lambda_5 u^5, \ldots, \lambda_{N+3} u^{N+3}][\lambda_2 u^2, \lambda_4 u^4, \lambda_5 u^5, \ldots, \lambda_{N+3} u^{N+3}]}{[\lambda_1 u^1, \lambda_4 u^4, \lambda_5 u^5, \ldots, \lambda_{N+3} u^{N+3}][\lambda_2 u^2, \lambda_3 u^3, \lambda_5 u^5, \ldots, \lambda_{N+3} u^{N+3}]}\\ &= \frac{\lambda_1\lambda_2\lambda_3\lambda_4\lambda_5^2 \ldots \lambda_{N+3}^2 [u^1, u^3, u^5, \ldots, u^{N+3}][u^2, u^4, u^5, \ldots, u^{N+3}]}{\lambda_1\lambda_2\lambda_3\lambda_4\lambda_5^2 \ldots \lambda_{N+3}^2 [u^1, u^4, u^5, \ldots, u^{N+3}][u^2, u^3, u^5, \ldots, u^{N+3}]}\\ &= \frac{[u^1, u^3, u^5, \ldots, u^{N+3}][u^2, u^4, u^5, \ldots, u^{N+3}]}{[u^1, u^4, u^5, \ldots, u^{N+3}][u^2, u^3, u^5, \ldots, u^{N+3}]} = (u^1, u^2, \ldots, u^{N+2}, u^{N+3}).\end{aligned}$$

□

By the same reasoning it is easy to see that the cross ratio N-tuple is also independent of the representative chosen.

We next show that this cross ratio is invariant under associated matrices of linear fractional maps.

THEOREM 3.7. *Let m_ϕ be the associated matrix to an invertible linear fractional map ϕ; that is,* $\det m_\phi \neq 0$. *Then the cross ratio is invariant under m_ϕ.*

PROOF. Note since $[Av_1, Av_2, \ldots, Av_N] = \det(A)[v_1, v_2, \ldots, v_N]$, we have

$$[m_\phi u^1, m_\phi u^2, \ldots, m_\phi u^k] = \det(m_\phi)[u^1, u^2, \ldots, u^k].$$

Thus

$$\begin{aligned}(m_\phi u^1, m_\phi u^2, \ldots, m_\phi u^{N+3})\\ &= \frac{[m_\phi u^1, m_\phi u^3, m_\phi u^5, \ldots, m_\phi u^{N+3}][m_\phi u^2, m_\phi u^4, m_\phi u^5, \ldots, m_\phi u^{N+3}]}{[m_\phi u^1, m_\phi u^4, m_\phi u^5, \ldots, m_\phi u^{N+3}][m_\phi u^2, m_\phi u^3, m_\phi u^5, \ldots, m_\phi u^{N+3}]}\\ &= \frac{\det(m_\phi)^2 [u^1, u^3, u^5, \ldots, u^{N+3}][u^2, u^4, u^5, \ldots, u^{N+3}]}{\det(m_\phi)^2 [u^1, u^4, u^5, \ldots, u^{N+3}][u^2, u^3, u^5, \ldots, u^{N+3}]}\\ &= \frac{[u^1, u^3, u^5, \ldots, u^{N+3}][u^2, u^4, u^5, \ldots, u^{N+3}]}{[u^1, u^4, u^5, \ldots, u^{N+3}][u^2, u^3, u^5, \ldots, u^{N+3}]} = (u^1, u^2, \ldots, u^{N+2}, u^{N+3})\end{aligned}$$

where we can divide by $\det(m_\phi)$ since we presume m_ϕ is invertible. □

4. Transitivity of linear fractional maps in several complex variables

It is well known that linear fractional maps acting on $\mathbb{C}$ are 3-transitive but not 4-transitive. That is, for six points z^1, z^2, z^3, w^1, w^2, and w^3 in $\mathbb{C}$ where the z^i's are pairwise distinct as are the w^i's, there is a unique linear fractional map χ such that $\chi(z^i) = w^i$ for $i = 1, 2, 3$. The usual way to achieve this result is to send the triple (z^1, z^2, z^3) to the standardized points $(1, 0, \infty)$ where ∞ represents the "north pole" on the Riemann sphere. In homogeneous coordinates, these standardized points are associated with $\begin{pmatrix}1\\1\end{pmatrix}$, $\begin{pmatrix}0\\1\end{pmatrix}$, and $\begin{pmatrix}1\\0\end{pmatrix}$, respectively. Thus, for the cross ratio as a function of z with associated point u given by $\phi(z) = \frac{(z-z^2)(z^1-z^3)}{(z-z^3)(z^1-z^2)}$, we may write the point associated with $\phi(z)$ as

$$m_\phi u = \begin{pmatrix} (z-z^2)(z^1-z^3) \\ (z-z^3)(z^1-z^2) \end{pmatrix}.$$

It is clear that this sends our triple (z^1, z^2, z^3) to the points $(1, 0, \infty)$, respectively. Note that taking two distinct points z^1 and z^2 in $\mathbb{C}$ implies that the associated points in $\mathbf{CP}^1$, when seen as vectors in $\mathbb{C}^2$, are linearly independent. However, it is not true that taking three distinct points in $\mathbb{C}^N$ implies the associated points, when seen as vectors in $\mathbb{C}^{N+1}$, will be linearly independent. They could be collinear, for example. This potential obstruction will necessitate an additional assumption of linear independence that was not needed in $\mathbb{C}$.

We would like our cross ratio N-tuple to be a generalized formula for a multi-dimensional linear fractional map that maps $N+2$ points to certain pre-specified points analogous to the triple $(1, 0, \infty)$. To achieve this, we consider an $N+2$-tuple in which $N+1$ of the points have associated points in $\mathbf{CP}^N$ which, when seen as vectors in $\mathbb{C}^{N+1}$, are linearly independent. We then send these points to the standard basis vectors $\{e_i\}_{i=1}^{N+1}$ and the remaining point is sent to $\sum_{i=1}^{N+1} e_i$. To begin, we see how to do this in $\mathbb{C}^2$ and generalize.

Let $u^i \in \mathbf{CP}^2$ be associated with $z^i \in \mathbb{C}^2$ for $i = 2, \ldots, 5$ and u associated with z. Define the linear fractional map $\phi : \mathbb{C}^2 \to \mathbb{C}^2$ by

$$\begin{aligned} \phi(z) &= (z, z^2, z^3, z^4, z^5)_2 = (u, u^2, u^3, u^4, u^5)_2 \\ &= \left(\frac{[u, u^3, u^5][u^2, u^4, u^5][u^2, u^3, u^4]}{[u, u^4, u^5][u^2, u^3, u^4][u^2, u^3, u^5]}, \frac{[u, u^3, u^4][u^2, u^3, u^5][u^2, u^4, u^5]}{[u, u^4, u^5][u^2, u^3, u^4][u^2, u^3, u^5]} \right). \end{aligned}$$

If the point $u \in \mathbf{CP}^2$ is associated with $z \in \mathbb{C}^2$, then the point $v \in \mathbf{CP}^2$ associated with $w = \phi(z) \in \mathbb{C}^2$ is given by

$$v = m_\phi u = \begin{pmatrix} [u, u^3, u^5][u^2, u^4, u^5][u^2, u^3, u^4] \\ [u, u^3, u^4][u^2, u^3, u^5][u^2, u^4, u^5] \\ [u, u^4, u^5][u^2, u^3, u^4][u^2, u^3, u^5] \end{pmatrix}. \tag{4.1}$$

We recall that $[u^i, u^j, u^k] = 0$ when any of i, j, k are equal or, more generally, when there is linear dependence. To keep the previous cross ratio map well-defined, we introduce the following hypothesis.

DEFINITION 4.1. Let the set $\{u^i\}_{i=3}^{N+3}$ in $\mathbb{C}^{N+1}$ form a linearly independent set. We make the assumption that $u^2 = \sum_{i=0}^{N} \alpha_i u^{i+3}$ where $\{\alpha_i\}_{i=0}^{N}$ are all non-zero complex numbers. We will call this *the independence hypothesis.*

Considering the N-dimensional space $\left(\{\alpha_i\}_{i=0}^{N}\right)$, the independence hypothesis asks only that we omit a set of measure zero. Additionally, If a set of points in $\mathbf{CP}^N$, when viewed as vectors in $\mathbb{C}^{N+1}$ satisfies the independence hypothesis, we say the set of points in $\mathbf{CP}^N$ satisfies the independence hypothesis.

With this in mind, it is clear to see that we have $m_\phi u^2 = (1, 1, 1)$, $m_\phi u^3 = (0, 0, 1)$, $m_\phi u^4 = (1, 0, 0)$, and $m_\phi u^5 = (0, 1, 0)$ as desired where $(1, 1, 1)$ is associated with $(1, 1)$, $(0, 0, 1)$ with $(0, 0)$, and vectors such as $(1, 0, 0)$, which we will denote by $e_{1,\infty}$, and $(0,1,0)$, which we will denote by $e_{2,\infty}$, correspond to the points in the hyperplane at infinity that are tangent to $(0,1)$ and $(1,0)$, respectively. In general, our goal will then be to map $N+2$ points associated with $N+2$ points in $\mathbf{CP}^N$ which satisfy the independence hypothesis to a set of $N+2$ standardized points in $\mathbb{C}^N$ with associated points $\sum_{k=1}^{N+1} e_k$ and $\{e_k\}_{k=1}^{N+1}$ where e_i is the vector that is zero on every coordinate except the ith coordinate where it takes the value of 1.

THEOREM 4.2. *Given $N+2$ distinct elements z^2, z^3,...,z^{N+3} in $\mathbb{C}^N$ with respective associated points u^2, u^3,...,u^{N+3} in $\mathbf{CP}^N$ that satisfy the independence hypothesis, there exists a unique linear fractional map ϕ such that $m_\phi u^2 = (1,1,\ldots,1) = \sum_{k=1}^{N+1} e_k$, $m_\phi u^3 = e_{N+1}$, $m_\phi u^4 = e_1$, $m_\phi u^5 = e_2$, and $m_\phi u^i = e_{i-3}$ for $i > 5$.*

PROOF. Define the linear fractional map ϕ by

$$\begin{aligned}\phi(z) &= (z^1, z^2, \ldots, z^{N+2}, z^{N+3})_N \\ &= (u^1, u^2, \ldots, u^{N+2}, u^{N+3})_N = \left(\left\{\frac{[u^2,u^i]_N^c[u,u^3]_N^c}{[u^2,u^3]_N^c[u,u^i]_N^c}\right\}_{i=4,\ldots,N+3}\right).\end{aligned}$$

We may then write the point associated with $\phi(z)$ as

$$v = m_\phi u = \begin{pmatrix} \frac{[u^2,u^4]_N^c}{[u,u^4]_N^c} & \frac{[u^2,u^5]_N^c}{[u,u^5]_N^c} & \cdots & \frac{[u^2,u^3]_N^c}{[u\cdot u^3]_N^c}\end{pmatrix}^T.$$

We recall that $[u^i,u^j]_N^c = [u,\ldots,u^{i-1},u^{i+1},\ldots,u^{j-1},u^{j+1},\ldots u^{N+3}] = 0$ when $u = u^k$ for some integer $1 < k < N+3$, $k \neq i,j$. In particular, note that none of the denominators in the entries of v are zero. Thus we have $m_\phi u^2 = \sum_{k=1}^{N+1} e_k$, $m_\phi u^3 = e_{N+1}$, $m_\phi u^4 = e_1$, $m_\phi u^5 = e_2$, and $m_\phi u^k = e_{k-3}$.

We next demonstrate uniqueness. Suppose there are two linear fractional maps ϕ and ψ such that the associated matrices m_ϕ and m_ψ send each u^i as previously defined. Then the associated matrix $M = m_\phi m_\psi^{-1}$ fixes e_i and $p = \sum_{i=1}^N e_i$. Denote M by

$$M = \begin{pmatrix} a_{11} & a_{12} & a_{13} & \cdots & a_{1N} \\ a_{21} & a_{22} & a_{23} & \cdots & a_{2N} \\ \vdots & \vdots & \vdots & \ddots & \vdots \\ a_{N1} & a_{N2} & a_{N3} & \cdots & a_{NN}\end{pmatrix}.$$

Then $Me_i = e_i$ implies $a_{ij} = 0$ for $i \neq j$ and $Mp = p$ implies $a_{ii} = a_{jj}$ for all $1 \leq i,j \leq N$. We conclude that $M = a_{11}I$ where I is the identity which implies $m_\phi = a_{11}m_\psi$ and thus $\phi = \psi$. □

To generalize in several variables, we use transitivity in a more restricted sense. In particular, we make the additional assumption that the initial data points in $\mathbb{C}^N$ have associated points in $\mathbf{CP}^N$ that satisfy the independence hypothesis. We next show that, in this sense of transitivity, linear fractional maps in $\mathbb{C}^N$ are $(N+2)$-transitive but not $(N+3)$-transitive.

THEOREM 4.3. *Given $N+2$ pairwise distinct points $\{z^i\}_{i=2}^{N+3}$ in $\mathbb{C}^N$ with associated points $\{u^i\}_{i=2}^{N+3}$ and $N+2$ pairwise distinct points $\{w^i\}_{i=2}^{N+3}$ in $\mathbb{C}^N$ with associated points $\{v^i\}_{i=2}^{N+3}$ such that the sets $\{u^i\}_{i=2}^{N+3}$ and $\{v^i\}_{i=2}^{N+3}$ satisfy the independence hypothesis, there exists a unique linear fractional map $\chi(z)$ such that $\chi(z^i) = w^i$ for $i = 2, \ldots, N+3$.*

PROOF. By Theorem 4.2, there exist maps ϕ and ψ such that $m_\phi u^2 = m_\psi v^2 = \sum_{k=1}^{N+1} e_k$, $m_\phi u^3 = m_\psi v^3 = e_{N+1}$, $m_\phi u^4 = m_\psi v^4 = e_1$, $m_\phi u^5 = m_\psi v^5 = e_2$, and $m_\phi u^k = m_\psi v^k = e_{k-3}$ otherwise. Define $m_\gamma = m_\psi^{-1} m_\phi$. This sends u^i to v^i for $i = 2, \ldots, N+3$. To show uniqueness, suppose m_σ maps u^i to v^i for $i = 2, \ldots, N+3$. Then m_ϕ and $m_\psi m_\sigma$ each send u^i to the same values for $i = 2, \ldots, N+3$. Thus by Theorem 4, we have $m_\phi = m_\psi m_\sigma$ which implies $m_\gamma = m_\psi^{-1} m_\phi = m_\sigma$ from which we conclude that $\gamma = \sigma$ as desired. □

THEOREM 4.4. *If m_ϕ fixes $N+2$ distinct points $\{u^i\}_1^{N+2}$ satisfying the independence hypothesis, then ϕ is the identity.*

PROOF. Suppose that m_ϕ fixes $N+2$ distinct points $\{u^i\}_1^{N+2}$ satisfying the independence hypothesis, then, since the identity also fixes these points, by Theorem 5 we have that m_ϕ is equal to the identity and thus ϕ is the identity. □

We can thus conclude that linear fractional maps in $\mathbb{C}^N$ are $(N+2)$-transitive but they are not $(N+3)$-transitive.

5. More to explore

The author invites the reader to investigate this generalized definition further. For example, in one variable, we have a well-known criterion to determine when the cross ratio is real.

THEOREM 5.1. *The cross ratio (z^1, z^2, z^3, z^4) is real if and only if the four points lie on a circle (where we let a line be a circle of infinite radius).*

PROOF. Consider the cross ratio as a function of the first argument, $\phi(z) = (z, z^1, z^2, z^3)$. We saw that ϕ maps the triple (z^1, z^2, z^3) to $(1, 0, \infty)$, respectively. Since linear fractional maps send circles to circles, we conclude that ϕ sends our circle containing the triple (z^1, z^2, z^3) to the extended real line. Now consider the map $\phi^{-1}(\phi(z)) = z$. It is known that linear fractional maps are bijective and thus $\phi(z)$ is real if and only if $z = \phi^{-1}(\phi(z))$ lies on the circle containing our triple (z^1, z^2, z^3) which is the image of the extended real line under our map ϕ^{-1}. □

In order for this theorem to extend to our generalized cross ratio N-tuple, it will take an upgrade of some geometric facts about these generalized maps. We appeal to results by Cowen and MacCluer (see [**1**], Section 3 for more details.). In particular, one finds more flexibility when working in higher dimensions. The appropriate generalization of circles in this situation is to consider *ellipsoids*, where an ellipsoid is a translate of the image of the unit ball under an invertible complex linear transformation. We recall Theorem 6 from [**1**].

THEOREM 5.2. *If ϕ is a one-to-one linear fractional map defined on a ball $\overline{\mathbb{B}}$ in $\mathbb{C}^N$, then $\phi(\mathbb{B})$ is an ellipsoid.*

PROOF. See [**1**], Section 3. □

This brings us to the following open question, left for the reader to demonstrate.

QUESTION 5.3. Is there an analogue to Theorem 5.1 for the generalized cross ratio if we relax the presumption of points lying on the boundary of a sphere and instead use the boundary of an ellipsoid?

In some form or another, the cross ratio and its special invariant properties have been examined and celebrated since antiquity. Now that we have a new perspective to talk about this quantity in higher dimensions, the author challenges the reader to explore what other similarities or differences the generalized version of this classical quantity has.

References

[1] C. C. Cowen and B. D. MacCluer, *Linear fractional maps of the ball and their composition operators*, Acta Sci. Math. (Szeged) **66** (2000), no. 1-2, 351–376. MR1768872

[2] S. Gong and C. H. FitzGerald, *The Schwarzian derivative in several complex variables*, Sci. China Ser. A **36** (1993), no. 5, 513–523. MR1228850

[3] F. Klein, *A comparative review of recent researches in geometry*, Bull. Amer. Math. Soc. **2** (1893), no. 10, 215–249, DOI 10.1090/S0002-9904-1893-00147-X. MR1557253

[4] T. Needham, *Visual complex analysis*, The Clarendon Press, Oxford University Press, New York, 1997. MR1446490

[5] C. L. Siegel, *Symplectic geometry*, Amer. J. Math. **65** (1943), 1–86, DOI 10.2307/2371774. MR8094

DEPARTMENT OF MATHEMATICAL SCIENCES, BALL STATE UNIVERSITY, MUNCIE, INDIANA 47405

Email address: michael.pilla@bsu.edu

Contemporary Mathematics
Volume **799**, 2024
https://doi.org/10.1090/conm/799/16020

Weighted estimates of the Bergman projection with matrix weights

Zhenghui Huo and Brett D. Wick

ABSTRACT. We establish a weighted inequality for the Bergman projection with matrix weights for a class of pseudoconvex domains. We extend a result of Aleman and Constantin and obtain the following estimate for the weighted norm of P:

$$\|P\|_{L^2(\Omega,W)} \leq C(\mathcal{B}_2(W))^2.$$

Here $\mathcal{B}_2(W)$ is the Bekollé–Bonami constant for the matrix weight W and C is a constant that is independent of the weight W but depends upon the dimension and the domain.

1. Introduction

The purpose of this paper is to investigate weighted inequalities with matrix weights for the Bergman projection on certain class of pseudoconvex domains.

In harmonic analysis, weighted inequalities characterize weighted spaces on which singular integral operators remain bounded and describe the norm dependence on the weights. For the case of the scalar-valued function with scalar weights, it is well-known that for $1 < p < \infty$, a Calderon–Zygmund singular integral operator is bounded over the weighted space $L^p(\mathbb{R}^N, W)$ with the weight function W satisfying the Muckenhoupt $\mathcal{A}_p$ condition, i.e., the $\mathcal{A}_p$ constant

$$\mathcal{A}_p(W) := \sup_B \langle W\rangle_B^{\frac{1}{p}} \langle W^{-\frac{1}{p-1}}\rangle_B^{\frac{1}{p'}} < \infty.$$

Here the supremum is taken over all Euclidean balls B in $\mathbb{R}^N$ and $\langle W\rangle_B$ denotes the average of W over the set B with respect to the Lebesgue measure. See for example [**HMW73**].

As to the dependence of the norm of the operator on $\mathcal{A}_p(W)$, extensive studies have been made in recent decades. For $p = 2$, the $\mathcal{A}_2$ conjecture speculates that the dependence is linear. The conjecture was solved by Wittwer for the martingale transform [**Wit00**], by Petermichl–Volberg for the Beurling operator, by Petermichl for the Hilbert transform in [**Pet07**], and then the general Calderon–Zygmund

2020 *Mathematics Subject Classification.* Primary 32A25, 32A36, 32A50, 42B20, 42B35.
Key words and phrases. Bergman projection, Bergman kernel, weighted inequality.
The first author's research was partially supported by National Science Foundation of China NSFC Grant # 12201265. The second author's research was partially supported by National Science Foundation grants DMS # 1800057, # 2054863, and # 2000510 and Australian Research Council DP 220100285.

operators in [**Hyt12**]. More elementary proofs were also obtained by Lerner [**Ler13**] and Lacey [**Lac17**].

One direction to extend weighted theory considers the setting of weighted L^p space of vector-valued functions with matrix-valued weight W. Let Ω be a domain. Let W be a locally integrable function on Ω with its range in the set of positive-semidefinite $d \times d$ matrices. The weighted space $L^2(\Omega, W)$ is the space of vector-valued measurable functions $f : \Omega \to \mathbb{C}^d$ for which

$$\|f\|^2_{L^2(\Omega,W)} := \int_\Omega \langle W(z)f(z), f(z)\rangle dV(z) < \infty;$$

where $\langle \cdot, \cdot \rangle$ is the usual inner product of $\mathbb{C}^d$.

For the vector-valued function setting with matrix weights, Treil and Volberg [**TV97a, TV97b**] first gave a matrix $\mathcal{A}_2$ condition and showed it is a necessary and sufficient for the boundedness of the Hilbert transform [**TV97a, TV97b**]. Then Nazarov–Treil [**NT96**] and Volberg [**Vol97**] separately generalized this result and established the boundedness of classical Calderon–Zygmund operators for matrix $\mathcal{A}_p$ weights. Christ–Goldberg [**CG01**] and Goldberg [**Gol03**] studied a class of weighted, vector analogues of the Hardy–Littlewood maximal function and used them to establish the boundedness of a class of singular integral operators on $L^p(W)$, where W is a matrix $\mathcal{A}_p$ weight. Recently, progress has also been made on norm dependence. Bickel–Wick [**BW16**] and Isralowitz–Kwon–Pott [**IKP17**] separately showed the following estimates for a sparse operator S:

$$\|S\|_{L^2(W)} \lesssim \mathcal{A}_2(W)^{\frac{3}{2}}.$$

Bickel–Petermichl–Wick [**BPW16**] obtained weighted L^2 estimates for the Hilbert transform with the norm bound $\mathcal{A}_2(W)^{\frac{3}{2}} \log(1 + \mathcal{A}_2(W))$. Nazarov–Petermichl–Treil–Volberg [**NPTV17**] then improved their result and obtained a better bound $\mathcal{A}_2(W)^{\frac{3}{2}}$ for Calderon-Zygmund operators as well as Haar shifts and paraproducts. The sharp bound was established to some particular cases such as sparse operators with a simple sparse family by Nazarov–Petermichl–Treil–Volberg [**NPTV17**] and the dyadic square function by Hytönen–Petermichl–Volberg [**HPV19**]. It is worth noting that a crucial ingredient in some of these results is the reverse Hölder inequality for a scalar $\mathcal{A}_2$ weight.

Let $\Omega \subseteq \mathbb{C}^n$ be a bounded domain. Let dV denote the Lebesgue measure. The Bergman projection P is the orthogonal projection from $L^2(\Omega)$ onto the Bergman space $A^2(\Omega)$, the space of all square-integrable holomorphic functions. Associated with P, there is a unique function K_Ω on $\Omega \times \Omega$ such that for any $f \in L^2(\Omega)$:

$$P(f)(z) = \int_\Omega K_\Omega(z; \bar{w}) f(w) dV(w). \tag{1.1}$$

Let P^+ denote the positive Bergman operator defined by:

$$P^+(f)(z) := \int_\Omega |K_\Omega(z; \bar{w})| f(w) dV(w). \tag{1.2}$$

In [**BB78, Bek82**], Bekollé and Bonami established the analogue of Muckenhoupt $\mathcal{A}_p$ condition in the Bergman setting:

THEOREM 1.1 (Bekollé–Bonami). *Let T_z denote the Carleson tent over z in $\mathbb{B}_n$ defined as:*

- $T_z := \left\{ w \in \mathbb{B}_n : \left|1 - \bar{w}\frac{z}{|z|}\right| < 1 - |z| \right\}$ *for $z \neq 0$, and*
- $T_z := \mathbb{B}_n$ *for $z = 0$.*

Let the weight W be a positive, locally integrable function on $\mathbb{B}_n$. Let $1 < p < \infty$. Then the following conditions are equivalent:

(1) $P : L^p(\mathbb{B}_n, W) \to L^p(\mathbb{B}_n, W)$ *is bounded;*
(2) $P^+ : L^p(\mathbb{B}_n, W) \to L^p(\mathbb{B}_n, W)$ *is bounded;*
(3) *The Bekollé–Bonami constant $\mathcal{B}_p(W)$ is finite where:*

$$\mathcal{B}_p(W) := \sup_{z\in\mathbb{B}_n} \langle W\rangle_{T_z}^{\frac{1}{p}} \langle W^{-\frac{1}{p-1}}\rangle_{T_z}^{\frac{1}{p'}}.$$

People have made progress on the dependence of the operator norm $\|P\|_{L^p(\mathbb{B}_n,W)}$ on $\mathcal{B}_p(W)$. In [**PR13**], Pott and Reguera gave a weighted L^p estimate for the Bergman projection on the upper half plane. Their estimates are in terms of the Bekollé–Bonami constant and the upper bound is sharp. Later, Rahm, Tchoundja, and Wick [**RTW17**] generalized the results of Pott and Reguera to the unit ball case, and also obtained estimates for the Berezin transform. Sharp weighted norm estimates of the Bergman projection have been obtained [**HW20**] on the Hartogs triangle and a broad class of pseudoconvex domains [**HWW21a, HWW21b, GHK22**].

Unlike Muckenhoupt $\mathcal{A}_2$ weights, the reverse Hölder inequality is generally not available for Békollé–Bonami weights, thereby making weighted inequalities for the Bergman projection harder to prove in the matrix weight setting. Nevertheless, Aleman and Constantin [**AC12**] extended Bekollé and Bonami's result on the unit disc to operator-valued weights for $p = 2$. They showed that a $\mathcal{B}_2$ condition for operator weights determines the boundedness of the Bergman projection in vector-valued L^2-spaces with operator-valued weights as opposed to just matrix valued weights (i.e., they consider the action on an infinite dimensional Hilbert space as opposed to just $\mathbb{C}^d$). They also obtained a weighted norm estimate for the projection:

$$\|P\|_{L^2(\mathbb{D},W)} \lesssim \mathcal{B}_2^{5/2}(W).$$

In their approach, they related the norms of analytic functions in weighted Bergman spaces to weighted norms of their derivatives, which relied on various properties of the unit disc $\mathbb{D}$.

In this paper, we consider the weighted estimates of the Bergman projection in the matrix weight case. We follow the settings from [**HWW21a, HWW21b**] and the domain Ω we consider belong to one of the following classes:

- a bounded, smooth, pseudoconvex domain of finite type in $\mathbb{C}^2$,
- a bounded, smooth, strictly pseudoconvex domain in $\mathbb{C}^n$,
- a bounded, smooth, convex domain of finite type in $\mathbb{C}^n$, or
- a bounded, smooth, decoupled domain of finite type in $\mathbb{C}^n$.

Such a Ω is referred to as a "simple domain" by McNeal in [**McN03**]. We will use the same terminology in this paper.

Let Ω be a simple domain. The main result can be summarized as follows:

THEOREM 1.2. *Let W be a matrix $\mathcal{B}_2$ weight on Ω. Let $\mathcal{B}_2(W)$ denote the matrix $\mathcal{B}_2$ constant given by*

$$\mathcal{B}_2(W) := \sup_{B\in\mathscr{B}} \left|\langle W\rangle_B^{1/2}\langle W^{-1}\rangle_B^{1/2}\right|^2,$$

where $|\cdot|$ *denotes the norm of the matrix acting on* $\mathbb{C}^d$, $\langle W\rangle_B := \frac{\int_B W dV}{\int_B dV}$, *and* $\mathscr{B}$ *is a collection of tents in* Ω *that touch its boundary. Then we have*

$$\|P\|_{L^2(\Omega,W)} \le C(\mathcal{B}_2(W))^2.$$

The constant C *in the inequality only depends on the domain* Ω *and the dimension* d, *but not the weight* W.

This theorem extends and improves the estimates by Aleman and Constantin [**AC12**] by lowering the constant power of $\mathcal{B}_2(W)$ in the upper bound and showing it works for more domains. See Section 2.2 for the detailed definition of the tents of Ω and the collection $\mathscr{B}$.

Our approach is motivated by ideas from [**AC12, NPTV17, APR19**] and can be outlined as follows: Given a matrix $\mathcal{B}_2$ weight W, we construct finitely many step averaging weights $\mathcal{W}_l$ according to dyadic systems $\mathcal{T}_l$ and their sum $\tilde{\mathcal{W}} = \sum \mathcal{W}_l$ with the following three properties:

(1) $\sup_{\hat{K}_j^k\in\mathcal{T}_l} |\langle \mathcal{W}_l^{-1}\rangle_{\hat{K}_j^k}^{1/2}\langle \mathcal{W}_l\rangle_{\hat{K}_j^k}^{1/2}|^2 \lesssim \mathcal{B}_2(W)$ (Lemma 4.2),

(2) for $v \in \mathbb{C}^d$ and each $\mathcal{W}_l$, the scalar weight $\langle \mathcal{W}_l v, v\rangle$ satisfies the reverse Hölder inequality (Lemma 4.4),

(3) $\|P\|_{L^2(W)} \lesssim \mathcal{B}_2^{1/2}(W)\|P\|_{L^2(\tilde{\mathcal{W}})}$ (Theorem 4.5).

The construction of the step averaging weight $\mathcal{W}_l$ is inspired by the work of [**APR19**]. Property (3) uses a duality argument from [**AC12**] and relates the weighted norm of P on $L^2(W)$ to the ones on $L^2(\mathcal{W}_l)$. Using the known estimates about the Bergman kernel on simple domains, we show that the Bergman projection belongs to a convex body-valued dyadic operator [**NPTV17**]. Since the reverse Hölder is available for $\langle \mathcal{W}_l v, v\rangle$, we are able to establish weighted estimates using some scalar dyadic square operators.

After the first draft of this paper was posted, we received a draft by Limani and Pott [**LP21**] where they considered the Bergman projection on the upper half plane and obtained weighted estimates of the projection with operator-valued weights. Their results sharpen the exponent of 2 in the estimate of Theorem 1.2 to $\frac{3}{2}$.

Our paper is organized as follows: In Section 2, we recall the definitions and known results about the dyadic tent structure of Ω, estimates for the Bergman kernel function. In Section 3, we give the definition of the convex body-valued sparse operator and show the convex body domination for the Bergman projection. In Section 4, we introduce the weight $\mathcal{W}$ and go over its properties. In Section 5, we prove Theorem 1.2. We make several remarks for our results in Section 6.

Throughout the paper, n will be the dimension of the complex Euclidean space $\mathbb{C}^n$ that contains the domain Ω of vector-valued functions and d will be the dimension of $\mathbb{C}^d$ in which vector-valued functions take their range. Hence the matrix weights will have range in the set of $d \times d$ matrices. We use the notation $\|\cdot\|$ for the norm of operators/functions on function spaces, and use $|\cdot|$ for the norm of the matrix acting on $\mathbb{C}^d$ or the length of a vector in $\mathbb{C}^d$. Given functions of several variables f and g, we use $f \lesssim g$ to denote that $f \le Cg$ for a constant C. If $f \lesssim g$ and $g \lesssim f$, then we say f is comparable to g and write $f \approx g$.

2. Preliminaries

2.1. Balls on the boundary of a simple domain. Let Ω be a bounded domain in $\mathbb{C}^n$ with smooth boundary $\mathbf{b}\Omega$. Then there is a smooth function ρ satisfying $\Omega = \{z \in \mathbb{C}^n : \rho(z) < 0\}$ and $\nabla\rho \neq 0$ on $\mathbf{b}\Omega$. We call such a function ρ a *defining function* of Ω. Ω is a *pseudoconvex domain* if for any $p \in \mathbf{b}\Omega$, the complex Hessian $(\rho_{i\bar{j}})$ of the defining function is positive semi-definite on holomorphic tangent space $T_p^{(1,0)}(\mathbf{b}\Omega)$ at p. For $m > 0$, a point $p \in \mathbf{b}\Omega$ is of *finite type* in the sense of D'Angelo [**D'A82**] if the maximum order of contact of one-dimensional complex analytic varieties with $\mathbf{\Omega}$ at p equals m. We say a domain Ω is of *finite type* if every boundary point p is of finite type.

When Ω is a simple domain in $\mathbb{C}^n$, non-isotropic sets can be constructed using a special coordinate system near the boundary of Ω. See results of McNeal [**McN94b, McN91, McN03**]. Let $p \in \mathbf{b}\Omega$ be a point of finite type m. For a small neighborhood U of the point p, there exists a holomorphic coordinate system $z = (z_1, \dots, z_n)$ centered at a point $q_0 \in U$ and defined on U and quantities $\tau_1(q,\delta), \tau_2(q,\delta), \dots, \tau_n(q,\delta)$ for each $q \in U$ such that

$$\tau_1(q,\delta) = \delta \quad \text{and} \quad \delta^{1/2} \lesssim \tau_j(q,\delta) \lesssim \delta^{1/m} \quad \text{for} \quad j = 2,3,\dots,n. \tag{2.1}$$

Moreover, the polydisc $D(q,\delta)$ defined by:

$$D(q,\delta) = \{z \in \mathbb{C}^n : |z_j| < \tau_j(q,\delta), j = 1,\dots,n\} \tag{2.2}$$

is the largest one centered at q on which the defining function ρ changes by no more than δ from its value at q, i.e., if $z \in D(q,\delta)$, then $|\rho(z) - \rho(q)| \lesssim \delta$.

The polydisc $D(q,\delta)$ is known to satisfy several "covering properties" [**McN94a**]:

(1) There exists a constant $C > 0$, such that for points $q_1, q_2 \in U \cap \Omega$ with $D(q_1,\delta) \cap D(q_2,\delta) \neq \varnothing$, we have

$$D(q_2,\delta) \subseteq CD(q_1,\delta) \text{ and } D(q_1,\delta) \subseteq CD(q_2,\delta). \tag{2.3}$$

(2) There exists a constant $c > 0$ such that for $q \in U \cap \Omega$ and $\delta > 0$, we have

$$D(q,2\delta) \subseteq cD(q,\delta). \tag{2.4}$$

It was also shown in [**McN94a**] that $D(p,\delta)$ induces a global quasi-metric on Ω. Here we will use it to define a quasi-metric on $\mathbf{b}\Omega$.

For $q \in \mathbf{b}\Omega$ and $\delta > 0$, we define the non-isotropic ball of radius δ to be the set

$$B(q,\delta) = D(q,\delta) \cap \mathbf{b}\Omega.$$

Set containments (2.3), (2.4), and the compactness and smoothness of $\mathbf{b}\Omega$ imply the following properties for the balls:

(1) There exists a constant C such that for $q_1, q_2 \in U \cap \mathbf{b}\Omega$ with $B(q_1,\delta) \cap B(q_2,\delta) \neq \varnothing$,

$$B(q_2,\delta) \subseteq CB(q_1,\delta) \text{ and } B(q_1,\delta) \subseteq CB(q_2,\delta). \tag{2.5}$$

(2) Let μ denote the Lebesgue surface area on $\mathbf{b}\Omega$. There exists a constant $c > 0$ such that for $q \in U \cap \mathbf{b}\Omega$ and $\delta > 0$, we have

$$B(q,\delta) \subseteq cB(q,\delta/2) \quad \text{and} \quad \mu(B(q,\delta)) \approx \prod_{j=2}^{n} \tau_j^2(q,\delta). \tag{2.6}$$

The balls B induce a quasi-metric on $\mathbf{b}\Omega \cap U$. For $q, p \in \mathbf{b}\Omega \cap U$, we set $\tilde{d}(q,p) = \inf\{\delta > 0 : p \in B(q,\delta)\}$. Note that $\mathbf{b}\Omega$ is compact. To extend this quasi-metric $\tilde{d}(\cdot,\cdot)$ to a global quasi-metric $d(\cdot,\cdot)$ defined on $\mathbf{b}\Omega \times \mathbf{b}\Omega$, one can just patch the local metrics together in an appropriate way. The resulting quasi-metric is not continuous, but satisfies all the relevant properties. We refer the reader to [**McN94a**] for more details on this matter. Since $d(\cdot,\cdot)$ and $\tilde{d}(\cdot,\cdot)$ are equivalent, we may abuse the notation B for the ball on the boundary induced by d. Then (2.5) and (2.6) still hold true for B.

2.2. Dyadic tents on Ω and the matrix $\mathcal{B}_2$ constant. The non-isotropic ball $B(p,\delta)$ on the boundary $\mathbf{b}\Omega$ induces "tents" in the domain Ω. To define what "tents" are we need the orthogonal projection map near the boundary. Let $\operatorname{dist}(\cdot,\cdot)$ denote the Euclidean distance in $\mathbb{C}^n$. For small $\epsilon > 0$, set

$$N_\epsilon(\mathbf{b}\Omega) = \{w \in \mathbb{C}^n : \operatorname{dist}(w, \mathbf{b}\Omega) < \epsilon\}.$$

LEMMA 2.1. *For sufficiently small $\epsilon_0 > 0$, there exists a map $\pi : N_{\epsilon_0}(\mathbf{b}\Omega) \to \mathbf{b}\Omega$ such that*

(1) *For each point $z \in N_{\epsilon_0}(\mathbf{b}\Omega)$ there exists a unique point $\pi(z) \in \mathbf{b}\Omega$ such that*
$$\operatorname{dist}(z, \pi(z)) = \operatorname{dist}(z, \mathbf{b}\Omega).$$
(2) *For $p \in \mathbf{b}\Omega$, the fiber $\pi^{-1}(p) = \{p - \epsilon n(p) : -\epsilon_0 < \epsilon < \epsilon_0\}$ where $n(p)$ is the outer unit normal vector of $\mathbf{b}\Omega$ at point p.*
(3) *The map π is smooth on $N_{\epsilon_0}(\mathbf{b}\Omega)$.*
(4) *If the defining function ρ is the signed distance function to the boundary, the gradient $\nabla\rho$ satisfies*
$$\nabla\rho(z) = n(\pi(z)) \quad \textit{for} \quad z \in N_{\epsilon_0}(\mathbf{b}\Omega).$$

A proof of Lemma 2.1 can be found in [**BB00**].

DEFINITION 2.2. Let ϵ_0 and π be as in Lemma 2.1. For $z \in \mathbf{b}\Omega$ and sufficiently small $\delta > 0$, the "tent" $B^\#(z,\delta)$ over the ball $B(z,\delta)$ is defined to be the subset of $N_{\epsilon_0}(\mathbf{b}\Omega)$ as follows: When Ω is a simple domain in $\mathbb{C}^n$,

$$B^\#(z,\delta) := \{w \in \Omega : \pi(w) \in B(z,\delta), \operatorname{dist}(\pi(w), w) < \delta\}.$$

For $\delta \gtrsim 1$ and any $z \in \mathbf{b}\Omega$, we set $B^\#(z,\delta) = \Omega$.

For the "tent" $B^\#(z,\delta)$ to be within $N_{\epsilon_0}(\mathbf{b}\Omega)$, the constant δ in Definition 2.2 needs to satisfy $\delta < \epsilon_0$.

Given a subset $U \in \mathbb{C}^n$, let $|U|$ denote the Lebesgue measure of U. By the definitions of the tents $B^\#(z,\delta)$ we have:

$$|B^\#(z,\delta)| \approx \delta^2 \prod_{j=2}^{n} \tau_j^2(z,\delta) \tag{2.7}$$

and hence also the "doubling property":

$$|B^\#(z,\delta)| \approx |B^\#(z,\delta/2)|. \tag{2.8}$$

Now we are in the position of constructing dyadic systems on $\mathbf{b}\Omega$ and Ω. Note that the ball $B(\cdot,\delta)$ on $\mathbf{b}\Omega$ satisfies the "doubling property" as in (2.6). By (2.5), the surface area $\mu(B(q_1,\delta)) \approx \mu(B(q_2,\delta))$ for any $q_1, q_2 \in \mathbf{b}\Omega$ satisfying $d(q_1,q_2) \leq \delta$. Combining these facts yields that the metric $d(\cdot,\cdot)$ is a doubling metric, i.e., for every

$q \in \mathbf{b}\Omega$ and $\delta > 0$, the ball $B(q,\delta)$ can be covered by at most M balls $B(x_i,\delta/2)$. Results of Hytönen and Kairema in [**HK12**] then give the following lemmas:

LEMMA 2.3. *Let δ be a positive constant that is sufficiently small and let $s > 1$ be a parameter. There exist reference points $\{p_j^{(k)}\}$ on the boundary $\mathbf{b}\Omega$ and an associated collection of subsets $\mathcal{Q} = \{Q_j^k\}$ of $\mathbf{b}\Omega$ with $p_j^{(k)} \in Q_j^k$ such that the following properties hold:*

(1) *For each fixed k, $\{p_j^{(k)}\}$ is a largest set of points on $\mathbf{b}\Omega$ satisfying $d_1(p_j^{(k)}, p_i^{(k)}) > s^{-k}\delta$ for all i,j. In other words, if $p \in \mathbf{b}\Omega$ is a point that is not in $\{p_j^{(k)}\}$, then there exists an index j_o such that $d_1(p, p_{j_o}^{(k)}) \le s^{-k}\delta$.*
(2) *For each fixed k, $\bigcup_j Q_j^k = \mathbf{b}\Omega$ and $Q_j^k \cap Q_i^k = \varnothing$ when $i \neq j$.*
(3) *For $k < l$ and any i,j, either $Q_j^k \supseteq Q_i^l$ or $Q_j^k \cap Q_i^l = \varnothing$.*
(4) *There exist positive constants c and C such that for all j and k,*

$$B(p_j^{(k)}, cs^{-k}\delta) \subseteq Q_j^k \subseteq B(p_j^{(k)}, Cs^{-k}\delta).$$

(5) *Each Q_j^k contains at most N numbers of Q_i^{k+1}. Here N does not depend on k,j.*

The points $p_j^{(k)}$ are dyadic points in $\mathbf{b}\Omega$.

LEMMA 2.4. *Let δ and $\{p_j^{(k)}\}$ be as in Lemma 2.3. There are finitely many collections $\{\mathcal{Q}_l\}_{l=1}^N$ such that the following hold:*

(1) *Each collection $\mathcal{Q}_l$ is associated to some dyadic points $\{z_j^{(k)}\}$ and satisfies all the properties in Lemma 2.3.*
(2) *For any $z \in \mathbf{b}\Omega$ and small $r > 0$, there exist $Q_{j_1}^{k_1} \in \mathcal{Q}_{l_1}$ and $Q_{j_2}^{k_2} \in \mathcal{Q}_{l_2}$ such that*

$$Q_{j_1}^{k_1} \subseteq B(z,r) \subseteq Q_{j_2}^{k_2} \quad \text{and} \quad \mu(B(z,r)) \approx \mu(Q_{j_1}^{k_1}) \approx \mu(Q_{j_2}^{k_2}).$$

Letting the sets Q_j^k in Lemma 2.3 serve as these bases, we construct dyadic tents in Ω as follows:

DEFINITION 2.5. Let δ, $\{p_j^{(k)}\}$ and $\mathcal{Q} = \{Q_j^k\}$ be as in Lemma 2.3. We define the collection $\mathcal{T} = \{\hat{K}_j^k\}$ of dyadic tents in the domain Ω as

$$\hat{K}_j^k := \{z \in \Omega : \pi(z) \in Q_j^k \text{ and } \operatorname{dist}(\pi(z), z) < s^{-k}\delta\}.$$

LEMMA 2.6. *Let $\mathcal{T} = \{\hat{K}_j^k\}$ be a collection of dyadic tents in Definition 2.5 and let $\{\mathcal{Q}_l\}_{l=1}^N$ be collections in Lemma 2.4. The following statements hold true:*

(1) *For any $\hat{K}_j^k$, $\hat{K}_i^{k+1}$ in $\mathcal{T}$, either $\hat{K}_j^k \supseteq \hat{K}_i^{k+1}$ or $\hat{K}_j^k \cap \hat{K}_i^{k+1} = \varnothing$.*
(2) *For any $z \in \mathbf{b}\Omega$ and small $r > 0$, there exist $Q_{j_1}^{k_1} \in \mathcal{Q}_{l_1}$ and $Q_{j_2}^{k_2} \in \mathcal{Q}_{l_2}$ such that*

$$\hat{K}_{j_1}^{k_1} \subseteq B^{\#}(z,r) \subseteq \hat{K}_{j_2}^{k_2} \quad \text{and} \quad |B^{\#}(z,r)| \approx |\hat{K}_{j_1}^{k_1}| \approx |\hat{K}_{j_2}^{k_2}|.$$

Set $\mathscr{B} := \{B^{\#}(z,r)\}_{z \in \mathbf{b}\Omega, r \le \epsilon_0} \cup \{\Omega\}$. Recall the matrix $\mathcal{B}_2$ constant given by

$$\mathcal{B}_2(W) := \sup_{B \in \mathscr{B}} \left| \langle W \rangle_B^{1/2} \langle W^{-1} \rangle_B^{1/2} \right|^2,$$

where $|\cdot|$ denotes the norm of the matrix acting on $\mathbb{C}^d$ and $\langle W\rangle_B := \frac{\int_B W dV}{\int_B dV}$. Let $\{\mathcal{T}_l\}_{l=1}^N$ be collections of dyadic tents induced by collections $\{\mathcal{Q}_l\}_{l=1}^N$ in Lemma 2.4 respectively. Set $\mathscr{T} := \{\hat{K}_j^k : \hat{K}_j^k \in \mathcal{T}_l \text{ for some } l = 1, \dots, N\} \cup \{\Omega\}$. Then Lemma 2.6 implies that

$$\mathcal{B}_2(W) \approx \sup_{\hat{K}_j^k \in \mathscr{T}} \left|\langle W\rangle_{\hat{K}_j^k}^{1/2} \langle W^{-1}\rangle_{\hat{K}_j^k}^{1/2}\right|^2.$$

From now on, we will abuse the notation of $\mathcal{B}_2(W)$ to represent both the supremum in $\mathscr{B}$ and $\mathscr{T}$.

2.3. Dyadic kubes on Ω.

DEFINITION 2.7. For a collection $\mathcal{T}$ of dyadic tents, we define the center $\alpha_j^{(k)}$ of each tent $\hat{K}_j^k$ to be the point satisfying

- $\pi(\alpha_j^{(k)}) = p_j^{(k)}$; and
- $\operatorname{dist}(p_j^{(k)}, \alpha_j^{(k)}) = \frac{1}{2} \sup_{\pi(p) = p_j^{(k)}} \operatorname{dist}(p, \mathbf{b}\Omega)$.

We set $K^{-1} = \Omega \backslash \left(\bigcup_j \hat{K}_j^0\right)$, and for each point $\alpha_j^{(k)}$ or its corresponding tent $\hat{K}_j^k$, we define the "kube" $K_j^k := \hat{K}_j^k \backslash \left(\bigcup_l \hat{K}_l^{k+1}\right)$, where l is any index with $p_l^{(k+1)} \in \hat{K}_j^k$.

By the definition of dyadic kubes, the following lemma for dyadic kubes holds true:

LEMMA 2.8. *Let $\mathcal{T} = \{\hat{K}_j^k\}$ be the system of tents induced by $\mathcal{Q}$ in Definition 2.5. Let K_j^k be the kubes of $\hat{K}_j^k$. Then*

(1) $\pi(K_j^k) = Q_j^k \in \mathcal{Q}$;
(2) *K_j^k's are pairwise disjoint and $\bigcup_{j,k} K_j^k = \Omega$;*
(3) *the following volume estimates hold:*

$$|K_j^k| \approx |\hat{K}_j^k| \approx s^{-2k}\delta^{-2} \prod_{j=2}^n \tau_j^2(p_j^{(k)}, s^{-k}\delta). \tag{2.9}$$

The next lemma is crucial in the proof of Lemma 4.4 which relates the norm of a holomorphic vector-valued function on $L^2(\tilde{\mathcal{W}})$.

LEMMA 2.9. *There exists finitely many collections of $\{\mathcal{T}_l\}_{l=1}^M$ such that for any $z \in \Omega$, we can find a tent $\hat{K}_j^k \in \mathcal{T}_l$ for some l such that the following holds:*

(1) $z \in K_j^k$,
(2) *If $k \geq 0$, then $K_j^k \supseteq D(z, cs^{-k})$ for some constant $c > 0$. If $k = -1$, then K^{-1} contains a Euclidean ball centered at z with radius $r \approx 1$.*

PROOF. Let $\{\mathcal{T}_l\}_{l=1}^N$ be collections of dyadic tents induced by collections $\{\mathcal{Q}_l\}_{l=1}^N$ in Lemma 2.4 respectively with parameter δ and reference points $p_j^{(k)}$. Now let $\{\mathcal{T}_l'\}_{l=1}^N$ and $\{\mathcal{T}_l''\}_{l=1}^N$ be collections also induced by $\{\mathcal{Q}_l\}_{l=1}^N$ in Lemma 2.4, but each $\hat{K}_j'^k \in \mathcal{T}_l'$ and $\hat{K}_j''^k \in \mathcal{T}_l'$ are defined by $Q_j^k \in \mathcal{Q}_l$ as follows:

$$\hat{K}_j'^k := \left\{z \in \Omega : \pi(z) \in Q_j^k \text{ and } \operatorname{dist}(\pi(z), z) < \frac{s+2}{3}\delta s^{-k}\right\},$$

$$\hat{K}_j''^k := \left\{z \in \Omega : \pi(z) \in Q_j^k \text{ and } \operatorname{dist}(\pi(z), z) < \frac{2s+1}{3}\delta s^{-k-1}\right\},$$

i.e., $\hat{K}'^k_j \in \mathcal{T}'_l$ and $\hat{K}''^k_j \in \mathcal{T}''_l$ have the same "base" Q^k_j as the tent $\hat{K}^k_j$ in $\mathcal{T}_l$ but different "height". Thus $\mathcal{T}'_l$ and $\mathcal{T}''_l$ still satisfies Lemma 2.6. We call the tents $\hat{K}'^k_j$ and $\hat{K}''^k_j$ *cousins* of $\hat{K}^k_j$ and use K'^k_j and K''^k_j to denote the dyadic kubes induced by $\hat{K}'^k_j$ and $\hat{K}''^k_j$ respectively as in Definition 2.7. We claim $\{\mathcal{T}_l\}_{l=1}^N \cup \{\mathcal{T}'_l\}_{l=1}^N \cup \{\mathcal{T}''_l\}_{l=1}^N$ are the desired collections in Lemma 2.9. Hence, $M = 3N$.

When the point $z \in \Omega$ is far away from the boundary $\mathbf{b}\Omega$, say $z \in K'^{-1}$. Then $\text{dist}(z, \pi(z)) \geq \frac{2s+1}{3}\delta$. Therefore, $\text{dist}(z, \mathbf{b}K^{-1}) \geq \frac{2s-2}{3}\delta$ and K^{-1} contains a ball centered at z with radius $\frac{2s-2}{3}\delta \approx 1$.

When the point $z \in \Omega \backslash K'^{-1}$. Lemma 2.6 implies that there exists a tent $\hat{K}^k_j \in \mathcal{T}_l$ with $z \in K^k_j$ and a constant $c_1 > 0$ such that $\pi(\hat{K}^k_j) \supseteq B(\pi(z), c_1 s^{-k})$. We first show that there exists a constant c_2 such that the polydisc $D(z, c_2 s^{-k})$ satisfies $\pi(D(z, c_2 s^{-k})) \subseteq B(\pi(z), c_1 s^{-k})$. Suppose $\zeta \in D(z, c_2 s^{-k})$. Let ζ' be the projection of ζ on the hypersurface $\{w : \rho(w) = \rho(z)\}$. Then we have $d(\zeta, z) \lesssim c_2 s^{-k}$ and

$$d(\zeta', z) \lesssim d(\zeta', \zeta) + d(z, \zeta) \lesssim c_2 s^{-k}.$$

Since the parameter δ in Lemma 2.3 is chosen to be sufficiently small and the gradient $\nabla\rho$ is continuous near $\mathbf{b}\Omega$, we may assume that $|\nabla\rho(x) - \nabla\rho(y)| < \epsilon$ for all $y \in \mathbf{b}\Omega$ and $x \in B(y, \delta)$. Then, from $\pi(\zeta) = \zeta' - \rho(z)\nabla\rho(\zeta)$ and $\pi(z) = z - \rho(z)(\nabla\rho(z))$, we obtain

$$\begin{aligned} d(\pi(\zeta), \pi(z)) &\lesssim d(\pi(\zeta), \zeta' - \rho(z)\nabla\rho(z)) + d(\pi(z), \zeta' - \rho(z)\nabla\rho(z)) \\ &\lesssim -\epsilon\rho(z) + d(\pi(z), \zeta' - \rho(z)\nabla\rho(z)). \end{aligned}$$

By (2.3) and (2.4), there exists a constant $C > 0$ such that $D(z, \delta s^{-k}) \subseteq CD(\pi(z), \delta s^{-k})$. Thus, $D(z, \delta s^{-k}) - \rho(z)\nabla\rho(z) \subseteq CD(\pi(z), \delta s^{-k})$. Since the sets $D(z, \delta s^{-k}) - \rho(z)\nabla\rho(z)$ and $CD(\pi(z), \delta s^{-k})$ have the common center $\pi(z)$, the containment still holds after applying a dilation to both sets. Therefore, there exists a constant C_2 such that

$$C^{-1}C_2^{-1}D(z, \delta s^{-k}) - \rho(z)\nabla\rho(z) \subseteq C^{-1}D(\pi(z), \delta s^{-k}) \subseteq D(\pi(z), c_1 s^{-k}).$$

We may choose c_2 to be small so that

$$D(z, c_2 s^{-k}) \subseteq C^{-1}C_2^{-1}D(z, \delta s^{-k}).$$

Then we obtain $D(z, c_2 s^{-k}) - \rho(z)\nabla\rho(z) \subseteq D(\pi(z), c_1 s^{-k})$ which also implies that

$$d(\pi(z), \zeta' - \rho(z)\nabla\rho(z)) \lesssim c_1 s^{-k}.$$

Hence $d(\pi(\zeta), \pi(z)) \lesssim c_1 s^{-k}$ and $\pi(D(z, c_2 s^{-k})) \subseteq B(\pi(z), c_1 s^{-k})$ for some c_2.

Now we are ready to show that $D(z, cs^{-k})$ is in K^k_j, K'^k_j, or K''^k_j for some $c > 0$. Since $Q^k_j \supseteq \pi(D(z, c_2 s^{-k}))$, it suffices to check the distance from z to the boundary $\mathbf{b}\Omega$. Note that

$$\begin{aligned} \{-\rho(\zeta) : \zeta \in K^k_j\} &= \left[\delta s^{-k-1}, \delta s^{-k}\right), \\ \{-\rho(\zeta) : \zeta \in K'^k_j\} &= \left[\frac{s+2}{3}\delta s^{-k-1}, \frac{s+2}{3}\delta s^{-k}\right), \\ \{-\rho(\zeta) : \zeta \in K''^k_j\} &= \left[\frac{2s+1}{3}\delta s^{-k-2}, \frac{2s+1}{3}\delta s^{-k-1}\right). \end{aligned}$$

If $z \in K^k_j \backslash K''^k_j$, then the distance from z to the boundary of K'^k_j along the normal direction $\pm\nabla\rho(z)$ is greater than $\frac{s-1}{3}\delta s^{-k-1}$. If $z \in K^k_j \backslash K'^k_j$, the distance from z to the boundary of $K_j''^k$ along the normal direction $\pm\nabla\rho(z)$ is greater than $\frac{s-1}{3}\delta s^{-k-2}$. Otherwise, $z \in K_j''^k \cap K_j'^k$ and the distance from z to the boundary of K_j^k along the normal direction $\pm\nabla\rho(z)$ is greater than $\frac{s-1}{3}\delta s^{-k-1}$. Hence by choosing $c = \min\{c_2, \frac{s-1}{3}\delta s^{-2}\}$, we have $D(z, cs^{-k})$ is contained in K_j^k, $K_j'^k$, or $K_j''^k$. $\square$

2.4. Estimates for the Bergman kernel. When Ω is a simple domain in $\mathbb{C}^n$, estimates of the Bergman kernel function on Ω were obtained in [**McN94b, McN91, McN03**].

THEOREM 2.10. *Let Ω be a simple domain in $\mathbb{C}^n$. Let p be a boundary point of Ω. There exists a neighborhood U of p so that for all $q_1, q_2 \in U \cap \Omega$,*

$$|K_\Omega(q_1; \bar{q}_2)| \lesssim t^{-2} \prod_{j=2}^{n} \tau_j(q_1, t)^{-2}, \tag{2.10}$$

where $t = |\rho(q_1)| + |\rho(q_2)| + \inf\{\epsilon > 0 : q_2 \in D(q_1, \epsilon)\}$.

We can reformulate Theorem 2.10 as Theorem 2.11.

THEOREM 2.11. *Let Ω be a simple domain in $\mathbb{C}^n$. Let p be a boundary point of Ω. There exists a neighborhood U of p so that for all $q_1, q_2 \in U \cap \Omega$,*

$$|K_\Omega(q_1; \bar{q}_2)| \lesssim |B^\#(\pi(q_1), t)|^{-1}, \tag{2.11}$$

where $t = |\rho(q_1)| + |\rho(q_2)| + \inf\{\epsilon > 0 : q_2 \in D(q_1, \epsilon)\}$. Moreover, there exists a constant c such that $q_1, q_2 \in B^\#(\pi(q_1), c\delta)$.

Here the estimate (2.11) follows from (2.7). Recall that the polydisc $D(q, \delta)$ induces a global quasi-metric [**McN94a**] on Ω. Then a triangle inequality argument using this quasi-metric yields the containment $q_2 \in B^\#(\pi(q_1), c\delta)$.

3. Convex body domination

For a function $f \in L^1(Q)$ on some set $Q \subseteq \mathbb{C}^n$ with values in $\mathbb{C}^d$, we recall the definition of the convex body average $\langle\!\langle f \rangle\!\rangle_Q$ in [**NPTV17**]:

$$\langle\!\langle f \rangle\!\rangle_Q := \{\langle \varphi f \rangle_Q : \varphi \text{ is any complex valued function on } Q \text{ with } \|\varphi\|_\infty \leq 1\}.$$

By its definition, the set $\langle\!\langle f \rangle\!\rangle_Q$ is symmetric, convex, and compact. For a collection of dyadic tents $\{\mathcal{T}_l\}_{l=1}^N$, we define the dyadic operator L by

$$Lf = \sum_{l=1}^{N} \sum_{\hat{K}_j^k \in \mathcal{T}_l} \langle\!\langle f \rangle\!\rangle_{\hat{K}_j^k} 1_{\hat{K}_j^k}.$$

This operator takes a vector-valued function and returns a set-valued function.

LEMMA 3.1 ([**NPTV17**, Lemma 2.7]). *Let $f \in L^1(Q, \mathbb{C}^d)$ and let $g(x) \in \langle\!\langle f \rangle\!\rangle_Q$ a.e. on Q. Then there exists a measurable function $H : Q \times Q \to \mathbb{C}$ with $\|H\|_\infty \leq |Q|^{-1}$ such that*

$$g(z) = \int_Q H(z, w) f(w) dV(w).$$

Thus to estimate L, it suffices to estimate all operators of the form

$$f \mapsto \sum_{\hat{K}_j^k \in \mathcal{T}} \int_{\hat{K}_j^k} H_{\hat{K}_j^k}(z, w) f(w) dV(w),$$

where $H_{\hat{K}_j^k}$ is supported on $\hat{K}_j^k \times \hat{K}_j^k$ and satisfies $\|H_{\hat{K}_j^k}\|_\infty \leq |\hat{K}_j^k|^{-1}$.

THEOREM 3.2. *There exists a constant C such that for any $f \in L^1(\Omega, \mathbb{C}^d)$ and $z \in \Omega$,*

$$P(f)(z) \in CLf(z).$$

PROOF. Let $\{\mathcal{T}_l\}_{l=1}^N$ be collections of dyadic tents induced by $\{\mathcal{Q}_l\}_{l=1}^N$ in Lemma 2.6. If $z \in \Omega \backslash N_{\epsilon_0}(\mathbf{b}\Omega)$, then $|K_\Omega(z;\cdot)| \lesssim 1$ and

$$P(f)(z) = \int_\Omega K_\Omega(z;\bar{w}) f(w) dV(w) \in C\langle\!\langle f \rangle\!\rangle_\Omega \subseteq CLf(z),$$

for some constant C. If $z \in \Omega \cap N_{\epsilon_0}(\mathbf{b}\Omega)$, then we can set $k_0 := \max\{k \in \mathbb{N} : z \in B^\#(\pi(z), s^{-k}\delta)\}$. Theorem 2.11 yields that for $l \in \{1, \dots, k_0 - 1\}$ and point $w \in B^\#(\pi(z), s^{-l}\delta) \backslash B^\#(\pi(z), s^{-l-1}\delta)$,

$$|K_\Omega(z;\bar{w})| \lesssim |B^\#(\pi(z), s^{-l}\delta)|^{-1}.$$

By Lemma 2.6, we can find k_0 tents $\{\hat{K}_j\}_{j=1}^{k_0}$ from $\{\mathcal{T}_l\}_{l=1}^N$ such that for each j,

$$B^\#(\pi(z), s^{-j}\delta) \subseteq \hat{K}_j \quad \text{and} \quad |\hat{K}_j| \approx |B^\#(\pi(z), s^{-j}\delta)|.$$

Set

$$\begin{aligned} U_0 &= \Omega \backslash B^\#(\pi(z), \delta) \\ U_j &= B^\#(\pi(z), s^{-j}\delta) \backslash B^\#(\pi(z), s^{-j-1}\delta) \quad \text{for} \quad j = 1, \dots, k_0 - 1. \end{aligned}$$

Then there is a constant C independent of j and z such that

$$\int_{U_j} K_\Omega(z;\bar{w}) f(w) dV(w) \in C\langle\!\langle f \rangle\!\rangle_{\hat{K}_j}.$$

This containment implies

$$\begin{aligned} P(f)(z) &= \int_\Omega K_\Omega(z;\bar{w}) f(w) dV(w) \\ &= \sum_{j=0}^{k_0-1} \int_{U_j} K_\Omega(z;\bar{w}) f(w) dV(w) \in CLf(z), \end{aligned} \tag{3.1}$$

which completes the proof. □

4. A step averaging weight $\mathcal{W}$ and its properties

In this section, we introduce the step averaging weight which will play a crucial rule in the proof of Theorem 1.2.

DEFINITION 4.1. Let W be a matrix weight over Ω. For a collection $\mathcal{T}$ of dyadic tents, we define a step weight $\mathcal{W}$ to be

$$\mathcal{W}(z) := \sum_{\hat{K}_j^k \in \mathcal{T}} \langle W \rangle_{K_j^k} 1_{K_j^k}(z).$$

For collections $\{\mathcal{T}_l\}_{l=1}^N$ of tents that satisfies both Lemmas 2.6 and 2.9, we define $\tilde{\mathcal{W}}$ to be

$$\tilde{\mathcal{W}}(z) := \sum_{l=1}^N \mathcal{W}_l(z) = \sum_{l=1}^N \sum_{\hat{K}_j^k \in \mathcal{T}_l} \langle W \rangle_{K_j^k} 1_{K_j^k}(z).$$

Since the K_j^k's are disjoint, it is easy to see that

$$\mathcal{W}^{-1}(z) = \sum_{\hat{K}_j^k \in \mathcal{T}} \langle W \rangle_{K_j^k}^{-1} 1_{K_j^k}(z).$$

LEMMA 4.2. *If W is a matrix $\mathcal{B}_2$ weight, then $\mathcal{W}$ is also a matrix $\mathcal{B}_2$ weight with*

$$\sup_{\hat{K}_j^k\in\mathcal{T}} \left|\langle\mathcal{W}^{-1}\rangle_{\hat{K}_j^k}^{1/2}\langle\mathcal{W}\rangle_{\hat{K}_j^k}^{1/2}\right|^2 \leq \mathcal{B}_2(W). \tag{4.1}$$

PROOF. By its definition, $\langle\mathcal{W}\rangle_{\hat{K}_j^k} = \langle W\rangle_{\hat{K}_j^k}$. Given a unit vector $v\in\mathbb{C}^d$,

$$\begin{aligned}
|\langle\mathcal{W}^{-1}\rangle_{\hat{K}_j^k}^{1/2}\langle\mathcal{W}\rangle_{\hat{K}_j^k}^{1/2}v|^2 &= \langle\langle\mathcal{W}\rangle_{\hat{K}_j^k}^{1/2}\langle\mathcal{W}^{-1}\rangle_{\hat{K}_j^k}\langle\mathcal{W}\rangle_{\hat{K}_j^k}^{1/2}v,v\rangle\\
&= |\hat{K}_j^k|^{-1}\sum_{K_m^l\subseteq\hat{K}_j^k}|K_m^l|\langle\langle W\rangle_{\hat{K}_j^k}^{1/2}\langle W\rangle_{K_m^l}^{-1}\langle W\rangle_{\hat{K}_j^k}^{1/2}v,v\rangle\\
&= |\hat{K}_j^k|^{-1}\sum_{K_m^l\subseteq\hat{K}_j^k}|K_m^l||\langle W\rangle_{K_m^l}^{-1/2}\langle W\rangle_{\hat{K}_j^k}^{1/2}v|^2\\
&= |\hat{K}_j^k|^{-1}\sum_{K_m^l\subseteq\hat{K}_j^k}|K_m^l||\langle W\rangle_{K_m^l}^{-1/2}\langle W^{-1}\rangle_{K_m^l}^{-1/2}\langle W^{-1}\rangle_{K_m^l}^{1/2}\langle W\rangle_{\hat{K}_j^k}^{1/2}v|^2\\
&\leq |\hat{K}_j^k|^{-1}\sum_{K_m^l\subseteq\hat{K}_j^k}|K_m^l||\langle W^{-1}\rangle_{K_m^l}^{1/2}\langle W\rangle_{\hat{K}_j^k}^{1/2}v|^2 = |\langle W^{-1}\rangle_{\hat{K}_j^k}^{1/2}\langle W\rangle_{\hat{K}_j^k}^{1/2}v|^2. \qquad (4.2)
\end{aligned}$$

The last inequality, (4.2), uses the fact that $\langle W^{-1}\rangle_{\hat{K}_j^k}^{1/2}\langle W\rangle_{\hat{K}_j^k}^{1/2}$ is expanding. See [**TV97b**, Corollary 3.3]. □

By the technique of [**TV97b**, Lemma 3.5], the weight $\langle\mathcal{W}^{-1}\rangle_{\hat{K}_j^k}^{1/2}\mathcal{W}(z)\langle\mathcal{W}^{-1}\rangle_{\hat{K}_j^k}^{1/2}$ is also a matrix weight satisfying (4.1). Moreover, for a unit vector $v\in\mathbb{C}^d$, the scalar function

$$\omega(v;z) := \left\langle\langle\mathcal{W}^{-1}\rangle_{\hat{K}_j^k}^{1/2}\mathcal{W}(z)\langle\mathcal{W}^{-1}\rangle_{\hat{K}_j^k}^{1/2}v,v\right\rangle,$$

is a scalar weight satisfying

$$\sup_{\hat{K}_j^k\in\mathcal{T}}\langle\omega(v;\cdot)^{-1}\rangle_{\hat{K}_j^k}\langle\omega(v;\cdot)\rangle_{\hat{K}_j^k} \lesssim \mathcal{B}_2(W).$$

Note that $\omega(v;z)$ is a fixed constant over each kube K_j^k. The next lemma shows the $\omega(v;z)$ satisfies the reverse Hölder inequality. In the case $\Omega=\mathbb{D}$, the lemma follows from [**APR19**, Theorem 1.7].

LEMMA 4.3. *The function $\omega(v;z)$ satisfies the reverse Hölder inequality over the tent $\hat{K}_j^k\in\mathcal{T}$, i.e., there exists a constant $r>1$ such that*

$$\langle\omega(v;z)^r\rangle_{\hat{K}_j^k}^{1/r} \lesssim \langle\omega(v;z)\rangle_{\hat{K}_j^k}. \tag{4.3}$$

Moreover, the constant r can be chosen such that $r-1\approx\mathcal{B}_2^{-1}(W)$.

PROOF. On the fixed tent $\hat{K}_j^k$ and a constant $R>1$ to be determined, we apply a corona decomposition of $\hat{K}_j^k$ as follows:

(1) First, we define the set of stopping children $\mathcal{L}(\hat{K}_m^l)$ of a tent $\hat{K}_m^l\in\mathcal{T}$:

$$\mathcal{L}(\hat{K}_m^l) := \left\{\text{maximal tents } \hat{K}_q^p \text{ such that } \frac{\int_{\hat{K}_q^p}\omega(v;z)dV}{|\hat{K}_q^p|} > R\frac{\int_{\hat{K}_m^l}\omega(v;z)dV}{|\hat{K}_m^l|}\right\}.$$

(2) Set $\mathcal{L}_1 := \mathcal{L}(\hat{K}_j^k)$ and for $i\geq 2$, set $\mathcal{L}_i := \cup_{\hat{K}_m^l\in\mathcal{L}_{i-1}}\mathcal{L}(\hat{K}_m^l)$ and $\mathcal{L} := \cup_{i\geq 1}\mathcal{L}_i$.

Then we have for $\hat{K}^l_m \in \mathcal{L}_i$,

$$\sum_{\hat{K}^p_q \subseteq \hat{K}^l_m, \hat{K}^p_q \in \mathcal{L}_{m+1}} |\hat{K}^p_q| \leq R^{-1} |\hat{K}^l_m|. \tag{4.4}$$

Since tents in $\mathcal{L}_m$ are maximal, we also have for $\hat{K}^l_m \in \mathcal{L}_i$,

$$R^i \frac{\int_{\hat{K}^k_j} \omega(v;z) dV}{|\hat{K}^k_j|} \leq \frac{\int_{\hat{K}^l_m} \omega(v;z) dV}{|\hat{K}^l_m|} \leq (cR)^i \frac{\int_{\hat{K}^k_j} \omega(v;z) dV}{|\hat{K}^k_j|}. \tag{4.5}$$

Here c can be chosen to be the supremum of $|\hat{K}^l_m|/|\hat{K}^{l+1}_{m'}|$ over all tents $\hat{K}^l_m$ and $\hat{K}^{l+1}_{m'}$ satisfying $\hat{K}^{l+1}_{m'} \subseteq \hat{K}^l_m$. By Lemma 2.8, c is finite. Recall the maximal operator $M_{\mathcal{T}}$ given by

$$M_{\mathcal{T}} f(z) = \sup_{\hat{K}^k_j \in \mathcal{T}} \langle |f| \rangle_{\hat{K}^k_j} 1_{\hat{K}^k_j}(z).$$

Then (4.5) implies

$$M_{\mathcal{T}}(1_{\hat{K}^k_j} \omega(v;z)) \lesssim \left(1 + \sum_i \sum_{\hat{K}^l_m \in \mathcal{L}_i} (cR)^i 1_{\hat{K}^l_m} \right) \frac{\int_{\hat{K}^k_j} \omega(v;z) dV}{|\hat{K}^k_j|}.$$

Since $\omega(v;z)$ is a fixed constant over each kube K^l_m, we have for $z \in K^l_m$,

$$\omega(v;z) = \frac{\int_{K^l_m} \omega(v;z) dV}{|K^l_m|} \lesssim \frac{\int_{\hat{K}^l_m} \omega(v;z) dV}{|\hat{K}^l_m|}.$$

Thus $\omega(v;z) \lesssim M_{\mathcal{T}}(\omega(v;\cdot) 1_{\hat{K}^k_j})(z)$. By applying a scalar multiplication to $\omega(v;z)$, we may assume that $\langle \omega(v;\cdot) \rangle_{\hat{K}^k_j} = 1$. Then, for a point $z \in \hat{K}^k_j \backslash \bigcup_{\hat{K}^{l'}_{m'} \in \mathcal{L}_i} \hat{K}^{l'}_{m'}$, we have

$$M_{\mathcal{T}}(\omega(v;\cdot) 1_{\hat{K}^k_j})(z) \approx \langle \omega(v;\cdot) \rangle_{\hat{K}^l_m} \approx c^i \langle \omega(v;\cdot) \rangle_{\hat{K}^k_j} = c^i, \tag{4.6}$$

which also implies

$$\log(e + \omega(v;z)) \lesssim i. \tag{4.7}$$

Since $\omega(v;z)$ is a scalar $\mathcal{B}_2$ weight, the proof of [**Gra14**, Lemma 7.1.9(b)] implies that

$$\|M_{\mathcal{T}}\|_{L^2(\omega^{-1}(v;\cdot))} \lesssim \mathcal{B}_2(\omega(v;\cdot)) \leq \mathcal{B}_2(W).$$

Then Hölder's inequality implies

$$\begin{aligned} \int_{\hat{K}^k_j} M_{\mathcal{T}}(\omega(v;\cdot) 1_{\hat{K}^k_j}) dV &= \int_{\hat{K}^k_j} M_{\mathcal{T}}(\omega(v;\cdot) 1_{\hat{K}^k_j}) \omega^{-1/2}(v;\cdot) \omega^{1/2}(v;\cdot) dV \\ &\leq \left(\int_{\hat{K}^k_j} M^2_{\mathcal{T}}(\omega(v;\cdot) 1_{\hat{K}^k_j}) \omega^{-1}(v;\cdot) dV \right)^{1/2} \left(\int_{\hat{K}^k_j} \omega(v;\cdot) dV \right)^{1/2} \\ &\leq \mathcal{B}_2(\omega(v,\cdot)) \left(\int_{\hat{K}^k_j} \omega^2(v;\cdot) \omega^{-1}(v;\cdot) dV \right)^{1/2} \left(\int_{\hat{K}^k_j} \omega(v;\cdot) dV \right)^{1/2} \\ &\leq \mathcal{B}_2(W) \int_{\hat{K}^k_j} \omega(v;z) dV(z). \end{aligned} \tag{4.8}$$

This together with (4.6) and (4.7) implies,

$$
\begin{aligned}
\mathcal{B}_2(W)\int_{\hat{K}_j^k}\omega(v;z)dV(z) &\gtrsim \int_{\hat{K}_j^k} M_{\mathcal{T}}(\omega(v;\cdot)1_{\hat{K}_j^k})dV\\
&\gtrsim \int_{\hat{K}_j^k}\sum_{i\ge 1}\sum_{\hat{K}_m^l\in\mathcal{L}_i}\langle\omega(v;\cdot)\rangle_{\hat{K}_m^l}1_{\hat{K}_m^l}dV\\
&= \int_{\hat{K}_j^k}\omega(v;\cdot)\sum_{i\ge 1} i\sum_{\hat{K}_m^l\in\mathcal{L}_i}1_{\hat{K}_j^k\setminus\cup_{\hat{K}_{m'}^{l'}\in\mathcal{L}_i}\hat{K}_{m'}^{l'}}+1dV\\
&\approx \int_{\hat{K}_j^k}\omega(v;z)\log(e+M_{\mathcal{T}}(\omega(v;\cdot))(z))dV\\
&\gtrsim \int_{\hat{K}_j^k}\omega(v;z)\log(e+\omega(v;z))dV. \qquad (4.9)
\end{aligned}
$$

Then an argument as in the proof of [**DMRO16**, Theorem 4.1] implies for $0<\beta<1$, there exists a constant $\alpha\approx c(\beta)\exp\{-\mathcal{B}_2(W)\}$ such that for $\hat{K}_m^l\subseteq\hat{K}_j^k$ with

$$\int_{\hat{K}_m^l}\omega(v;z)dV\le\beta\int_{\hat{K}_j^k}\omega(v;z)dV,$$

we have

$$|\hat{K}_m^l|\le\alpha|\hat{K}_j^k|.$$

Now we are ready to show (4.3). Choose $\beta=(4c)^{-1}$ where c is the constant in (4.5). Set $R=\alpha^{-1}\approx\exp\{\mathcal{B}_2(W)\}$. Then we have

$$
\begin{aligned}
\langle\omega(v;\cdot)^r\rangle_{\hat{K}_j^k} &\lesssim \langle(M_{\mathcal{T}}(\omega(v;\cdot)1_{\hat{K}_j^k})(z))^r\rangle_{\hat{K}_j^k}\\
&\lesssim \frac{\left(\int_{\hat{K}_j^k}\omega(v;z)dV\right)^r}{|\hat{K}_j^k|^{1+r}}\int_{\hat{K}_j^k}\left(1+\sum_i\sum_{\hat{K}_m^l\in\mathcal{L}_i}(cR)^i1_{\hat{K}_m^l}\right)^r dV\\
&\approx \frac{\left(\int_{\hat{K}_j^k}\omega(v;z)dV\right)^r}{|\hat{K}_j^k|^{1+r}}\int_{\hat{K}_j^k}\left(1+\sum_{i\ge1}\sum_{\hat{K}_m^l\in\mathcal{L}_i}(cR)^{ir}1_{\hat{K}_m^l}\right)dV\\
&= \left(\frac{\int_{\hat{K}_j^k}\omega(v;z)dV}{|\hat{K}_j^k|}\right)^r+\frac{\left(\int_{\hat{K}_j^k}\omega(v;z)dV\right)^r}{|\hat{K}_j^k|^{1+r}}\sum_{i\ge1}\sum_{\hat{K}_m^l\in\mathcal{L}_i}(cR)^{ir}|\hat{K}_m^l|\\
&\le \left(\frac{\int_{\hat{K}_j^k}\omega(v;z)dV}{|\hat{K}_j^k|}\right)^r\\
&\quad+\frac{\left(\int_{\hat{K}_j^k}\omega(v;z)dV\right)^{r-1}}{|\hat{K}_j^k|^r}\sum_{i\ge1}c^{ir}R^{i(r-1)}\sum_{\hat{K}_m^l\in\mathcal{L}_i}\int_{\hat{K}_m^l}\omega(v;z)dV\\
&\le \left(\langle\omega(v;)\rangle_{\hat{K}_j^k}\right)^r(1+\sum_{i\ge1}c^{ir}R^{i(r-1)}(4c)^{-i})\\
&= \langle\omega(v;)\rangle_{\hat{K}_j^k}^r(1+\sum_{i\ge1}(cR)^{i(r-1)}\frac{1}{4^i}). \qquad (4.10)
\end{aligned}
$$

By choosing r such that $(cR)^{(r-1)}=2$, we obtain (4.3). Applying the natural logarithm it yields:

$$(r-1)\ln(cR)=\ln 2.$$

Since $R\approx\exp\{\mathcal{B}_2(W)\}$, we obtain $r-1\approx\mathcal{B}_2^{-1}(W)$. □

LEMMA 4.4. *For any holomorphic vector-valued function f on Ω,*

$$\|f\|_{L^2(W)} \lesssim \|f\|_{L^2(\tilde{\mathcal{W}})}.$$

PROOF. Recall K'^{-1} and $D(z, cs^{-k})$ from Lemma 2.9. For each $z \in \Omega$, we define a z-dependent subset $D^\star(z)$ as follows: For $z \in K'^{-1}$, we set $D^\star(z)$ to be the maximal Euclidean ball in K^{-1} centered at z. For $z \in \Omega \backslash K'^{-1}$, we set $D^\star(z)$ to be the polydisc $D(z, cs^{-k})$ from Lemma 2.9. Then for all $z \in \Omega$

$$\langle \tilde{\mathcal{W}}(z) f(z), f(z)\rangle \gtrsim \langle \langle W\rangle_{D^\star(z)} f(z), f(z)\rangle.$$

We claim that there is a constant $c < 1$ such that for all $\zeta \in cD^\star(z)$, there exists a polydisc (or ball) $D'(\zeta)$ centered ζ satisfying the following properties: (1) $z \in D'(\zeta)$; (2) $D'(\zeta) \subseteq D^\star(z)$; (3) $|D'(\zeta)| \approx |D^\star(z)|$. When $D^\star(z)$ is a Euclidean ball, this is obvious. When $D^\star(z)$ is a polydisc, the claim follows from (2.3) and (2.4). Then Fubini's theorem implies that

$$\begin{aligned} \int_\Omega \langle \tilde{\mathcal{W}} f(z), f(z)\rangle dV(z) &\gtrsim \int_\Omega |D^\star(z)|^{-1} \int_{D^\star(z)} \langle W(\zeta) f(z), f(z)\rangle dV(\zeta) dV(z) \\ &\gtrsim \int_\Omega |D'(\zeta)|^{-1} \int_{D'(\zeta)} \langle W(\zeta) f(z), f(z)\rangle dV(z) dV(\zeta). \end{aligned} \tag{4.11}$$

Note that $\langle W(\zeta) f(z), f(z)\rangle$ is subharmonic in z, we have

$$\int_\Omega |D'(\zeta)|^{-1} \int_{D'(\zeta)} \langle W(\zeta) f(z), f(z)\rangle dV(z) dV(\zeta) \geq \int_\Omega \langle W(\zeta) f(\zeta), f(\zeta)\rangle dV(\zeta).$$

This implies $\|f\|_{L^2(W)} \lesssim \|f\|_{L^2(\tilde{\mathcal{W}})}$. □

As a consequence of Lemma 4.4, the norm of P on the weighted space $L^2(W)$ can be dominated by the norm of P on the weighted space $L^2(\tilde{\mathcal{W}})$.

THEOREM 4.5. *Let W be a matrix $\mathcal{B}_2$ weight on Ω. Then $\|P\|_{L^2(W)} \lesssim \mathcal{B}_2^{1/2}(W)\|P\|_{L^2(\tilde{\mathcal{W}})}$.*

PROOF. By Lemma 4.4 and a duality argument, we have

$$\|P\|_{L^2(W)} \lesssim \|P : L^2(W) \to L^2(\tilde{\mathcal{W}})\| = \|P : L^2(\tilde{\mathcal{W}}^{-1}) \to L^2(W^{-1})\|.$$

Set

$$\tilde{\mathcal{W}}_{-1}(z) := \sum_{l=1}^{N} \sum_{\hat{K}_j^k \in \mathcal{T}_l} \langle W^{-1}\rangle_{K_j^k} 1_{K_j^k}(z).$$

Applying Lemma 4.4 again to the weight W^{-1} yields

$$\|P\|_{L^2(W)} \lesssim \|P : L^2(\tilde{\mathcal{W}}^{-1}) \to L^2(W^{-1})\| \lesssim \|P : L^2(\tilde{\mathcal{W}}^{-1}) \to L^2(\tilde{\mathcal{W}}_{-1})\|.$$

For any $z \in \Omega$, we consider those kubes K_m^l of $\hat{K}_j^k \in \bigcup_{l=1}^N \mathcal{T}_l$ that contain the point z. If $K_m^l \ni z$, then $\operatorname{dist}(K_m^l, \mathbf{b}\Omega) \approx \operatorname{dist}(z, \mathbf{b}\Omega)$. By Lemma 2.4 and Definition 2.7, there exists a constant r such that $\pi(K_m^l) \subseteq B(\pi(z), r)$ and $|K_m^l| \approx |B^\#(\pi(z), r)|$. Then Lemma 2.6 implies that there is a dyadic tent $\hat{K}_j^k \in \bigcup_{l=1}^N \mathcal{T}_l$ such that $\hat{K}_j^k \supseteq B^\#(\pi(z), r)$ with $|\hat{K}_j^k| \approx |B^\#(\pi(z), r)|$. Thus

$$\tilde{\mathcal{W}}(z) = \sum_{\hat{K}_m^l \in \bigcup_{l=1}^N \mathcal{T}_l} \langle W\rangle_{K_m^l} 1_{K_m^l}(z) \lesssim N \langle W\rangle_{\hat{K}_j^k}, \tag{4.12}$$

$$\tilde{\mathcal{W}}_{-1}(z) = \sum_{\hat{K}_m^l \in \bigcup_{l=1}^N \mathcal{T}_l} \langle W^{-1}\rangle_{K_m^l} 1_{K_m^l}(z) \lesssim N \langle W^{-1}\rangle_{\hat{K}_j^k}. \tag{4.13}$$

Inequalities (4.12) and (4.13) yield

$$|\tilde{\mathcal{W}}_{-1}^{1/2}(z)\tilde{\mathcal{W}}^{1/2}(z)|^2 \lesssim |\langle W^{-1}\rangle_{\hat{K}_j^k}^{1/2}\langle W\rangle_{\hat{K}_j^k}^{1/2}|^2 \le \mathcal{B}_2(W).$$

Hence, we have $|\tilde{\mathcal{W}}_{-1}^{1/2}(z)f(z)| \le \mathcal{B}_2^{1/2}(W)|\tilde{\mathcal{W}}^{-1/2}(z)f(z)|$, which implies

$$\|P\|_{L^2(W)} \lesssim \|P : L^2(\tilde{\mathcal{W}}^{-1}) \to L^2(\tilde{\mathcal{W}}_{-1})\| \lesssim \mathcal{B}_2^{1/2}(W)\|P\|_{L^2(\tilde{\mathcal{W}}^{-1})} = \mathcal{B}_2^{1/2}(W)\|P\|_{L^2(\tilde{\mathcal{W}})}.$$

□

5. Proof of Theorem 1.2

By Theorem 4.5 and the fact that $\tilde{\mathcal{W}} = \sum_{l=1}^N \mathcal{W}_l$, it suffices to show that

$$\|P\|_{L^2(\mathcal{W})} \lesssim \mathcal{B}_2^{3/2}(W),$$

where $\mathcal{W}$ is a weight defined as in Definition 4.1.

We begin with the following lemma which can be viewed as a special case of the Carleson Embedding Theorem for $\mathcal{T}$. We provide the proof of the lemma for the completeness of the paper.

LEMMA 5.1. *Let $\mathcal{T}$ be a collection of dyadic tents as in Lemma 2.8. Then for any scalar-valued measurable function f on Ω and for any $p \in (1,\infty)$,*

$$\sum_{\hat{K}_j^k \in \mathcal{T}} \langle f\rangle_{\hat{K}_j^k}^p |\hat{K}_j^k| \lesssim (p')^p \|f\|_{L^p(\Omega)}^p.$$

PROOF. Recall the maximal operator $M_{\mathcal{T}}$ given by

$$M_{\mathcal{T}} f(z) = \sup_{\hat{K}_j^k \in \mathcal{T}} \langle |f|\rangle_{\hat{K}_j^k} 1_{\hat{K}_j^k}(z).$$

For $p \in (1,\infty)$, a standard argument yields that $\|M_{\mathcal{T}}\|_{L^p(\Omega)} \le p'$. Recall also that for each tent $\hat{K}_j^k$, its induced kube K_j^k satisfying $|\hat{K}_j^k| \approx |K_j^k|$. Then

$$\sum_{\hat{K}_j^k \in \mathcal{T}} \langle |f|\rangle_{\hat{K}_j^k}^p |\hat{K}_j^k| \lesssim \sum_{\hat{K}_j^k \in \mathcal{T}} \langle |f|\rangle_{\hat{K}_j^k}^p |K_j^k| \le \int_\Omega (M_{\mathcal{T}} f)^p dV \le (p')^p \|f\|_{L^p(\Omega)}^p.$$

□

By Lemma 3.1 and Theorem 3.2, we need to estimate the norm of the operator

$$\sum_{\hat{K}_j^k \in \mathcal{T}} \int_{\hat{K}_j^k} H_{\hat{K}_j^k}(z,w) f(w) dV(w) 1_{\hat{K}_j^k}(z),$$

on the weighted space $L^2(\mathcal{W})$. Here the kernel function $H_{\hat{K}_j^k}$ satisfies $|H_{\hat{K}_j^k}(z,w)| \le |\hat{K}_j^k|^{-1}$. Thus it is equivalent to estimate the operator

$$Q_W(f)(z) = \sum_{\hat{K}_j^k \in \mathcal{T}} \int_{\hat{K}_j^k} H_{\hat{K}_j^k}(z,w) \mathcal{W}^{1/2}(z) \mathcal{W}^{-1/2}(w) f(w) dV(w) 1_{\hat{K}_j^k}(z),$$

over the unweighted space $L^2(\Omega, \mathbb{C}^d)$.

Set

$$\mathcal{W}_{-1}(z) = \sum_{\hat{K}_j^k \in \mathcal{T}} \langle W^{-1}\rangle_{K_j^k} 1_{K_j^k}(z).$$

For vector-valued functions $f,g\in L^2(\Omega,\mathbb{C}^d)$,

$$
\begin{aligned}
&|\langle Q_W(f),g\rangle|\\
&\le \sum_{\hat K_j^k\in\mathcal{T}}\left|\int_\Omega\left\langle\int_{\hat K_j^k}H_{\hat K_j^k}(z,w)\mathcal{W}^{-1/2}(w)f(w)dV(w)1_{\hat K_j^k}(z),\right.\right.\\
&\qquad\qquad\left.\left.\mathcal{W}^{1/2}(z)g(z)\right\rangle dV(z)\right|\\
&=\sum_{\hat K_j^k\in\mathcal{T}}\left|\int_\Omega\int_{\hat K_j^k}\left\langle H_{\hat K_j^k}(z,w)\langle\mathcal{W}^{-1}\rangle_{\hat K_j^k}^{1/2}\langle\mathcal{W}^{-1}\rangle_{\hat K_j^k}^{-1/2}\mathcal{W}^{-1/2}(w)f(w),\right.\right.\\
&\qquad\qquad\left.\left.\mathcal{W}^{1/2}(z)g(z)1_{\hat K_j^k}(z)\right\rangle dV(w)dV(z)\right|\\
&=\sum_{\hat K_j^k\in\mathcal{T}}\left|\int_{\hat K_j^k}\int_{\hat K_j^k}\left\langle\langle\mathcal{W}^{-1}\rangle_{\hat K_j^k}^{-1/2}\mathcal{W}^{-1/2}(w)f(w),\right.\right.\\
&\qquad\qquad\left.\left.\langle\mathcal{W}^{-1}\rangle_{\hat K_j^k}^{1/2}\mathcal{W}^{1/2}(z)g(z)H_{\hat K_j^k}(z,w)\right\rangle dV(w)dV(z)\right|\\
&=\sum_{\hat K_j^k\in\mathcal{T}}\left|\int_{\hat K_j^k}\left\langle\langle\mathcal{W}^{-1}\rangle_{\hat K_j^k}^{-1/2}\mathcal{W}^{-1/2}(w)f(w),\right.\right.\\
&\qquad\qquad\left.\left.\int_{\hat K_j^k}\langle\mathcal{W}^{-1}\rangle_{\hat K_j^k}^{1/2}\mathcal{W}^{1/2}(z)g(z)H_{\hat K_j^k}(z,w)dV(z)\right\rangle dV(w)\right|\\
&\le\sum_{\hat K_j^k\in\mathcal{T}}\int_{\hat K_j^k}\left|\langle\mathcal{W}^{-1}\rangle_{\hat K_j^k}^{-1/2}\mathcal{W}^{-1/2}(w)f(w)\right|\\
&\qquad\qquad\cdot\left|\int_{\hat K_j^k}\langle\mathcal{W}^{-1}\rangle_{\hat K_j^k}^{1/2}\mathcal{W}^{1/2}(z)g(z)H_{\hat K_j^k}(z,w)dV(z)\right|dV(w)\\
&\le\sum_{\hat K_j^k\in\mathcal{T}}\int_{\hat K_j^k}\left|\langle\mathcal{W}^{-1}\rangle_{\hat K_j^k}^{-1/2}\mathcal{W}^{-1/2}(w)f(w)\right|\\
&\qquad\qquad\cdot\int_{\hat K_j^k}\left|\langle\mathcal{W}^{-1}\rangle_{\hat K_j^k}^{1/2}\mathcal{W}^{1/2}(z)g(z)\right||\hat K_j^k|^{-1}dV(z)dV(w)\\
&\lesssim\sum_{\hat K_j^k\in\mathcal{T}}\left\langle\left|\langle\mathcal{W}^{-1}\rangle_{\hat K_j^k}^{-1/2}\mathcal{W}^{-1/2}f\right|\right\rangle_{\hat K_j^k}\left\langle\left|\langle\mathcal{W}^{-1}\rangle_{\hat K_j^k}^{1/2}\mathcal{W}^{1/2}g\right|\right\rangle_{\hat K_j^k}|K_j^k|
\end{aligned}
$$

$$
\le\left(\sum_{\hat K_j^k\in\mathcal{T}}\left\langle\left|\langle\mathcal{W}^{-1}\rangle_{\hat K_j^k}^{-1/2}\mathcal{W}^{-1/2}f\right|\right\rangle_{\hat K_j^k}^2|K_j^k|\right)^{\frac12}\left(\sum_{\hat K_j^k\in\mathcal{T}}\left\langle\left|\langle\mathcal{W}^{-1}\rangle_{\hat K_j^k}^{1/2}\mathcal{W}^{1/2}g\right|\right\rangle_{\hat K_j^k}^2|K_j^k|\right)^{\frac12}. \tag{5.1}
$$

Set operators $S_{1,W}$ and $S_{2,W}$ to be as follows:

$$
S_{1,W}(f)(z):=\sum_{\hat K_j^k\in\mathcal{T}}\left\langle\left|\langle\mathcal{W}^{-1}\rangle_{\hat K_j^k}^{-1/2}\mathcal{W}^{-1/2}f\right|\right\rangle_{\hat K_j^k}^2 1_{K_j^k}(z),
$$

$$
S_{2,W}(g)(z):=\sum_{\hat K_j^k\in\mathcal{T}}\left\langle\left|\langle\mathcal{W}^{-1}\rangle_{\hat K_j^k}^{1/2}\mathcal{W}^{1/2}g\right|\right\rangle_{\hat K_j^k}^2 1_{K_j^k}(z).
$$

For $S_{2,W}(g)$, Hölder's inequality implies

$$
\langle|\langle\mathcal{W}^{-1/2}\rangle_{\hat K_j^k}^{1/2}\mathcal{W}^{1/2}g|\rangle_{\hat K_j}\le\langle|\langle\mathcal{W}^{-1/2}\rangle_{\hat K_j^k}^{1/2}\mathcal{W}^{1/2}|^{2r}\rangle_{\hat K_j}^{1/2r}\langle|g|^{(2r)'}\rangle_{\hat K_j}^{1/(2r)'}. \tag{5.2}
$$

Choosing r as in Lemma 4.3 and applying the lemma, we have

$$
\begin{aligned}
\langle |\langle \mathcal{W}^{-1}\rangle_{\hat{K}_j^k}^{\frac{1}{2}} \mathcal{W}^{\frac{1}{2}}|^{2r}\rangle_{\hat{K}_j^k} &= |\hat{K}_j^k|^{-1}\int_{\hat{K}_j^k} |\langle \mathcal{W}^{-1}\rangle_{\hat{K}_j^k}^{\frac{1}{2}} \mathcal{W}(w)\langle \mathcal{W}^{-1}\rangle_{\hat{K}_j^k}^{\frac{1}{2}}|^{r} dV(w)\\
&\lesssim |\hat{K}_j^k|^{-1}\int_{\hat{K}_j^k} \left(\operatorname{Tr}\left(\langle \mathcal{W}^{-1}\rangle_{\hat{K}_j^k}^{\frac{1}{2}} \mathcal{W}(w)\langle \mathcal{W}^{-1}\rangle_{\hat{K}_j^k}^{\frac{1}{2}}\right)\right)^{r} dV(w)\\
&\lesssim |\hat{K}_j^k|^{-1}\int_{\hat{K}_j^k} \max_{1\le k\le d}\left\langle \langle \mathcal{W}^{-1}\rangle_{\hat{K}_j^k}^{\frac{1}{2}} \mathcal{W}(w)\langle \mathcal{W}^{-1}\rangle_{\hat{K}_j^k}^{\frac{1}{2}}e_k, e_k\right\rangle_{\mathbb{C}^d}^{r} dV(w)\\
&\le \sum_{k=1}^{d}|\hat{K}_j^k|^{-1}\int_{\hat{K}_j^k} \left\langle \langle \mathcal{W}^{-1}\rangle_{\hat{K}_j^k}^{\frac{1}{2}} \mathcal{W}(w)\langle \mathcal{W}^{-1}\rangle_{\hat{K}_j^k}^{\frac{1}{2}}e_k, e_k\right\rangle_{\mathbb{C}^d}^{r} dV(w)\\
&\lesssim \sum_{k=1}^{d}\left(|\hat{K}_j^k|^{-1}\int_{\hat{K}_j^k} \left\langle \langle \mathcal{W}^{-1}\rangle_{\hat{K}_j^k}^{\frac{1}{2}} \mathcal{W}(w)\langle \mathcal{W}^{-1}\rangle_{\hat{K}_j^k}^{\frac{1}{2}}e_k, e_k\right\rangle_{\mathbb{C}^d} dV(w)\right)^{r}\\
&\lesssim (\mathcal{B}_2(W))^r. \qquad (5.3)
\end{aligned}
$$

Applying this inequality to (5.2) gives

$$S_{2,W}(g)(z) \lesssim (\mathcal{B}_2(W))^{\frac{1}{2}} \sum_{\hat{K}_j^k\in\mathcal{T}} \langle |g|^{(2r)'}\rangle_{\hat{K}_j^k}^{\frac{1}{(2r)'}} 1_{K_j^k}(z).$$

Set $h = \|g\|^{(2r)'}$ and $p = 2/(2r)'$. Then

$$
\begin{aligned}
\|S_{2,W}(g)\|_{L^2(\Omega)}^2 &\lesssim \mathcal{B}_2(W) \sum_{\hat{K}_j^k\in\mathcal{T}} \langle h\rangle_{\hat{K}_j^k}^{p}|\hat{K}_j^k| \lesssim \mathcal{B}_2(W)(p')^p\|h\|_{L^p(\Omega)}^p\\
&= \mathcal{B}_2(W)(p')^p\|g\|_{L^2(\Omega)}^2,
\end{aligned}
$$

where the second inequality follows from Lemma 5.1. Since $p' = 2+\frac{1}{\epsilon} \lesssim \frac{1}{\epsilon}$ and $p\le 1+\epsilon$, we have $(p')^p \lesssim 1/\epsilon = (r-1)^{-1}$. From the choice of r in Lemma 4.3,

$$(r-1)^{-1} \approx \mathcal{B}_2(W).$$

Therefore,

$$\|S_{2,W}g(z)\|_{L^2(\Omega)}^2 \lesssim \mathcal{B}_2^2(W)\|g\|_{L^2(\Omega)}^2.$$

We turn to estimate the norm of $S_{1,W}f$. Note that

$$\int_{\hat{K}_j^k} \langle \mathcal{W}^{-1}\rangle_{\hat{K}_j^k}^{-\frac{1}{2}} \mathcal{W}^{-1}(w)\langle \mathcal{W}^{-1}\rangle_{\hat{K}_j^k}^{-\frac{1}{2}} dV(w) = \mathbf{I}_d.$$

Using this fact and going through the same argument as $S_{2,W}g$ yield that

$$\|S_{1,W}f(z)\|_{L^2(\Omega)}^2 \lesssim \mathcal{B}_2(W)\|f\|_{L^2(\Omega)}^2.$$

Combining these inequalities, we obtain

$$\langle Q_w f, g\rangle \lesssim (\mathcal{B}_2(W))^{\frac{3}{2}}\|f\|_{L^2(\Omega)}\|g\|_{L^2(\Omega)},$$

which shows the desired estimate $\|P\|_{L^2(\Omega,\mathcal{W})} \le \|Q_W\|_{L^2(\Omega)} \lesssim (\mathcal{B}_2(W))^{\frac{3}{2}}$.

6. Remarks

(1) By the proof of Theorem 3.2, it's not hard to see that the positive Bergman operator P^+ also belongs to a convex body valued sparse operator. The same argument in Section 5 then yields the following corollary for the weighted norm of P^+ on the weighted space $L^2(\tilde{\mathcal{W}})$:

COROLLARY 6.1. *Let Ω be a simple domain. Let W be a matrix $\mathcal{B}_2$ weight. Let $\tilde{\mathcal{W}}$ be a weight constructed based on W as in Definition* 4.1. *Then $\|P^+\|_{L^2(\tilde{\mathcal{W}})} \lesssim \mathcal{B}_2^{3/2}(W)$.*

It is worth noting that Limani and Pott [**LP21**] obtained a similar result for $\|P^+\|_{L^2(W)}$ on the unit disc, which then yields a sharper estimate of $\|P\|_{L^2(W)}$.

(2) As pointed out in [**HWW21b**] for the scalar-valued case, the products of averages of W and W^{-1} over the whole domain and over the small tents will have different impacts on the estimate for the weighted norm of the projection P. However, our estimate in Theorem 1.2 is less likely to be sharp. Hence we didn't consider this difference in here for the simplicity of the argument.

References

[AC12] A. Aleman and O. Constantin, *The Bergman projection on vector-valued L^2-spaces with operator-valued weights*, J. Funct. Anal. **262** (2012), no. 5, 2359–2378, DOI 10.1016/j.jfa.2011.12.005. MR3103020

[APR19] A. Aleman, S. Pott, and M. C. Reguera, *Characterizations of a limiting class B_∞ of Békollé-Bonami weights*, Rev. Mat. Iberoam. **35** (2019), no. 6, 1677–1692, DOI 10.4171/rmi/1097. MR4029779

[BB78] D. Bekollé and A. Bonami, *Inégalités à poids pour le noyau de Bergman* (French, with English summary), C. R. Acad. Sci. Paris Sér. A-B **286** (1978), no. 18, A775–A778. MR497663

[BB00] Z. M. Balogh and M. Bonk, *Gromov hyperbolicity and the Kobayashi metric on strictly pseudoconvex domains*, Comment. Math. Helv. **75** (2000), no. 3, 504–533, DOI 10.1007/s000140050138. MR1793800

[Bek82] D. Bekollé, *Inégalité à poids pour le projecteur de Bergman dans la boule unité de $\mathbf{C}^n$* (French), Studia Math. **71** (1981/82), no. 3, 305–323, DOI 10.4064/sm-71-3-305-323. MR667319

[BPW16] K. Bickel, S. Petermichl, and B. D. Wick, *Bounds for the Hilbert transform with matrix A_2 weights*, J. Funct. Anal. **270** (2016), no. 5, 1719–1743, DOI 10.1016/j.jfa.2015.12.006. MR3452715

[BW16] K. Bickel and B. D. Wick, *A study of the matrix Carleson embedding theorem with applications to sparse operators*, J. Math. Anal. Appl. **435** (2016), no. 1, 229–243, DOI 10.1016/j.jmaa.2015.10.023. MR3423392

[CG01] M. Christ and M. Goldberg, *Vector A_2 weights and a Hardy-Littlewood maximal function*, Trans. Amer. Math. Soc. **353** (2001), no. 5, 1995–2002, DOI 10.1090/S0002-9947-01-02759-3. MR1813604

[D'A82] J. P. D'Angelo, *Real hypersurfaces, orders of contact, and applications*, Ann. of Math. (2) **115** (1982), no. 3, 615–637, DOI 10.2307/2007015. MR657241

[DMRO16] J. Duoandikoetxea, F. J. Martín-Reyes, and S. Ombrosi, *On the A_∞ conditions for general bases*, Math. Z. **282** (2016), no. 3-4, 955–972, DOI 10.1007/s00209-015-1572-y. MR3473651

[GHK22] C. Gan, B. Hu, and I. Khan, *Dyadic decomposition of convex domains of finite type and applications*, Math. Z. **301** (2022), no. 2, 1939–1962, DOI 10.1007/s00209-022-02984-y. MR4418342

[Gol03] M. Goldberg, *Matrix A_p weights via maximal functions*, Pacific J. Math. **211** (2003), no. 2, 201–220, DOI 10.2140/pjm.2003.211.201. MR2015733

[Gra14] L. Grafakos, *Classical Fourier analysis*, 3rd ed., Graduate Texts in Mathematics, vol. 249, Springer, New York, 2014, DOI 10.1007/978-1-4939-1194-3. MR3243734

[HK12] T. Hytönen and A. Kairema, *Systems of dyadic cubes in a doubling metric space*, Colloq. Math. **126** (2012), no. 1, 1–33, DOI 10.4064/cm126-1-1. MR2901199

[HMW73] R. Hunt, B. Muckenhoupt, and R. Wheeden, *Weighted norm inequalities for the conjugate function and Hilbert transform*, Trans. Amer. Math. Soc. **176** (1973), 227–251, DOI 10.2307/1996205. MR312139

[HPV19] T. Hytönen, S. Petermichl, and A. Volberg, *The sharp square function estimate with matrix weight*, Discrete Anal., posted on 2019, Paper No. 2, 8, DOI 10.19086/da. MR3936542

[HW20] Z. Huo and B. D. Wick, *Weighted estimates for the Bergman projection on the Hartogs triangle*, J. Funct. Anal. **279** (2020), no. 9, 108727, 34, DOI 10.1016/j.jfa.2020.108727. MR4134894

[HWW21a] Z. Huo, N. A. Wagner, and B. D. Wick, *A Békollè-Bonami class of weights for certain pseudoconvex domains*, J. Geom. Anal. **31** (2021), no. 6, 6042–6066, DOI 10.1007/s12220-020-00516-w. MR4267636

[HWW21b] Z. Huo, N. A. Wagner, and B. D. Wick, *Bekollé-Bonami estimates on some pseudoconvex domains*, Bull. Sci. Math. **170** (2021), Paper No. 102993, 36, DOI 10.1016/j.bulsci.2021.102993. MR4263006

[Hyt12] T. P. Hytönen, *The sharp weighted bound for general Calderón-Zygmund operators*, Ann. of Math. (2) **175** (2012), no. 3, 1473–1506, DOI 10.4007/annals.2012.175.3.9. MR2912709

[IKP17] J. Isralowitz, H.-K. Kwon, and S. Pott, *Matrix weighted norm inequalities for commutators and paraproducts with matrix symbols*, J. Lond. Math. Soc. (2) **96** (2017), no. 1, 243–270, DOI 10.1112/jlms.12053. MR3687948

[Lac17] M. T. Lacey, *An elementary proof of the A_2 bound*, Israel J. Math. **217** (2017), no. 1, 181–195, DOI 10.1007/s11856-017-1442-x. MR3625108

[Ler13] A. K. Lerner, *A simple proof of the A_2 conjecture*, Int. Math. Res. Not. IMRN **14** (2013), 3159–3170, DOI 10.1093/imrn/rns145. MR3085756

[LP21] A. Limani and S. Pott, *Sparse Lerner operators in infinite dimensions*, Preprint, `arXiv:2103.17005`, 2022.

[McN91] J. D. McNeal, *Local geometry of decoupled pseudoconvex domains*, Complex analysis (Wuppertal, 1991), Aspects Math., E17, Friedr. Vieweg, Braunschweig, 1991, pp. 223–230. MR1122183

[McN94a] J. D. McNeal, *The Bergman projection as a singular integral operator*, J. Geom. Anal. **4** (1994), no. 1, 91–103, DOI 10.1007/BF02921594. MR1274139

[McN94b] J. D. McNeal, *Estimates on the Bergman kernels of convex domains*, Adv. Math. **109** (1994), no. 1, 108–139, DOI 10.1006/aima.1994.1082. MR1302759

[McN03] J. D. McNeal, *Subelliptic estimates and scaling in the $\overline{\partial}$-Neumann problem*, Explorations in complex and Riemannian geometry, Contemp. Math., vol. 332, Amer. Math. Soc., Providence, RI, 2003, pp. 197–217, DOI 10.1090/conm/332/05937. MR2018340

[NPTV17] F. Nazarov, S. Petermichl, S. Treil, and A. Volberg, *Convex body domination and weighted estimates with matrix weights*, Adv. Math. **318** (2017), 279–306, DOI 10.1016/j.aim.2017.08.001. MR3689742

[NT96] F. L. Nazarov and S. R. Treĭl′, *The hunt for a Bellman function: applications to estimates for singular integral operators and to other classical problems of harmonic analysis* (Russian, with Russian summary), Algebra i Analiz **8** (1996), no. 5, 32–162; English transl., St. Petersburg Math. J. **8** (1997), no. 5, 721–824. MR1428988

[Pet07] S. Petermichl, *The sharp bound for the Hilbert transform on weighted Lebesgue spaces in terms of the classical A_p characteristic*, Amer. J. Math. **129** (2007), no. 5, 1355–1375, DOI 10.1353/ajm.2007.0036. MR2354322

[PR13] S. Pott and M. C. Reguera, *Sharp Békollé estimates for the Bergman projection*, J. Funct. Anal. **265** (2013), no. 12, 3233–3244, DOI 10.1016/j.jfa.2013.08.018. MR3110501

[RTW17] R. Rahm, E. Tchoundja, and B. D. Wick, *Weighted estimates for the Berezin transform and Bergman projection on the unit ball*, Math. Z. **286** (2017), no. 3-4, 1465–1478, DOI 10.1007/s00209-016-1809-4. MR3671584

[TV97a] S. Treil and A. Volberg, *Continuous frame decomposition and a vector Hunt-Muckenhoupt-Wheeden theorem*, Ark. Mat. **35** (1997), no. 2, 363–386, DOI 10.1007/BF02559975. MR1478786

[TV97b] S. Treil and A. Volberg, *Wavelets and the angle between past and future*, J. Funct. Anal. **143** (1997), no. 2, 269–308, DOI 10.1006/jfan.1996.2986. MR1428818

[Vol97] A. Volberg, *Matrix A_p weights via S-functions*, J. Amer. Math. Soc. **10** (1997), no. 2, 445–466, DOI 10.1090/S0894-0347-97-00233-6. MR1423034

[Wit00] J. Wittwer, *A sharp estimate on the norm of the martingale transform*, Math. Res. Lett. **7** (2000), no. 1, 1–12, DOI 10.4310/MRL.2000.v7.n1.a1. MR1748283

Zu Chongzhi Center for Mathematics and Computational Sciences, Division of Natural and Applied Sciences, Duke Kunshan University, Kunshan, Jiangsu 215316, People's Republic of China

Email address: zhenghui.huo@duke.edu

Department of Mathematics and Statistics, Washington University in Saint Louis, Saint Louis, Missouri 63130-4899

Email address: bwick@wustl.edu

Contemporary Mathematics
Volume **799**, 2024
https://doi.org/10.1090/conm/799/16021

h^1 boundedness of localized operators and commutators with bmo and lmo

Galia Dafni and Chun Ho Lau

ABSTRACT. We first consider two types of localizations of singular integral operators of convolution type, and show, under mild decay and smoothness conditions on the auxiliary functions, that their boundedness on the local Hardy space $h^1(\mathbb{R}^n)$ is equivalent. We then study the boundedness on $h^1(\mathbb{R}^n)$ of the commutator $[b, T]$ of an inhomogeneous singular integral operator with b in bmo$(\mathbb{R}^n)$, the nonhomogeneous space of functions of bounded mean oscillation. We define local analogues of the atomic space $H^1_b(\mathbb{R}^n)$ introduced by Pérez in the case of the homogeneous Hardy space and BMO, including a variation involving atoms with approximate cancellation conditions. For such an atom a, we prove integrability of the associated commutator maximal function and of $[b, T](a)$. For b in lmo$(\mathbb{R}^n)$, this gives h^1 to L^1 boundedness of $[b, T]$. Finally, under additional approximate cancellation conditions on T, we show boundedness from h^1 to h^1.

1. Introduction

The real Hardy spaces $H^p(\mathbb{R}^n)$ are a class of function spaces that has been extensively studied in harmonic analysis for over 50 years. In particular, while the boundedness of singular integral operators on $L^p(\mathbb{R}^n)$ for $1 < p < \infty$ does not extend to $0 < p \leq 1$, under sufficient smoothness and cancellation conditions such operators are bounded on $H^p(\mathbb{R}^n)$, $0 < p \leq 1$. At the other end of the scale, $p = \infty$, the space BMO$(\mathbb{R}^n)$ of functions of bounded mean oscillation introduced by John and Nirenberg [**32**] replaces L^∞ and is the dual of H^1—see the fundamental paper of Fefferman and Stein [**20**] and more recent exposition in [**46**].

The focus of our paper is on a nonhomogeneous version of these spaces, introduced by Goldberg [**23, 24**] under the name "local Hardy spaces", denoted $h^p(\mathbb{R}^n)$. Membership in $h^p(\mathbb{R}^n)$ does not require the global vanishing moment conditions necessary for $f \in H^p(\mathbb{R}^n)$ and so, unlike the case of $H^p(\mathbb{R}^n)$, important classes of smooth functions such as the Schwartz spaces $\mathcal{S}(\mathbb{R}^n)$ are contained (and are dense) in $h^p(\mathbb{R}^n)$. Moreover, $h^p(\mathbb{R}^n)$ is closed under multiplication by smooth cut-off functions, which is why these are sometimes called "localizable" Hardy spaces, and are more suitable to analysis in different settings such as domains, manifolds and spaces of homogeneous type, as well as the study of partial differential equations and pseudo-differential operators.

The authors were partially supported by the Natural Sciences and Engineering Research Council (NSERC) of Canada and the Centre de Recherches Mathématiques (CRM).

We restrict ourselves to $h^1(\mathbb{R}^n)$, which satisfies $H^1(\mathbb{R}^n) \subsetneq h^1(\mathbb{R}^n) \subsetneq L^1(\mathbb{R}^n)$, and the nonhomogeneous version of $\mathrm{BMO}(\mathbb{R}^n)$, denoted $\mathrm{bmo}(\mathbb{R}^n)$ and identified by Goldberg as the dual of $h^1(\mathbb{R}^n)$. In addition to characterizations via maximal functions and atomic decomposition, Goldberg also identified $h^1(\mathbb{R}^n)$ as the space of functions in $L^1(\mathbb{R}^n)$ whose "local Riesz transforms" r_j, $j = 1, \ldots, n$, are in $L^1(\mathbb{R}^n)$. The r_j are obtained by localizing the multipliers of the Riesz transforms R_j on the Fourier transform side via multiplication with a Schwartz function vanishing near the origin. We consider more general localizations of Calderón–Zygmund convolution-type singular integral operators, obtained either by multiplying the kernel by an appropriate function (as in the "truncated" Hilbert transform defined in [**23**]), or localizing on the Fourier transform side. We show, under mild decay and smoothness assumptions on the auxiliary functions, that the boundedness of different localizations of the same kernel on $h^1(\mathbb{R}^n)$ is equivalent (Theorem 3.4), allowing us to characterize h^1 by a broader class of localized Riesz transforms (Corollary 3.5).

We then turn to another important class of operators, the commutators of singular integral operators with multiplication operators: $[b, T]f := bT(f) - T(bf)$. Coifman, Rochberg and Weiss [**10**] proved that $[b, T]$, where $b \in BMO(\mathbb{R}^n)$ and T is a Calderón–Zygmund operator, is bounded on $L^p(\mathbb{R}^n)$ for all $1 < p < \infty$. They also proved a converse: for the Riesz transforms R_j, if $[b, R_j]$, $j = 1, \ldots, n$, are all bounded on $L^p(\mathbb{R}^n)$ for some $1 < p < \infty$, then $b \in BMO(\mathbb{R}^n)$. Subsequently, Uchiyama [**49**] proved that $[b, T]$ is compact on $L^p(\mathbb{R}^n)$, $1 < p < \infty$, if and only if $b \in \mathrm{CMO}(\mathbb{R}^n)$, where $\mathrm{CMO}(\mathbb{R}^n)$ is the $\mathrm{BMO}(\mathbb{R}^n)$-closure of $C_c^\infty(\mathbb{R}^n)$. Janson [**31**] extended the boundedness of the commutator to Orlicz-type spaces. Recently, Hytönen [**30**] completely characterized $L^p \to L^q$ boundedness.

The above results are all for L^p spaces with $1 < p < \infty$. We are interested in the endpoint case $p = 1$. For a Calderón–Zygmund operator T, we have weak-type $(1, 1)$ boundedness as well as from $H^1(\mathbb{R}^n)$ to $L^1(\mathbb{R}^n)$. Does the commutator $[b, T]$ also satisfy this type of boundedness? The answer turns out to be no. Harboure, Segovia and Torrea [**26**] proved that there is no way to get boundedness of $[b, H]$ from $H^1(\mathbb{R})$ to $L^1(\mathbb{R})$, where H be the Hilbert transform, unless b is a constant function, and this also holds in the case of boundedness of $[b, H]$ from $L_c^\infty(\mathbb{R})$ to $BMO(\mathbb{R})$.

This means we need to seek a smaller subspace of H^1 to get the integrability of the commutator. In addition to proving weak-type $L \operatorname{Log} L$ inequalities for the commutator $[b, T]$, Pérez [**39**] introduced a special type of H^1 atom that has extra cancellation against b, and showed that the commutator $[b, T](a) \in L^1(\mathbb{R}^n)$ for all such a. He then defined the space $H_b^1(\mathbb{R}^n)$ via atomic decompositions. The boundedness on infinite linear combinations of atoms, however, does not follow from the boundedness on one atom (see [**7**]), so another characterization of the space was needed.

Ky [**34**] introduced another space $H_b^1(\mathbb{R}^n)$ through a specially defined commutator maximal function, and proved that it is the largest subspace of $H^1(\mathbb{R}^n)$ such that the commutator $[b, T]$ is bounded from this space to $L^1(\mathbb{R}^n)$. As Pérez's atomic space is continuously contained in Ky's space, the boundedness on atoms could be extended to get boundedness on the whole atomic space. The reverse inclusion, namely whether every function in Ky's maximal $H_b^1(\mathbb{R}^n)$ space has an atomic decomposition into Pérez atoms, was not shown. Ky also showed that under suitable assumptions, $[b, T]$ is bounded from H^1 to $h^1(\mathbb{R}^n)$ and also to $H^1(\mathbb{R}^n)$, and these results were extended to metric measure space of homogeneous type by Fu, D. Yang and S. Yang [**21**].

Unlike the case for $H^1(\mathbb{R}^n)$, commutators of the form $[b, T]$ with nonconstant b can be bounded on $h^1(\mathbb{R}^n)$. Hung and Ky [**28**] proved such boundedness for b with logarithmically vanishing mean oscillation and T in a class of pseudo-differential operators.

In this paper we introduce an $h^1(\mathbb{R}^n)$ version of the Pérez atomic space, as well as another atomic space which we call $h^1_{\text{atom},b}(\mathbb{R}^n)$, containing it, defined using new h^1_b atoms which only satisfy approximate cancellation conditions. We then study the boundedness of $[b, T]$ for $b \in \text{bmo}(\mathbb{R}^n)$ and T an inhomogeneous singular integral operator, and show that such commutators take h^1_b atoms to $L^1(\mathbb{R}^n)$ (Theorem 5.1). Moreover, imposing nonhomogeneous cancellation conditions on T by considering T^*1 and T^*b, we are able to show boundedness to $h^1(\mathbb{R}^n)$ (Theorem 5.4). However, as in the case of the Pérez atomic space H^1_b, it is not possible to extend this boundedness to all of $h^1_{\text{atom},b}(\mathbb{R}^n)$ without a different characterization of the space. While we do not have a maximal characterization, we do show that a local analogue of Ky's commutator maximal function takes h^1_b atoms to $L^1(\mathbb{R}^n)$.

Finally, restricting b to $\text{lmo}(\mathbb{R}^n)$, the functions of logarithmically vanishing mean oscillation in $\text{bmo}(\mathbb{R}^n)$, we are able to identify the space $h^1_{\text{atom},b}(\mathbb{R}^n)$ with $h^1(\mathbb{R}^n)$ (Theorem 4.10) and thus conclude the boundedness of the commutator on the whole space, generalizing the results of Hung and Ky [**28**].

Section 2 gives some background on h^1, with particular emphasis on atomic decompositions and approximate cancellation conditions, as well as on bmo and lmo. Section 3 deals with the various localizations of Calderón–Zygmund operators of convolution type and their boundedness on h^1, and ends with the definition of inhomogeneous singular integral operators. Section 4 introduces the commutator maximal functions and atomic spaces in the nonhomogeneous case, and proves relations between them. Finally, Section 5 contains the results on the boundedness of the commutators on the atoms.

2. Preliminaries

The Lebesgue measure of a set $A \subset \mathbb{R}^n$ is denoted by $|A|$, and the average of an integrable function f on a set E of positive finite measure by

$$f_E := \fint_E f := \frac{1}{|E|} \int_E f(x)dx.$$

Balls will be denoted by $B(x_0, r)$ and the radius of a ball B by $r(B)$. The dilated ball $B(x_0, \delta r)$, $\delta > 0$, will be denoted by δB.

The constant C used in inequalities will may vary from line to line. We write $\alpha \lesssim \beta$ if there exists a constant C (which may depend on other factors but not on α and β) such that $\alpha \leq C\beta$, and $\alpha \approx \beta$ if $\alpha \lesssim \beta$ and $\beta \lesssim \alpha$.

We denote the Schwartz space by $\mathcal{S}(\mathbb{R}^n)$ and the seminorms defining its topology by

$$\|\varphi\|_{\alpha,\beta} := \sup_{x\in\mathbb{R}^n} |x^\alpha \partial_x^\beta \varphi(x)|, \quad \alpha, \beta \in \mathbb{Z}^n_+.$$

Its dual, $\mathcal{S}'(\mathbb{R}^n)$, is the space of tempered distributions.

2.1. Local Hardy spaces. Goldberg [**23, 24**] introduced nonhomogeneous versions of the real Hardy spaces $H^p(\mathbb{R}^n)$, called "local Hardy spaces" and denoted by $h^p(\mathbb{R}^n)$, and characterized them by means of maximal functions and atomic decomposition.

DEFINITION 2.1 ([**23**]). For $0 < p \leq 1$, a (normalized) Goldberg h^p atom a is a function supported in a cube Q with

(1) $\|a\|_{L^\infty} \leq |Q|^{-1}$;
(2) if $|Q| < 1$, $\int a(x)x^\alpha dx = 0$ for $|\alpha| \leq \lfloor n(1/p - 1)\rfloor$.

An atomic decomposition for h^p where the atoms need only have approximate moment conditions was introduced in [**12**] and later refined in [**17, 18**] for the case $p = 1$ in the more general setting of spaces of homogeneous type (see also [**14, 15, 33**] for $p < 1$).

DEFINITION 2.2 ([**18**]). Fix $R > 0$, $1 < q \leq \infty$. We say a is an R-approximate $(1, q)$ atom if

- $\operatorname{supp}(a) \subset B(x_0, r)$ for some $x_0 \in \mathbb{R}^n$ and $r > 0$;
- $\|a\|_{L^q} \leq |B(x_0, r)|^{\frac{1}{q}-1}$;
- $|\int a| \leq [\log(1 + \frac{R}{r})]^{-1}$.

Any of the following equivalent characterizations can be taken as the definition of $h^1(\mathbb{R}^n)$.

PROPOSITION 2.3. *Let $f \in \mathcal{S}'(\mathbb{R}^n)$. Then the following are equivalent:*

(1) *For a fixed $\psi \in \mathcal{S}(\mathbb{R}^n)$ with $\int \psi \neq 0$,*

$$\|f\|_{h^1,\psi} := \|M_\psi(f)\|_{L^1(\mathbb{R}^n)} < \infty,$$

where

$$M_\psi(f)(x) = \sup_{0<t<T} |f * \psi_t(x)|, \quad \psi_t(y) = t^{-n}\psi(t^{-1}y).$$

(2) *For a fixed $T > 0$,*

$$\|f\|_{h^1,T} := \|\mathcal{M}(f)\|_{L^1(\mathbb{R}^n)} < \infty,$$

where

$$\mathcal{M}(f)(x) := \sup_\phi |\langle f, \phi\rangle|,$$

and the supremum is taken over all $\phi \in C^1(\mathbb{R}^n)$ with $\operatorname{supp}(\phi) \subset B(x,t)$, $0 < t < T$, $\|\phi\|_{L^\infty} \leq |B(x,t)|^{-1}$ and $\|\nabla\phi\|_{L^\infty} \leq [t|B(x,t)|]^{-1}$.

(3) *For N sufficient large*

$$\|f\|_{h^1,\max} := \|\mathfrak{M}(f)\|_{L^1(\mathbb{R}^n)} < \infty,$$

where

$$\mathfrak{M}f(x) := \sup\{|f * \varphi_t(x)| : 0 < t < 1, \varphi \in \mathcal{F}_N\}$$

and

$$\mathcal{F}_N = \{\varphi \in \mathcal{S}(\mathbb{R}^n) : \|\varphi\|_{\alpha,\beta} \leq 1 \;\forall\; |\alpha|, |\beta| \leq N\}.$$

(4)

$$\|f\|_{h^1,\mathrm{atom}} := \inf \sum |\lambda_j| < \infty,$$

where the infimum is taken over all decompositions $f = \sum \lambda_j a_j$ *in* $\mathcal{S}'(\mathbb{R}^n)$ *with* $\{\lambda_j\} \in \ell^1$ *and where the* a_j *are Goldberg* h^1 *atoms as in Definition* 2.1.

(5) *For a fixed* $R > 0$, $1 < q \leq \infty$,

$$\|f\|_{h^1,R} := \inf \sum |\lambda_j| < \infty,$$

where the infimum is taken over all decompositions $f = \sum \lambda_j a_j$ *in* $\mathcal{S}'(\mathbb{R}^n)$ *with* $\{\lambda_j\} \in \ell^1$ *and where the* a_j *are* R*-approximate* $(1, q)$ *atoms as in Definition* 2.2.

In case these conditions hold, we have

$$\|f\|_{h^1,\psi} \approx \|f\|_{h^1,T} \approx \|f\|_{h^1,\max} \approx \|f\|_{h^1,\mathrm{atom}} \approx \|f\|_{h^1,R},$$

where the constants depend on the choices of T, R *and* N.

The equivalence of conditions (1), (3) and (4) of Proposition 2.3 is contained in [**23**, Theorem 1 and Lemma 5]. The equivalence of conditions (2), (4) and (5) is shown in [**18**, Section 7] on a doubling metric-measure space, with C^1 replaced by Lipschitz, for arbitrary T, R and q (see also [**14**, Lemma 1] for a proof that (4) implies (2) in the more general case of all $p \leq 1$). The maximal functions $\mathcal{M}$ and $\mathfrak{M}$ are often given the name *grand maximal function.* It is also possible to define nontangential versions of these (see Definition 4.1 for the case of H^1) and get an equivalent characterization of h^1 (resp. H^1)—this is part of Goldberg's theorem (see [**46**, Section III.1] for the homogeneous Hardy spaces). Further characterizations of local Hardy spaces on spaces of homogeneous type were given in [**27**].

In what follows, we will use mostly condition (2) with $T = 1$, and abbreviate $\|f\|_{h^1,1}$ to $\|f\|_{h^1}$, and condition (5) with $R = 1$ and $q = 2$, in which case we will call the 1 approximate $(1, 2)$ atoms simply h^1 atoms.

We will also need the notion of a molecule, which is a generalization of an atom not requiring compact support. Molecules are useful in showing boundedness of operators on h^p since, typically, singular integral operators take atoms to molecules. If we know that molecules are in h^p with bounded norm, then we have shown the boundedness of the operator on atoms.

DEFINITION 2.4 ([**14**]). Let $1 < s < \infty$, $\lambda > n(s-1)$. A measurable function M on $\mathbb{R}^n$ is called a (s, λ) molecule (for $h^1(\mathbb{R}^n)$) if there exists a ball $B = B(x_0, r)$ such that

(M1) $\|M\|_{L^s(B)} \leq r^{n(\frac{1}{s}-1)}$

(M2) $\|M\,|\cdot - x_0|^{\frac{\lambda}{s}}\|_{L^s(B^c)} \leq r^{\frac{\lambda}{s}+n(\frac{1}{s}-1)}$

(M3) $\left|\int M\right| \leq [\ln(1 + r^{-1})]^{-1}$.

PROPOSITION 2.5 ([**14**]). *There exists a constant* $C_{n,s,\lambda}$ *such that* $\|M\|_{h^1} \leq C_{n,s,\lambda}$ *every* (s, λ) *molecule* M.

Conversely, for any $f \in h^1(\mathbb{R}^n)$, there exist a sequence $\{\lambda_j\} \in \ell^1$ and a sequence of h^1 molecules $\{M_j\}$ such that

$$f = \sum_{j=1}^{\infty} \lambda_j M_j \quad \text{in } h^1(\mathbb{R}^n).$$

Moreover,

$$\|f\|_{h^1} \approx \inf \sum |\lambda_j| < \infty,$$

where the infimum is taken over all such decompositions.

The following proposition (case $p = 1$ of [**15**, Proposition 1]) shows that the cancellation condition on atoms in Definition 2.2 is not only sufficient but also necessary.

PROPOSITION 2.6 ([**15**]). *Suppose $g \in h^1(\mathbb{R}^n)$ with* $\operatorname{supp}(g) \subset B(x_0, r)$. *Then*

$$\left| \int g \right| \lesssim \frac{\|g\|_{h^1}}{\log(1 + r^{-1})}.$$

2.2. Nonhomogeneous BMO. Goldberg [**24**, Corollary 1] identifies the dual of $h^1(\mathbb{R}^n)$ with the space $\operatorname{bmo}(\mathbb{R}^n)$, a nonhomogeneous version of the John–Nirenberg space $\operatorname{BMO}(\mathbb{R}^n)$ of functions of bounded mean oscillation. See Bourdaud [**6**] for an exposition of the properties of $\operatorname{bmo}(\mathbb{R}^n)$. In the following definition, we use the notation introduced in [**18**], where it is shown that the definition is independent of the choice of scale used to distinguish large and small balls. Note that unlike $\operatorname{BMO}(\mathbb{R}^n)$, we do not need to take the space modulo constants in order to get a Banach space.

DEFINITION 2.7. Let $b \in L^1_{\text{loc}}(\mathbb{R}^n)$. We say $b \in \operatorname{bmo}(\mathbb{R}^n)$ if

$$\|b\|_{\text{bmo}} := \sup_B \fint_B |b(x) - c_B| dx < \infty,$$

where the supremum is taken over all balls, and for each ball B we set

$$c_B := \begin{cases} b_B := \fint_B b, & \text{if } r(B) < 1, \\ 0, & \text{if } r(B) \geq 1. \end{cases} \tag{2.1}$$

Since there will only be one bmo function under discussion at any point in time, we will continue to use the notation c_B defined in (2.1) in the rest of the paper. We can then state a corresponding version of the John–Nirenberg inequality for $b \in \operatorname{bmo}(\mathbb{R}^n)$ (see [**18**, Theorem 3.1]): there exist $C, c > 0$ such that for any ball B,

$$|\{x \in B : |b(x) - c_B| > \lambda\}| \leq C|B|e^{-\lambda c/\|b\|_{\text{bmo}}}$$

for all $\lambda > 0$. As consequence, one gets that

$$\|b\|_{\text{bmo}} \approx \|b\|_{\text{bmo},p} := \left(\sup_B \fint_B |b(x) - c_B|^p dx \right)^{1/p} \tag{2.2}$$

for $1 \leq p < \infty$, with constants depending on p.

This gives us the following useful inequality (see the proof of [**10**, Lemma 2.4] for the case $\delta = 1$ and $b \in \operatorname{BMO}(\mathbb{R}^n)$).

LEMMA 2.8. *Let $b \in \mathrm{bmo}(\mathbb{R}^n)$. Then for any $\delta > 0$, $p \geq 1$, there exists $C_{n,p,\delta} > 0$, independent of b, x_0 and r, such that*

$$r^\delta \int_{|x-x_0|>r} \frac{|b(x) - c_{B(x_0,r)}|^p}{|x - x_0|^{n+\delta}} \leq C_{n,p,\delta}\|b\|^p_{\mathrm{bmo}}. \tag{2.3}$$

PROOF. By translation invariance of $\mathrm{bmo}(\mathbb{R}^n)$, we may assume $x_0 = 0$. Denote $B(0, 2^k r)$ by B_k, $k = 0, 1, 2, \ldots$. Then using (2.2) and the fact that for $2^j r < 1$

$$|c_{B_j} - c_{B_{j-1}}| = |b_{B_j} - b_{B_{j-1}}| \leq 2^n \|b\|_{\mathrm{bmo}} \tag{2.4}$$

and therefore $c_{B_0} \lesssim \|b\|_{\mathrm{bmo}} \log(1 + r^{-1})$ (see also [**18**, Lemma 6.1]), we have

$$\begin{aligned} r^\delta \int_{\mathbb{R}^n \setminus B_0} \frac{|b(x) - c_{B_0}|^p}{|x|^{n+\delta}} &\lesssim \sum_{k=0}^{\infty} 2^{-k\delta} |B_k|^{-1} \int_{B_{k+1}\setminus B_k} |b(x) - c_{B_{k+1}}|^p + |c_{B_{k+1}} - c_{B_0}|^p dx \\ &\lesssim \sum_{k=0}^{\infty} 2^{-k\delta} \|b\|^p_{\mathrm{bmo}} (1 + (\min\{k+1, \log_+ r^{-1}\})^p) dx \\ &\lesssim \|b\|^p_{\mathrm{bmo}}. \end{aligned}$$

□

While bmo is the dual of the local Hardy space h^1, it does not make sense to call it "local BMO" as it is actually strictly smaller. It is more appropriate to give that name to the space consisting of locally integrable functions b for which

$$\|b\|_{\mathrm{BMO}_{\mathrm{loc}}(\mathbb{R}^n)} := \sup_{r(B)<R} \fint_B |b(x) - b_B| dx < \infty,$$

where R is some fixed constant (see [**48**]). This space is strictly larger than $\mathrm{BMO}(\mathbb{R}^n)$ since it contains all Lipschitz functions, for example, and is independent of the choice of $R < \infty$. Taking those elements of $\mathrm{BMO}_{\mathrm{loc}}$ for which $\sup_{r(B)\geq R} \fint |b| < \infty$ gives $\mathrm{bmo}(\mathbb{R}^n)$ (in [**6**] it is shown that it suffices to take $\sup_{r(B)=R} \fint |b| < \infty$). In particular $\mathrm{BMO}_{\mathrm{loc}}(\mathbb{R}^n) \cap L^p(\mathbb{R}^n) \subset \mathrm{bmo}(\mathbb{R}^n)$ for $1 \leq p < \infty$.

Vanishing mean oscillation refers to the property that

$$\lim_{R\to 0} \sup_{r(B)<R} \fint_B |b(x) - b_B| dx = 0. \tag{2.5}$$

For $b \in \mathrm{BMO}(\mathbb{R}^n)$, this holds if and only if $b \in \mathrm{VMO}(\mathbb{R}^n)$, the closure of the uniformly continuous functions in $\mathrm{BMO}(\mathbb{R}^n)$, as shown by Sarason [**40**] for $n = 1$. Analogously, we can denote by $\mathrm{vmo}(\mathbb{R}^n)$ the space of functions in $\mathrm{bmo}(\mathbb{R}^n)$ which have vanishing mean oscillation, or alternatively the closure of the bounded uniformly continuous functions in $\mathrm{bmo}(\mathbb{R}^n)$. Finally, we use $\mathrm{cmo}(\mathbb{R}^n)$ to denote the closure of the continuous (or smooth) functions with compact support in $\mathrm{bmo}(\mathbb{R}^n)$. More about these spaces can be found in [**6**] and [**13**] (where the notation vmo is used for cmo). It was shown there that $h^1(\mathbb{R}^n)$ can be identified with the dual of $\mathrm{cmo}(\mathbb{R}^n)$ and that $b \in \mathrm{cmo}(\mathbb{R}^n)$ if and only if (2.5) holds and in addition

$$\lim_{R\to\infty} \sup_{\substack{|B|\geq 1 \\ B\subset (B(0,R))^c}} \fint_B |b| = 0. \tag{2.6}$$

One can specify the rate at which the oscillation vanishes as the radius goes to zero in (2.5) (see [**40**, Chapter 5], [**43**]). A special class are those functions of *logarithmic mean oscillation*, usually denoted LMO (not to be confused with *bounded lower oscillation* denoted by BLO). Bounded functions in this class have been identified as the pointwise multipliers of H^1 and BMO on the circle [**44**], on the sphere [**35**] and on spaces of homogeneous type [**9**]. The analogous result in the case of bmo was shown in [**5**]. These spaces have also proved useful in the study of PDE. In [**47**] a logarithmic mean oscillation condition is imposed on the coefficients of a parabolic equation, while in [**3, 4**] a range of such conditions is imposed on the initial vorticity in the Euler and Navier-Stokes equations. Note that the latter articles use slightly different notation (for example bmo there refers to "local BMO" rather than Goldberg's nonhomogeneous space) and an L^2 oscillation, which is equivalent to the L^1 one by the John–Nirenberg inequality for these spaces (see [**43**]). In [**28**], a space called LMO_∞ is introduced which consists of functions in $\mathrm{BMO}_{\mathrm{loc}}$ which have logarithmic vanishing mean oscillation on small balls while the mean oscillation on large balls can grow as a power of the radius. This space is contained in $\mathrm{LMO}_{\mathrm{loc}}(\mathbb{R}^n)$, defined as follows.

DEFINITION 2.9. Let $b \in L^1_{\mathrm{loc}}(\mathbb{R}^n)$. We say $b \in \mathrm{LMO}_{\mathrm{loc}}(\mathbb{R}^n)$ if

$$\|b\|_{\mathrm{LMO}_{\mathrm{loc}}(\mathbb{R}^n)} := \sup_{r(B)<1} \frac{[\log(1+r(B)^{-1})]}{|B|} \int_B |b(x) - b_B|dx < \infty.$$

We also define $\mathrm{lmo}(\mathbb{R}^n) := \mathrm{LMO}_{\mathrm{loc}}(\mathbb{R}^n) \cap \mathrm{bmo}(\mathbb{R}^n)$ and

$$\|b\|_{\mathrm{lmo}(\mathbb{R}^n)} := \|b\|_{\mathrm{LMO}_{\mathrm{loc}}(\mathbb{R}^n)} + \sup_{r(B)\geq 1} |b|_B < \infty.$$

Since $[\log(1+r(B)^{-1})]^{-1} \to 0$ as $r \to 0$, we have that functions in $\mathrm{LMO}_{\mathrm{loc}}(\mathbb{R}^n)$ satisfy the vanishing mean oscillation condition (2.5) and therefore $\mathrm{lmo}(\mathbb{R}^n) \subset \mathrm{vmo}(\mathbb{R}^n)$. If we take $b \in \mathrm{LMO}_{\mathrm{loc}}(\mathbb{R}^n) \cap L^p(\mathbb{R}^n)$, $1 \leq p < \infty$, then b will also satisfy condition (2.6) since if $|B| \geq 1$ and $B \subset \mathbb{R}^n \setminus B(0,R) \subset \mathbb{R}^n$, then $⨍_B |b| \leq \|b\|_{L^p(B(0,R)^c)} \to 0$ as $R \to \infty$. Thus $\mathrm{LMO}_{\mathrm{loc}}(\mathbb{R}^n) \cap L^p(\mathbb{R}^n) \subset \mathrm{cmo}(\mathbb{R}^n)$.

3. Localized convolution kernels

In addition to the characterizations via maximal functions and atomic decomposition, as in Proposition 2.3, Goldberg also characterized $h^1(\mathbb{R}^n)$ by singular integral operators in an analogous way to the characterization of $H^1(\mathbb{R}^n)$ via the Riesz transforms R_j, $j = 1, \dots, n$, given by their multipliers as $\widehat{R_j f}(\xi) = i\frac{\xi_j}{|\xi|}\hat{f}(\xi)$.

THEOREM 3.1. [**24**, Theorem 2] *Fix a function $\varphi \in \mathcal{S}(\mathbb{R}^n)$ with $\varphi \equiv 1$ in a neighborhood of the origin. Then a distribution f belongs to $h^1(\mathbb{R}^n)$ if and only if $f \in L^1(\mathbb{R}^n)$ and $r_j f \in L^1(\mathbb{R}^n)$, $j = 1, \dots, n$, where r_j is defined by*

$$\widehat{r_j f}(\xi) = (1 - \varphi(\xi))i\frac{\xi_j}{|\xi|}\hat{f}(\xi).$$

See [**38**] for a generalization of this result to the case $0 < p \leq 1$, using φ of compact support.

For $n = 1$, in [**23**], Goldberg also claims a characterization of $h^1(\mathbb{R})$ via a localized version of the Hilbert transform. This can also be described as a "truncated"

Hilbert transform (not to be confused with the usual truncation used to the defined the principal value integral). We set

$$\mathcal{H}_{\text{loc}}(f)(x) := p.v. \int_{\mathbb{R}} \frac{\eta(x-y)f(y)}{x-y} dy, \tag{3.1}$$

where $\eta \equiv 1$ on $(-1, 1)$ and is supported in $[-2, 2]$. In [**16**] it was shown that for smooth η, the local Hilbert transform $\mathcal{H}_{\text{loc}}$ characterizes $h^1(\mathbb{R})$.

Note that the two localizations are different: $\mathcal{H}_{\text{loc}}$ is obtained by localizing the kernel of the Hilbert transform near 0, while r_j is obtained by localizing the Fourier transform of R_j away from 0. This may cause some confusion. For example, in order to characterize weighted local Hardy spaces, higher-dimensional analogues of the local Hilbert transforms, namely "truncated" versions of the Riesz transforms where the kernel is multiplied by a smooth function of compacted support, were defined in [**48**, Section 8] using the notation R_j. However, the result referred to as [**48**, Theorem A], which is [**8**, Corollary to Theorem 3.1], actually uses the same r_j as in Theorem 3.1, localized on the Fourier transform side.

In this section we consider these two types of localization for a more general convolution kernel, with minimal smoothness and decay conditions on the auxiliary functions used for the localization, and prove that the boundedness of the associated singular integral operators on $h^1(\mathbb{R}^n)$ does not depend on which type of localization we choose.

3.1. Calderón–Zygmund singular integral operators of convolution type. We assume that our kernel is a δ-kernel, namely K is a measurable function on $\mathbb{R}^n \setminus \{0\}$ and there exist $\delta \in (0, 1]$ and $C > 0$ such that

$$|K(x)| \leq \frac{C}{|x|^n} \tag{3.2}$$

and

$$|K(x-y) - K(x)| \leq C\frac{|y|^\delta}{|x|^{n+\delta}} \quad \text{for } 2|y| \leq |x|. \tag{3.3}$$

We also assume that a cancellation condition holds in the following form:

$$\exists A < \infty \quad \sup_{0<r<R} \left| \int_{r<|x|<R} K(x)dx \right| \leq A. \tag{3.4}$$

We then define the associated convolution operator $T : \mathcal{S}(\mathbb{R}^n) \to \mathcal{S}'(\mathbb{R}^n)$ by the principal value integral

$$Tf(x) = \lim_{\varepsilon \to 0} \int_{|x-y| \geq \varepsilon} K(x-y)f(y)dy.$$

This extends to a bounded operator on $L^2(\mathbb{R}^n)$ (see [**46**, Section VII.3]). Furthermore, T is bounded on $L^p(\mathbb{R}^n)$, $1 < p < \infty$, and from $L^1(\mathbb{R}^n)$ to $L^{1,\infty}(\mathbb{R}^n)$. See [**45**, Chapter 2] for the classical theory of Calderón–Zygmund operators of convolution type and [**25**, Chapter 4] for more general kernels. A general Calderón–Zygmund operator can be written as a sum of an operator of this form plus an operator given by multiplication by a bounded function (see [**46**, Section I.7.4],[**25**, Proposition 4.1.11]), but since multiplication by a bounded function is typically not a bounded operator on h^1, here we only consider the singular integral part.

We define two types of operators via localizations of the kernel K. On the one hand, we look at operators of the form

$$T^{\psi}(f) := T(f - \psi * f)$$

for suitable functions ψ. These are modelled on Goldberg's localized Riesz transforms in Theorem 3.1, with $\psi = \check{\varphi}$.

On the other, based on the definition of the local Hilbert transform (3.1), we look at operators T_η, associated to the kernel $K\eta$, for a class of functions η.

3.2. The Operator T_η. We now show that under certain weak smoothness and decay conditions on η, the kernel $K\eta$ satisfies the same conditions as K and so we can associate to it an operator T_η in the same way that we associate T to K, and this operator enjoys the same boundedness properties as T.

LEMMA 3.2. *Suppose that η is a bounded function and there exists a constant C for which*

$$|\eta(x-y) - \eta(x)| \leq C\frac{|y|^\delta}{|x|^\delta} \qquad \text{for } 0 < 2|y| \leq |x|. \tag{3.5}$$

Then $K\eta$ satisfies conditions (3.2) *and* (3.3).

PROOF. We have that $|K\eta| \leq \|\eta\|_{L^\infty}|x|^{-n}$, and for $0 < 2|y| \leq |x|$, since this implies $|x-y| \geq |x|/2$, we get

$$\begin{aligned}
&|K(x-y)\eta(x-y) - K(x)\eta(x)| \\
&\quad \leq |K(x-y)||\eta(x-y) - \eta(x)| + |K(x-y) - K(x)||\eta(x)| \\
&\quad \lesssim \frac{1}{|x-y|^n}\frac{|y|^\delta}{|x|^\delta} + \frac{|y|^\delta}{|x|^{n+\delta}} \\
&\quad \lesssim \frac{|y|^\delta}{|x|^{n+\delta}}.
\end{aligned}$$

□

The condition on η is a local δ-Lipschitz condition such that the Lipschitz constant in $B(x, |x|/2)$ decays like $|x|^{-\delta}$. In particular, if η has compact support, then we just require $\eta \in \mathrm{Lip}_\delta$.

Next we show that $K\eta$ satisfies (3.4). For this we need further assumptions on η guaranteeing that η has a certain decay at infinity and $\eta - 1$ vanishes sufficiently fast at the origin. Note that we do not require that $\eta \equiv 1$ in a neighborhood of the origin. For example, if $\eta \in \mathrm{Lip}_\delta$ then it suffices that $\eta(0) = 1$.

LEMMA 3.3. *If η satisfies,*

$$\int_{0<|x|<1} \frac{|\eta(x) - 1|}{|x|^n}dx + \int_{|x|\geq 1} \frac{|\eta(x)|}{|x|^n}dx < \infty, \tag{3.6}$$

then (3.4) *holds for $K\eta$.*

PROOF. For $0 < r < R$, by the assumptions on K and η,

$$\begin{aligned}\left|\int_{r<|y|<R} K(y)\eta(y)dy\right| &\le \left|\int_{r<|y|<1} K(y)[\eta(y)-1]dy\right| + A + \left|\int_{1\le|y|<R} K(y)\eta(y)dy\right| \\ &\lesssim \int_{0<|y|<1} \frac{|\eta(y)-1|}{|y|^n}dy + \int_{|y|\ge 1} \frac{|\eta(y)|}{|y|^n}dy + 1 \le A',\end{aligned}$$

where the constant A' is independent of r and R. □

If η satisfies the hypotheses of both lemmas, we get the boundedness of T_η on $L^2(\mathbb{R}^n)$. Once we have L^2-boundedness, we can show that T_η is bounded on $L^p(\mathbb{R}^n)$ for $1 < p < \infty$ and from $L^1(\mathbb{R}^n)$ to weak-$L^1(\mathbb{R}^n)$.

3.3. Equivalent boundedness of T_η and T^ψ. For nice choices of ψ, one can study the boundedness of the localized operator T^ψ as a pseudo-differential operator. Alternatively, one can show that the kernel of T^ψ is sufficiently good. In particular, assuming that $\psi \in L^1$, T^ψ inherits from T the boundedness on L^p for $1 < p < \infty$ and the weak $(1,1)$ boundedness. We are interested in the boundedness on h^1.

Instead of working with T^ψ directly, we give sufficient conditions on ψ to be able to obtain the boundedness of T^ψ on h^1 from the boundedness of T_η. As a consequence of the following equivalence, it suffices to prove the boundedness for one good η (e.g. a Lip_δ function of compact support) in order to obtain it for the whole class of η and ψ.

THEOREM 3.4. *Suppose η satisfies the hypotheses of Lemmas 3.2 and 3.3, and ψ satisfies*

$$\psi \in L^1(\mathbb{R}^n) \cap L^2(\mathbb{R}^n), \qquad \int \psi = 1,$$

$$\int_{|x|\ge 1} \left[\frac{1}{|x|^n}\int_{|y|\ge|x|/2} |\psi(y)|dy + \frac{1}{|x|^{n+\delta}}\int_{|y|\le|x|/2} |y|^\delta|\psi(y)|dy\right] dx < \infty \tag{3.7}$$

and

$$\int_{|x|\ge 1} \left[\int_{|y|\le|x|/2} \frac{|\psi(x-y)-\psi(x)|}{|y|^n}dy\right] dx < \infty. \tag{3.8}$$

Then the operator T_η is bounded on $h^1(\mathbb{R}^n)$ if and only if T^ψ is bounded on $h^1(\mathbb{R}^n)$.

Note that condition (3.7) will hold, for example, if the function $|x|^\eta\psi(x)$ is integrable for some $\eta > 0$, while (3.8) will hold if $\psi \in \mathrm{Lip}_\alpha$ for $0 < \alpha \le 1$ and $|x|^\alpha\|\psi\|_{\mathrm{Lip}_\alpha(B(x,|x|/2))}$ is integrable.

PROOF. We will show that the operator $T_E := T_\eta - T^\psi$ is bounded on $L^1(\mathbb{R}^n)$, and from that deduce that T_E is bounded on $h^1(\mathbb{R}^n)$.

For $0 < \varepsilon < 1/2$, let $K_\varepsilon(x) = \chi_{|x|>\varepsilon}K(x)$ be the usual truncation of the kernel. Applying the same to $K\eta$, we consider $(K\eta)_\varepsilon - K_\varepsilon + K_\varepsilon * \psi$. Set

$$K_*(x) = \sup_{\varepsilon>0} |(K\eta)_\varepsilon(x) - K_\varepsilon(x) + K_\varepsilon * \psi(x)|.$$

We claim that $K_* \in L^1(\mathbb{R}^n)$.

For the local estimate, we write

$$K_*(x) \leq |K(x)(\eta(x)-1)| + T_*(\psi)(x),$$

where $T_*(\psi)(x) = \sup_{\varepsilon>0} |K_\varepsilon * \psi(x)|$. Recall that the maximal operator T_* has the same boundedness as T (see [**46**, Section I.7],[**25**, Section 4.2.2]). Thus we have

$$\int_{|x|<1} K_*(x)dx \lesssim \int_{|x|<1} \frac{|\eta(x)-1|}{|x|^n}dx + \|\psi\|_{L^2} < \infty$$

For $|x| \geq 1$, we can write

$$K_*(x) \leq |K\eta(x)| + \sup_{\varepsilon>0} |K(x) - K_\varepsilon * \psi(x)|.$$

The first term is integrable for $|x| \geq 1$ by condition (3.6) on η, so it remains to bound the integral of the second term. We do this by fixing ε and obtaining a pointwise estimate on $|K(x) - K_\varepsilon * \psi(x)|$ in terms of ψ.

Using the hypotheses on ψ, we can write, for $|x| \geq 1$ and $\varepsilon < 1/2$,

$$\begin{aligned} \left|K(x) - \int_{|x-y|>\varepsilon} K(x)\psi(y)dy\right| &\leq |K(x)| \int_{|x-y|\leq\varepsilon} |\psi(y)|dy \\ &\lesssim |x|^{-n} \int_{|y|\geq|x|/2} |\psi(y)|dy. \end{aligned} \tag{3.9}$$

Thus it remains to consider

$$\begin{aligned} &\int_{|x-y|>\varepsilon} [K(x) - K(x-y)]\psi(y)dy \\ &\quad = \iint_{\left\{\substack{|y|\leq|x|/2 \\ |x-y|\geq|x|/2}\right\}} + \iint_{\left\{\substack{|y|\geq|x|/2 \\ |x-y|\geq|x|/2}\right\}} + \int_{\varepsilon<|x-y|\leq|x|/2} [K(x) - K(x-y)]\psi(y)dy \\ &\quad =: I_1 + I_2 + I_3. \end{aligned}$$

To estimate I_1, note that $B(0, |x|/2) \subset B(x, |x|/2)^c$ so we can use the smoothness condition (3.3) on K to write

$$\begin{aligned} |I_1| &\leq \int_{|y|\leq|x|/2} |K(x) - K(x-y)||\psi(y)|dy \\ &\lesssim \frac{1}{|x|^{n+\delta}} \int_{|y|\leq|x|/2} |y|^\delta |\psi(y)|dy. \end{aligned}$$

For I_2, we use the decay bound (3.2) on K to get, as in (3.9),

$$\begin{aligned} |I_2| &\leq \iint_{\left\{\substack{|y|\geq|x|/2 \\ |x-y|\geq|x|/2}\right\}} (|K(x)||\psi(y)| + |K(x-y)||\psi(y)|)dy \\ &\lesssim \frac{1}{|x|^n} \int_{|y|\geq|x|/2} |\psi(y)|dy. \end{aligned}$$

Finally, for I_3, we again use the decay bound (3.2) as well as the cancellation condition (3.4) on K to obtain

$$\begin{aligned}
|I_3| &\leq \left| \int_{\varepsilon<|x-y|\leq|x|/2} [K(x)\psi(y) - K(x-y)(\psi(y)-\psi(x))]\,dy \right| \\
&\quad + \left| \int_{\varepsilon<|x-y|\leq|x|/2} K(x-y)\psi(x)dy \right| \\
&\lesssim \int_{|y|\geq|x|/2} \frac{|\psi(y)|}{|x|^n}dy + \int_{|x-y|\leq|x|/2} \frac{|\psi(y)-\psi(x)|}{|x-y|^n}dy + |\psi(x)| \\
&\lesssim \frac{1}{|x|^n}\int_{|y|\geq|x|/2} |\psi(y)|dy + \int_{|y'|\leq|x|/2} \frac{|\psi(x-y')-\psi(x)|}{|y'|^n}dy + |\psi(x)|.
\end{aligned}$$

Since all these estimates are independent of ε, we can combine (3.9), the bounds on I_1, I_2 and I_3, and the integrability conditions (3.7) and (3.8) on ψ to get that

$$\int_{|x|\geq 1} \sup_{\varepsilon>0} |K(x) - K_\varepsilon * \psi(x)| < \infty$$

and as a result, $K_* \in L^1(\mathbb{R}^n)$. Let

$$K_E(x) = \lim_{\varepsilon\to 0}((K\eta)_\varepsilon - K_\varepsilon + K_\varepsilon * \psi)(x) = K\eta(x) - K(x) + \lim_{\varepsilon\to 0} K_\varepsilon * \psi(x).$$

This limit exists for almost every $x \in \mathbb{R}^n$ by the properties of T_* (see [**45**, p. 45]). By the Dominated Convergence Theorem, $K_E \in L^1(\mathbb{R}^n)$ and

$$\lim_{\varepsilon\to 0}((K\eta)_\varepsilon - K_\varepsilon + K_\varepsilon * \psi) * f = K_E * f \quad \text{for } f \in L^1(\mathbb{R}^n).$$

Let $f \in h^1(\mathbb{R}^n)$. Then $T_E f \in L^1(\mathbb{R}^n)$ with

$$\|T_E f\|_{L^1(\mathbb{R}^n)} \leq \|K_E\|_{L^1(\mathbb{R}^n)} \|f\|_{L^1(\mathbb{R}^n)}.$$

Moreover, the localized Riesz transforms r_j, $j = 1, \ldots, n$ of Goldberg (see Theorem 3.1) commute with convolution with the L^1 function K_E, so

$$r_j(T_E(f)) = T_E(r_j(f)).$$

Since $r_j(f) \in L^1(\mathbb{R}^n)$, hence $T_E(r_j(f)) \in L^1(\mathbb{R}^n)$, we have

$$\begin{aligned}
\|T_E f\|_{h^1(\mathbb{R}^n)} &\lesssim \|T_E f\|_{L^1(\mathbb{R}^n)} + \sum_{j=1}^n \|r_j(T_E(f))\|_{L^1(\mathbb{R}^n)} \\
&\leq \|K_E\|_{L^1(\mathbb{R}^n)}(\|f\|_{L^1(\mathbb{R}^n)} + \sum_{j=1}^n \|r_j(f)\|_{L^1(\mathbb{R}^n)}) \\
&\lesssim \|f\|_{h^1(\mathbb{R}^n)}.
\end{aligned}$$

□

Starting with the kernels of the Riesz transforms R_j, $j = 1, \ldots, n$, which satisfy conditions (3.2) and (3.3) with $\delta = 1$ and (3.4) with $A = 0$, and noting that the function $\psi = \check{\varphi} \in \mathcal{S}(\mathbb{R}^n)$ for φ as in Theorem 3.1 satisfies the hypotheses of Theorem 3.4, we can use both theorems to get the following.

COROLLARY 3.5. *Suppose that η and ψ satisfy assumptions in Theorem 3.4. For $j = 1, \dots, n$, define, for $f \in \mathcal{S}(\mathbb{R}^n)$,*

$$\mathcal{R}_{j,\eta} f(x) := c_n \lim_{\varepsilon \to 0} \int_{|x-y| \geq \varepsilon} \frac{(x_j - y_j)\eta(x-y)}{|x-y|^{n+1}} f(y)dy \quad \text{and}$$
$$\mathcal{R}_j^{\psi} f(x) := R_j(f - \psi * f).$$

Then

(1) *$\mathcal{R}_{j,\eta}$ and $\mathcal{R}_j^{\psi}$ both map $h^1(\mathbb{R}^n)$ to $L^1(\mathbb{R}^n)$;*
(2) *for $f \in L^1(\mathbb{R}^n)$, $f \in h^1(\mathbb{R}^n) \iff \mathcal{R}_{j,\eta}(f) \in L^1(\mathbb{R}^n), j = 1, \dots, n \iff \mathcal{R}_j^{\psi}(f) \in L^1(\mathbb{R}^n), j = 1, \dots, n$.*

As seen above, because convolution operators commute with the r_j, the boundedness from $h^1(\mathbb{R}^n)$ to $L^1(\mathbb{R}^n)$ implies the boundedness on $h^1(\mathbb{R}^n)$, and extra cancellation conditions on the kernel are not needed (see also the remark following the proof of [**24**, Theorem 4]).

3.4. Inhomogeneous singular integral operators. We end this section by introducing a class of operators which are not necessarily of convolution type. As can be seen from the definition, however, for a convolution δ-kernel K and an η which decays sufficiently fast at infinity, the localized kernel $K\eta$ will be in this class.

Let $\Delta := \{(x, x) : x \in \mathbb{R}^n\}$ denote the diagonal in $\mathbb{R}^{2n}$.

DEFINITION 3.6. Let $\delta \in (0, 1]$. We say T is an he inhomogeneous singular integral operator associated with an inhomogeneous δ-kernel K if K is a locally integrable function on $\mathbb{R}^{2n} \setminus \Delta$ and there exist $\varepsilon > 0$ and C such that

(1) $|K(x, y)| \leq C \min\{|x-y|^{-n}, |x-y|^{-n-\varepsilon}\}$ for all $(x, y) \in \mathbb{R}^{2n} \setminus \Delta$;
(2) $|K(x, y) - K(x, z)| + |K(y, x) - K(z, x)| \leq C|y-z|^{\delta}|x-y|^{-n-\delta}$ for $2|y-z| \leq |x-y|$.

In addition, we assume that the operator T, defined by $Tf(x) := \int_{\mathbb{R}^n} K(x, y)f(y)dy$ for $f \in C_c^{\infty}(\mathbb{R}^n)$ and $x \notin \text{supp}(f)$, can be extended to a bounded operator on $L^2(\mathbb{R}^n)$.

This type of operator was introduced in [**19**] and a generalization was considered in [**14**], where it was shown that if certain cancellation conditions are imposed on T, then it is bounded from $h^p(\mathbb{R}^n)$ to $h^p(\mathbb{R}^n)$, $0 < p \leq 1$. If no cancellation conditions are imposed, we still have boundedness from $h^1(\mathbb{R}^n)$ to $L^1(\mathbb{R}^n)$, as can be seen by noting that h^1 atoms are mapped to "pre-molecules" which are bounded in $L^1(\mathbb{R}^n)$ (see [**15**, Definition 3]). As T is a singular integral operator, T is also bounded from $L^p(\mathbb{R}^n)$ to itself for $1 < p < \infty$. We will consider the commutators of these operators with bmo functions in Section 5.

4. Local Hardy spaces related to commutators with bmo functions

In this section we introduce some "local" analogues of commutator maximal function defined in [**34**], and the atoms in [**39**], and consider the relationships between them.

4.1. Some maximal functions. Recall that the Hardy-Littlewood Maximal function defined for $f \in L^1_{\text{loc}}(\mathbb{R}^n)$ by

$$Mf(x) := \sup_{B \ni x} |f|_B$$

is bounded on $L^p(\mathbb{R}^n)$ for $1 < p \leq \infty$ and also bounded from $L^1(\mathbb{R}^n)$ to $L^{1,\infty}(\mathbb{R}^n)$.

In connection with commutators, the following two maximal functions were introduced in [**10**, Section 6] in the one-dimensional case. Setting

$$M(b,f)(x) := \sup_{I \ni x} \big|(b(x) - b_I)f_I\big|,$$

it was shown in [**10**, Theorem IX] that $b \in \text{BMO}(\mathbb{R})$ if and only if $f \mapsto M(b,f)$ is bounded on $L^2(\mathbb{R})$. The proof uses [**10**, Theorem X] which states that for a different maximal function,

$$N(b,f)(x) := \sup_{I \ni x} |(bf)_I - b(x)f_I|,$$

$f \mapsto M(b,f)$ is bounded on $L^2(\mathbb{R})$ when $b \in \text{BMO}(\mathbb{R})$.

The commutator of M with multiplication by b, that is

$$[M,b](f)(x) := M(bf)(x) - b(x)Mf(x),$$

was studied in [**1**, **2**, **37**] and shown to be bounded on $L^p(\mathbb{R}^n)$ for $1 < p < \infty$ if and only if $b \in BMO(\mathbb{R}^n)$ and b is bounded below.

Finally, one has the following maximal function,

$$M_b f(x) = \sup_{B \ni x} \frac{1}{|B|} \int_B |b(x) - b(y)||f(y)|dy,$$

which was studied in [**1**, **22**, **42**] and is sometimes denoted by C_b. In this case, as in [**10**], one has that M_b is bounded on $L^p(\mathbb{R}^n)$ for $1 < p < \infty$ if and only if $b \in BMO(\mathbb{R}^n)$. Moreover, for $b \in BMO(\mathbb{R}^n)$, [**1**, Corollary 1.11] gives the pointwise domination $M_b f(x) \lesssim \|b\|_{\text{BMO}} M(Mf)(x)$ for $f \in L^1_{\text{loc}}(\mathbb{R}^n)$, and therefore

$$\|M_b\|_{L^p \to L^p} \lesssim \|b\|_{\text{BMO}}, \quad 1 < p < \infty. \tag{4.1}$$

Ky [**34**] introduced the following commutator maximal function definition of the space H^1_b.

Definition 4.1 ([**34**], Definition 2.2). Let $b \in \text{BMO}(\mathbb{R}^n)$ be non-constant. The space $H^1_b(\mathbb{R}^n)$ consists of $f \in H^1(\mathbb{R}^n)$ such $[b, \mathfrak{M}_{\text{nt}}]f \in L^1(\mathbb{R}^n)$, where $[b, \mathfrak{M}_{\text{nt}}]f(x) := \mathfrak{M}_{\text{nt}}[b(x)f(\cdot) - b(\cdot)f(\cdot)](x)$ and $\mathfrak{M}_{\text{nt}}$ is the non-tangential grand maximal function defined by

$$\mathfrak{M}_{\text{nt}} f(x) := \sup\{|f * \phi_t(y)| : |y - x| < t, \varphi \in \mathcal{A}\}$$

with

$$\mathcal{A} = \{\phi \in \mathcal{S}(\mathbb{R}^n) : \|\phi\|_\infty + \|\nabla\phi\|_\infty \leq (1 + |x|^2)^{-n-1}\}.$$

The norm on H^1_b is given by

$$\|f\|_{H^1_b} := \|f\|_{H^1}\|b\|_{\text{BMO}} + \|[b, \mathfrak{M}_{\text{nt}}]f\|_{L^1}.$$

Ky also defines a "local" version of the non-tangential grand maximal function $\mathfrak{M}_{\text{nt}}$ by restricting t to $(0,1)$. As pointed out following Proposition 2.3, this is equivalent to the maximal functions $\mathcal{M}$ and $\mathfrak{M}$ in terms of characterizing $h^1(\mathbb{R}^n)$. However, there is no definition of h^1_b given in [**34**].

We choose the grand maximal function $\mathcal{M}$, using test functions of compact support, to give the analogue of Ky's definition in the local case.

Definition 4.2. Given $b \in L^2_{loc}(\mathbb{R}^n)$, we define the maximal commutator function for $f \in L^2_{loc}(\mathbb{R}^n)$ to be

$$\mathcal{M}_b f(x) := [b, \mathcal{M}](x) := \mathcal{M}[b(x)f(\cdot) - b(\cdot)f(\cdot)](x).$$

Recall that in Proposition 2.3 (for $T = 1$) we defined, for $f \in \mathcal{S}'(\mathbb{R}^n)$,

$$\mathcal{M}(f)(x) := \sup_{\phi} |\langle f, \phi \rangle|,$$

where the supremum is taken over all $\phi \in C^1(\mathbb{R}^n)$ with $\operatorname{supp}(\phi) \subset B(x,t)$ for some $0 < t < 1$, $\|\phi\|_{L^\infty} \leq |B(x,t)|^{-1}$ and $\|\nabla\phi\|_{L^\infty} \leq [t|B(x,t)|]^{-1}$. We therefore have that for $b, f \in L^2_{loc}(\mathbb{R}^n)$,

$$\mathcal{M}_b f(x) = \sup_{\phi} \left|\langle b(x)f(\cdot) - b(\cdot)f(\cdot), \phi\rangle\right| = \sup_{\phi} \left|\int_{\mathbb{R}^n} [b(x) - b(y)]f(y)\phi(y)dy\right|.$$

Since $(b(x) - b)f \in L^1_{loc}(\mathbb{R}^n)$ for almost every $x \in \mathbb{R}^n$, the integral is well-defined almost everywhere, and by the size and support conditions on ϕ,

$$\mathcal{M}_b f(x) \lesssim M_b f(x). \tag{4.2}$$

The extension of this definition from $f \in L^2_{loc}(\mathbb{R}^n)$ to general $f \in h^1(\mathbb{R}^n)$, except in the special case where b is bounded, is not obvious. In fact, this is also the case for Ky's definition, Definition 4.1, and additional work, such as the sublinear decomposition [**34**, Theorem 3.1], is needed there.

4.2. Atomic commutator Hardy spaces. In [**39**], Pérez defined a restricted notion of atom with cancellation against a function $b \in \mathrm{BMO}$.

Definition 4.3 ([**39**], Definition 1.3). For $b \in \mathrm{BMO}(\mathbb{R}^n)$, a b-atom a is a measurable function such that for some cube Q the following hold:

(1) $\operatorname{supp}(a) \subset Q$,
(2) $\|a\|_{L^\infty} \leq |Q|^{-1}$,
(3) $\int a = 0$ and $\int ab = 0$.

Pérez then proceeded to define an atomic space, which he called H^1_b (Ky [**34**] uses $\mathcal{H}^1_b$ for this space) by taking all $f \in L^1$ which can be written as $f = \sum_{j=1}^\infty \lambda_j a_j$, where a_j are b-atoms and $\{\lambda_j\} \in \ell^1$. Note that since b-atoms are also H^1 atoms, such a decomposition will converge in the H^1 norm with $\|f\|_{H^1} \leq \sum |\lambda_j|$, and hence Pérez's space is contained in H^1.

We give the following straightforward adaptation of Pérez's definition to the local case.

Definition 4.4. Fix $b \in \mathrm{bmo}(\mathbb{R}^n)$. We say a measurable function a is a Pérez h^1_b atom if for some ball $B(x_0, r)$ we have

(1) $\operatorname{supp}(a) \subset B(x_0, r)$,
(2) $\|a\|_{L^2(\mathbb{R}^n)} \leq |B(x_0, r)|^{-\frac{1}{2}}$,
(3) $\displaystyle\int a = 0$ and $\displaystyle\int ab = 0$ whenever $r < 1$.

For $f \in L^1(\mathbb{R}^n)$, we say $f \in h^1_{\mathrm{Perez},b}(\mathbb{R}^n)$ if f can be written as $f = \sum_{j=1}^\infty \lambda_j a_j$, where a_j are Pérez h^1_b atoms and $\{\lambda_j\} \in \ell^1$. For such functions, we define

$$\|f\|_{h^1_{\mathrm{Perez},b}} := \inf \sum_j |\lambda_j|,$$

where the infimum is taken over all such decompositions.

As in the homogeneous case, we note that a Pérez h^1_b atom is a $(1,2)$ atom in h^1 and therefore the decomposition converges in h^1 and we have $\|f\|_{h^1} \le \sum|\lambda_j|$. Moreover, each Pérez h^1_b atom supported in a ball $B = B(x_0,r)$ also satisfies that $a(b-c_B) \in h^1$, where c_B is as in (2.1). Indeed, we have that $a(b-c_B)$ is also supported in B, has integral zero whenever $r<1$, and by (2.2)

$$\begin{aligned}\|a(b-c_B)\|_{L^s(\mathbb{R}^n)} &\le \|a\|_{L^2(\mathbb{R}^n)}\|b-c_B\|_{L^p(\mathbb{R}^n)}\\ &\le |B(x_0,r)|^{-\frac{1}{2}}\|b\|_{\mathrm{bmo},p}|B(x_0,r)|^{\frac{1}{p}} = \|b\|_{\mathrm{bmo},p}|B(x_0,r)|^{\frac{1}{s}-1}\end{aligned}\tag{4.3}$$

for $1 \le s < 2$ and $\frac{1}{2}+\frac{1}{p}=\frac{1}{s}$; taking $s>1$ makes $a(b-c_B)$ a multiple of a $(1,s)$ atom and we have

$$\|a(b-c_B)\|_{h^1} \lesssim \|b\|_{\mathrm{bmo},p}.$$

We now give a variation on this definition which involves approximate cancellation conditions.

Definition 4.5. Fix $b \in \mathrm{bmo}(\mathbb{R}^n)$ and let $C_b = \|b\|_{\mathrm{bmo},2}$ as defined in (2.2). We say a measurable function a is an approximate h^1_b atom if for some ball $B(x_0,r)$ we have

(1) $\mathrm{supp}(a) \subset B(x_0,r)$,
(2) $\|a\|_{L^2(\mathbb{R}^n)} \le |B(x_0,r)|^{-\frac{1}{2}}$,
(3) $\left|\int a\right| \le [\log(1+r^{-1})]^{-2}$ and $\left|\int a(b-c_B)\right| \le C_b[\log(1+r^{-1})]^{-1}$.

For $f \in L^1(\mathbb{R}^n)$, we say $f \in h^1_{\mathrm{atom},b}(\mathbb{R}^n)$ if f can be written as $f = \sum_{j=1}^\infty \lambda_j a_j$, where a_j are approximate h^1_b atoms and $\{\lambda_j\} \in \ell^1$. For such functions, we define

$$\|f\|_{h^1_{\mathrm{atom},b}} := \inf \sum_j |\lambda_j|,$$

where the infimum is taken over all such decompositions.

The choice of $C_b = \|b\|_{\mathrm{bmo},2}$ guarantees that, by Cauchy–Schwarz, condition (2) implies condition (3) when $r \ge 1$. Thus every Pérez h^1_b atom is an approximate h^1_b atom, and

$$h^1_{\mathrm{Perez},b}(\mathbb{R}^n) \subset h^1_{\mathrm{atom},b}(\mathbb{R}^n).$$

Moreover, comparing with Definition 2.2 for $R=1$ and $q=2$, we see, as in (4.3), that not only is every approximate h^1_b atom a an approximate $(1,2)$ atom (up to a factor of $\log 2$), but also $a(b-c_B)$ is a multiple of an approximate $(1,s)$ atom for $1<s<2$ and $\frac{1}{2}+\frac{1}{p}=\frac{1}{s}$, with

$$\|a(b-c_B)\|_{h^1} \lesssim \|b\|_{\mathrm{bmo},p}.\tag{4.4}$$

Recall that in h^1 (or h^p), every atom (or molecule) with approximate cancellation conditions can be written as an infinite linear combination of atoms with exact cancellation—see [**14, 18**]. Is it possible to decompose an approximate h^1_b atom into Pérez h^1_b atoms with exact cancellation? If by exact cancellation we mean the vanishing of both integrals in part (3) of Definition 4.4, the answer is not clear. In the following proposition we give a partial answer, showing that every approximate h^1_b atom can be written as a finite sum of atoms of integral zero.

Proposition 4.6. *Suppose A is an approximate h^1_b atom with respect to the ball $B(x_0,r)$, and $r<1$. Then $A = \sum \lambda_j a_j$ where the sum is* **finite**, *a_j are* *approximate* h^1_b *atoms supported in balls* B_j*, respectively, and in addition, when* $r(B_j) < 1$,

$$\int a_j = 0.$$

Moreover, for some constant depending only on n, $\sum |\lambda_j| \leq C_n$.

PROOF. Set $\alpha = \int A \neq 0$ (otherwise we are done). Denote $B(0, 2^j r)$ by B_j, $j = 0, 1, 2, \ldots, k$, where $2^{k-1}r < 1 \leq 2^k r$, and let $\eta_j = \frac{\chi_{B_j}}{|B_j|}$ be the normalized characteristic function of B_j. Then

$$A = A - \alpha\eta_0 + \sum_{j=1}^{k} \alpha(\eta_{j-1} - \eta_j) + \alpha\eta_k = \sum_{i=0}^{k+1} A_j$$

where

$$\begin{aligned} A_0 &:= A - \alpha\eta_0 \\ A_j &:= \alpha(\eta_{j-1} - \eta_j), \quad j = 1, \ldots, k, \quad \text{and} \\ A_{k+1} &:= \alpha\eta_k. \end{aligned}$$

First observe that $\operatorname{supp}(A_j) \subset B_j$ for $j = 1, \ldots, k$, and $\operatorname{supp}(A_{k+1}) \subset B_k$. Moreover, since $\int \eta_j = 1$ for all j, we have that $\int A_j = 0$ for $j = 0, \ldots, k$.

Write $A_j = \lambda_j a_j$ where

$$a_0 := \frac{1}{2} A_0, \quad \lambda_0 := 2, \quad a_{k+1} := \alpha^{-1} A_{k+1}, \quad \lambda_{k+1} := \alpha,$$

and

$$a_j := \frac{A_j}{\alpha 2^n \log(1 + [2^j r]^{-1})}, \quad \lambda_j := \alpha 2^n \log(1 + [2^j r]^{-1}), \quad j = 1, \ldots, k.$$

Since $|\alpha| \leq \|A\|_{L^1} \leq 1$, $\|\eta_j\|_{L^\infty} \leq |B_j|^{-1}$, and $2^n \log(1 + [2^j r]^{-1}) > 1$ for $j = 1, \ldots, k$, we have that $\|a_0\|_{L^2} \leq |B_0|^{-\frac{1}{2}}$, $\|a_j\|_{L^\infty} \leq |B_j|^{-1}$ for $j = 1, \ldots, k$, and $\|a_{k+1}\|_{L^\infty} \leq |B_k|^{-1}$. Thus, all the a_j satisfy Conditions (1) and (2) in Definition 4.5 and in addition $\int a_j = 0$ for $j = 0, \ldots, k$. Since $r(B_k) = 2^k r \geq 1$, the latter condition is not required for a_{k+1}.

To verify Condition (3) in Definition 4.5 , we need to check that the a_j satisfy the approximate cancellation condition against b. For $j = 0$, we get the approximate cancellation for a_0 from that of A:

$$\left| \int a_0 b \right| = \frac{1}{2} \left| \int Ab - \alpha \int b\eta_0 \right| = \frac{1}{2} \left| \int A[b - b_B] \right| \leq \frac{C_b}{\log(1 + r^{-1})}.$$

For $j = 1, \ldots, k$, we have, using (2.4),

$$\begin{aligned} \left| \int a_j b \right| &= \frac{1}{2^n \log(1 + [2^j r]^{-1})} \left| \int b(\eta_{j-1} - \eta_j) \right| \\ &= \frac{1}{2^n \log(1 + [2^j r]^{-1})} |b_{2^{j-1}B} - b_{2^j B}| \\ &\leq \frac{\|b\|_{\text{bmo}}}{\log(1 + [2^j r]^{-1})}, \end{aligned}$$

and for $j = k+1$, since $r(B_k) = 2^k r \geq 1$,

$$\left|\int a_{k+1} b\right| = |b_{B_k}| \leq \frac{\|b\|_{\text{bmo}}}{\log(1+(2^k r)^{-1})}.$$

If, as noted after Definition 4.5, we take $C_b = \|b\|_{\text{bmo},2} \geq \|b\|_{\text{bmo}}$, then we can conclude that a_j for $j = 0, \ldots, k+1$ are all approximate h^1_b atoms.

Finally, we have

$$\begin{aligned}
\sum |\lambda_j| &= 2 + \alpha 2^n \sum_{j=1}^{k} \log(1 + [2^j r]^{-1}) + \alpha \\
&\leq 3 + \frac{2^n}{[\log(1+r^{-1})]^2} \sum_{j=1}^{k} \log(1+[2^j r]^{-1}) \\
&\leq 3 + \frac{2^n k \log(1+r^{-1})}{[\log(1+r^{-1})]^2} \\
&\leq 3 + \frac{2^n(\log_2 r^{-1} + 1)}{[\log(1+r^{-1})]} \\
&\leq C_n.
\end{aligned}$$

□

4.3. Relations between the spaces. For the homogeneous case, Ky shows (see [**34**, Theorem 5.2]) that

$$\mathcal{H}^1_b(\mathbb{R}^n) \subset H^1_b(\mathbb{R}^n) \subset H^1(\mathbb{R}^n),$$

where $\mathcal{H}^1_b$ is Pérez atomic space and H^1_b is the maximal commutator space, but does not discuss whether the inclusion is proper. The result of [**26**] shows that the last inclusion is proper, at least in dimension 1, unless b is a constant.

For the nonhomogeneous case, the discussion following the definitions of the atomic spaces in the previous section gave us the inclusions

$$h^1_{\text{Perez},b}(\mathbb{R}^n) \subset h^1_{\text{atom},b}(\mathbb{R}^n) \subset h^1(\mathbb{R}^n).$$

In the trivial case, when b is a constant function, the cancellation conditions against b reduce to the conditions on the integral of the atoms. Thus every h^1 atom with exact cancellation is a Pérez h^1_b atom, so

$$h^1_{\text{Perez},b}(\mathbb{R}^n) = h^1_{\text{atom},b}(\mathbb{R}^n) = h^1(\mathbb{R}^n). \tag{4.5}$$

In the next section will see later that the second equality can hold for b in a nontrivial subspace of bmo, i.e. not just for constant functions.

While we have not defined a maximal space, we can consider the action of the commutator maximal function $\mathcal{M}_b$ on atoms.

PROPOSITION 4.7. *Let $b \in \text{bmo}(\mathbb{R}^n)$. Then for every h^1_b atom a,*

$$\|\mathcal{M}_b a\|_{L^1} \lesssim \|b\|_{\text{bmo}}.$$

PROOF. Let a be an approximate h^1_b atom with support in $B = B(x_0, r)$. We want to bound $\mathcal{M}_b(a)$.

For the local estimate, using (4.2), we can apply the L^p boundedness of the maximal commutator operator $M_b f$ defined in Section 4.1, with the bound (4.1). Thus, using the L^2 size condition on a we have

$$\int_{2B} \mathcal{M}_b(a) \leq |2B|^{1/2}\|M_b a\|_{L^2(2B)} \lesssim |B|^{\frac{1}{2}}\|a\|_{L^2}\|M_b\|_{L^2\to L^2} \lesssim \|b\|_{\mathrm{bmo}}.$$

Next we handle the integral on $(2B)^c$. Note that

$$\left|\int_{B(x,t)} \phi(y)[b(x)-b(y)]a(y)dy\right| \neq 0$$

implies that there exists $y \in B(x,t) \cap B(x_0,r)$, which in turn implies

$$r \leq \frac{|x-x_0|}{2} \leq |x-x_0| - |x_0-y| \leq |x-y| \leq t < 1. \tag{4.6}$$

This cannot happen when $r \geq 1$, so in that case the integral vanishes for all test functions ϕ and $\mathcal{M}_b(a) = 0$ on $(2B)^c$.

For $r<1$, we may assume, by Proposition 4.6, that $\int a = 0$. Thus we have, for almost all $x \in (2B)^c$,

$$\begin{aligned}
&\left|\int [b(x)-b(y)]a(y)\phi(y)dy\right| \\
&\quad\leq \left|\int [b(x)-b(y)]a(y)[\phi(y)-\phi(x_0)]dy\right| + \left|\phi(x_0)\int b(y)a(y)dy\right| \\
&\quad\leq |b(x)-b_B|\|\nabla\phi\|_{L^\infty} r\int |a(y)|dy + \|\nabla\phi\|_{L^\infty} r\int |b(y)-b_B||a(y)|dy \\
&\qquad + \|\phi\|_{L^\infty}\left|\int b(y)a(y)dy\right| \\
&\quad\lesssim \frac{|b(x)-b_B|r}{|x-x_0|^{n+1}} + \frac{r\|b\|_{\mathrm{bmo},2}}{|x-x_0|^{n+1}} + \frac{C_b}{|x-x_0|^n\log(1+r^{-1})}, \qquad (4.7)
\end{aligned}$$

where in the last step we used the conditions on ϕ, (4.6), the L^1 estimate on $a(b-b_B)$ (see (4.3) with $s=1$, $p=2$), and condition (3) in Definition 4.5. Since the estimate above is independent of ϕ, it holds for $\mathcal{M}_b(a)(x)$ and therefore, again using (4.6),

$$\begin{aligned}
\int_{(2B)^c} \mathcal{M}_b(a)(x)dx &= \int_{2r<|x-x_0|<2} \mathcal{M}_b(a)(x)dx \\
&\lesssim \int_{(2B)^c} \frac{|b(x)-b_B|r}{|x-x_0|^{n+1}} + \int_{(2B)^c} \frac{r\|b\|_{\mathrm{bmo},2}}{|x-x_0|^{n+1}} \qquad (4.8)\\
&\quad + \int_{2r<|x-x_0|<2} \frac{C_b}{|x-x_0|^n\log(1+r^{-1})} \\
&\lesssim \|b\|_{\mathrm{bmo}}.
\end{aligned}$$

Here we have used (2.3), (2.2), and the fact that $C_b = \|b\|_{\mathrm{bmo},2}$. □

If we take an h^1 atom a without assuming any cancellation against b, the boundedness of the maximal function $\mathcal{M}_b$ turns out to be equivalent to whether $a(b-c_B)$ belongs to h^1, where B is the ball containing the support of a. Such boundedness gives us automatically the approximate cancellation against b.

PROPOSITION 4.8. *Let $b \in \mathrm{bmo}(\mathbb{R}^n)$. Suppose a is an h^1 atom supported in a ball $B = B(x_0, r)$, and has vanishing integral when $r < 1$. Then*

$$\|\mathcal{M}_b a\|_{L^1} \lesssim \|a(b - c_B)\|_{h^1} + \|b\|_{\mathrm{bmo}} \lesssim \|\mathcal{M}_b a\|_{L^1} + \|b\|_{\mathrm{bmo}}.$$

Moreover,

$$\left|\int ab\right| \lesssim \min\left\{\frac{\|\mathcal{M}_b a\|_{L^1} + \|b\|_{\mathrm{bmo}}}{\log(1+r^{-1})}, \|b\|_{\mathrm{bmo}}\right\}.$$

PROOF. We assume that a satisfies an L^2 size condition. First note that by (4.3), $a(b - c_B) \in L^s(\mathbb{R}^n)$ for $1 < s < 2$, and as it has compact support, $a(b - c_B)$ is in $h^1(\mathbb{R}^n)$. By Definition 4.2, we have, for almost every $x \in \mathbb{R}^n$,

$$\mathcal{M}_b(a)(x) = \mathcal{M}([b(x) - b]a)(x) \leq |b(x) - c_B|\mathcal{M}(a)(x) + \mathcal{M}([b - c_B]a)(x),$$

hence

$$\begin{aligned}\|\mathcal{M}_b(a)\|_{L^1} &\leq \int |b(x) - c_B|\mathcal{M}(a)(x)dx + \|\mathcal{M}([b - c_B]a)\|_{L^1}\\ &\lesssim \int |b(x) - c_B|\mathcal{M}(a)(x)dx + \|a(b - c_B)\|_{h^1}.\end{aligned}$$

Conversely,

$$\mathcal{M}([b - c_B]a)(x) \leq \mathcal{M}_b(a)(x) + |b(x) - c_B|\mathcal{M}(a)(x),$$

so

$$\|a(b - c_B)\|_{h^1} \leq \|\mathcal{M}_b(a)\|_{L^1} + \int |b(x) - c_B|\mathcal{M}(a)(x)dx.$$

It thus remains to show that the integral of $|b - c_B|\mathcal{M}(a)$ is controlled by $\|b\|_{\mathrm{bmo}}$. The arguments are similar to those in the proof of Proposition 4.7. First, we have

$$\int_{2B} |b - c_B|\mathcal{M}(a) \leq \|b - c_B\|_{L^2(2B)}\|Ma\|_{L^2(2B)} \lesssim \|b\|_{\mathrm{bmo}}|B|^{\frac{1}{2}}\|a\|_{L^2} \lesssim \|b\|_{\mathrm{bmo}}.$$

Moreover, (4.6) implies that if $r \geq 1$, $\mathcal{M}(a)$ vanishes outside $2B$. For $r < 1$, we can use the cancellation condition on a to write, as in (4.7), for every test function ϕ, and for almost every $x \in (2B)^c$,

$$\begin{aligned}|b(x) - c_B|\left|\int a(y)\phi(y)dy\right| &\leq |b(x) - b_B|\|\nabla\phi\|_{L^\infty} r \int |a(y)|dy\\ &\lesssim \frac{|b(x) - b_B|r}{|x - x_0|^{n+1}}.\end{aligned}$$

The integral on $(2B)^c$ is then estimated as in the first term of (4.8).

By applying Proposition 2.6 to $a(b - c_B) \in h^1(\mathbb{R}^n)$, followed by the estimates above, we have

$$\left|\int ab\right| = \left|\int a(b - c_B)\right| \lesssim \frac{\|a(b - c_B)\|_{h^1}}{\log(1+r^{-1})} \lesssim \frac{(\|\mathcal{M}_b a\|_{L^1} + \|b\|_{\mathrm{bmo}})}{\log(1+r^{-1})}.$$

The other upper bound follows from the duality of bmo and h^1, or by Cauchy–Schwarz. □

It is interesting to note that if we had not assumed exact cancellation on a in the hypotheses of Proposition 4.8 when $r < 1$, we would have ended up having to estimate an extra term of the form

$$\int_{2r<|x-x_0|<2} \frac{|b(x)-b_B|}{|x-x_0|^n} dx \left| \int a \right|.$$

Looking at the proof of (2.3), if $\delta = 0$ and $p = 1$, the integral in x can be estimated by $\|b\|_{\mathrm{bmo}}(\log \frac{1}{r})^2$. Therefore, we would need a cancellation condition of the form $\left|\int a\right| \leq [\log(1+r^{-1})]^{-2}$, as in condition (3) of Definition 4.5.

Combining Propositions 4.6, 4.7 and 4.8, we get the following.

COROLLARY 4.9. *Let $f \in h^1(\mathbb{R}^n)$. Then the following are equivalent.*

(1) *The function $f \in h^1_{\mathrm{atom},b}(\mathbb{R}^n)$.*
(2) *There exists a sequence $\{\lambda_j\} \in \ell^1$ and a collection of h^1 atoms $\{a_j\}$, where B_j denotes the ball containing the support of a_j and a_j has vanishing integral when $r(B_j) < 1$, so that $f = \sum_j \lambda_j a_j$ and $\mathcal{M}_b a_j$ are uniformly bounded in $L^1(\mathbb{R}^n)$.*
(3) *There exists a sequence $\{\lambda_j\} \in \ell^1$ and a collection of h^1 atoms $\{a_j\}$, where B_j denotes the ball containing the support of a_j and a_j has vanishing integral when $r(B_j) < 1$, so that $f = \sum_j \lambda_j a_j$ and*

$$\sum_j \lambda_j a_j (b - c_{B_j}) \text{ converges absolutely in } h^1(\mathbb{R}^n).$$

4.4. The case of b in lmo. We now come to the promised result which gives us the second equality in (4.5) even when b is not constant. Recall that the pointwise multipliers of $\mathrm{bmo}(\mathbb{R}^n)$, and hence $h^1(\mathbb{R}^n)$, were identified as the elements of $L^\infty \cap \mathrm{lmo}(\mathbb{R}^n)$ in [**5**]. It is therefore not surprising that we have the following results.

THEOREM 4.10. *Let $b \in \mathrm{lmo}(\mathbb{R}^n)$. Then $h^1_{\mathrm{atom},b}(\mathbb{R}^n) = h^1(\mathbb{R}^n)$. Moreover,*

$$\|f\|_{h^1} \leq \|f\|_{h^1_{\mathrm{atom},b}} \lesssim \frac{\|b\|_{\mathrm{LMO}_{loc}}}{\|b\|_{\mathrm{BMO}_{loc}}} \|f\|_{h^1}.$$

PROOF. It was already observed following Definition 4.5 that for $b \in \mathrm{bmo}$, h^1_b atoms are approximate h^1 atoms (up to a factor of $\log 2$) and therefore $h^1_b(\mathbb{R}^n) \subset h^1(\mathbb{R}^n)$ with $\|f\|_{h^1} \lesssim \|f\|_{h^1_{\mathrm{atom},b}}$. We only need the lmo condition for the reverse inclusion.

Let a be an h^1 atom with support in the ball $B = B(x_0, r)$, with L^2 size condition and such that $\int a = 0$. This means that conditions (1), (2), and the first part of (3) in Definition 4.5 are satisfied, and we only need to check the approximate cancellation against b. Since this holds automatically for $r \geq 1$, we assume $r < 1$. By the L^2 size condition on a,

$$\begin{aligned}\left|\int_{\mathbb{R}^n} a(x)[b(x)-b_B]dx\right| &\leq \frac{1}{\log(1+r^{-1})} \frac{\log(1+r^{-1})}{|B|^{\frac{1}{2}}} \left(\int_B |b(x)-b_B|^2 dx\right)^{\frac{1}{2}} \\ &\leq \frac{\|b\|_{\mathrm{LMO}_{\mathrm{loc},2}}}{\log(1+r^{-1})},\end{aligned}$$

where

$$\|b\|_{\mathrm{LMO}_{\mathrm{loc},2}} := \sup_{r(B)<1} \log(1+r(B)^{-1})\left(\fint_B |b(x)-b_B|^2 dx\right)^{\frac{1}{2}}.$$

Since

$$\gamma := \frac{\log 2\|b\|_{\mathrm{BMO}_{\mathrm{loc},2}}}{\|b\|_{\mathrm{LMO}_{\mathrm{loc},2}}} \leq 1$$

and $\log 2\|b\|_{\mathrm{BMO}_{\mathrm{loc},2}} \leq \|b\|_{\mathrm{bmo},2} = C_b$, we have that γa is an h^1_b atom.

Finally, for $f \in h^1$ with an atomic decomposition $f = \sum_{j=1}^{\infty} \lambda_j a_j$, we can regard it as a decomposition using h^1_b atoms multiplied by γ^{-1}, so $f \in h^1_{\mathrm{atom},b}(\mathbb{R}^n)$ with

$$\|f\|_{h^1_{\mathrm{atom},b}} \leq \gamma^{-1} \sum_{j=1}^{\infty} |\lambda_j|.$$

Taking the infimum over all such decompositions, we get $\|f\|_{h^1_{\mathrm{atom},b}} \leq \gamma^{-1}\|f\|_{h^1}$. The inequality in the statement of the theorem is obtained by applying the John–Nirenberg inequality for LMO (see [**43**]). □

We give a partial converse to the theorem.

PROPOSITION 4.11. *Let $b \in L^1_{\mathrm{loc}}(\mathbb{R}^n)$. Suppose there exists a constant C_b such that every h^1 atom a satisfies*

$$\left|\int ab\right| \leq \min\left\{C_b, \frac{C_b \log(2)}{\log(1+r^{-1})}\right\},$$

where r is the radius of the ball B containing the support of a. Then $b \in \mathrm{lmo}(\mathbb{R}^n)$.

PROOF. We first handle a ball B with radius $r < 1$. We define $s(x) := \mathrm{sgn}[b(x) - b_B]$ and $a(x) := \frac{s(x)-s_B}{|B|}\chi_B(x)$. Then a is an h^1 atom. Moreover, we can write

$$\begin{aligned}
\log(1+r^{-1}) \fint_B |b(x)-b_B|dx &= \log(1+r^{-1}) \fint_B s(x)(b(x)-b_B)dx \\
&= \frac{\log(1+r^{-1})}{|B|} \int_B (s(x)-s_B)(b(x)-b_B)dx \\
&= \log(1+r^{-1}) \int_B a(x)b(x)dx \\
&= \log(1+r^{-1}) \left|\int_B ab\right| \\
&\leq C_b.
\end{aligned}$$

If $r \geq 1$, we define $a(x) := \frac{\mathrm{sgn}[b(x)]}{|B|}\chi_B(x)$. A similar argument shows that

$$\frac{1}{|B|}\int_B |b(x)|dx = \left|\int_B a(x)b(x)dx\right| \leq C_b.$$

Therefore, we can conclude that $b \in \mathrm{lmo}(\mathbb{R}^n)$ with $\|b\|_{\mathrm{lmo}} \leq C_b$. □

In a similar vein, we give a partial converse to Proposition 4.7

PROPOSITION 4.12. *Let $b \in \mathrm{bmo}(\mathbb{R}^n)$. Suppose there exists a constant κ_b such that every h^1 atom a satisfies*

$$\|\mathcal{M}_b a\|_{L^1} \leq \kappa_b.$$

Then b satisfies

$$\sup_{r(B)<1} \frac{1}{|B|} \||b - b_B|\chi_B\|_{h^1} + \sup_{r(B)\geq 1} \frac{1}{|B|} \||b|\chi_B\|_{h^1} < \infty.$$

PROOF. Fix a ball B with radius $r < 1$. As in the previous proof, set $s(x) := \mathrm{sgn}[b(x) - b_B]$ and $a(x) := \frac{s(x)-s_B}{|B|}\chi_B(x)$. Then a is an h^1 atom, and from Proposition 4.8,

$$\begin{aligned}
\||b - b_B|\frac{\chi_B}{|B|}\|_{h^1} &= \|(b - b_B)s(x)\frac{\chi_B}{|B|}\|_{h^1} \\
&\leq \|a(b - b_B)\|_{h^1} + |s_B| \left\| \frac{(b - b_B)\chi_B}{|B|} \right\|_{h^1} \\
&\lesssim \|\mathcal{M}_b(a)\|_{L^1} + \|b\|_{\mathrm{bmo}} + |s_B| \left\| \frac{(b - b_B)\chi_B}{|B|} \right\|_{h^1} \\
&\lesssim \|\mathcal{M}_b(a)\|_{L^1} + \|b\|_{\mathrm{bmo}}.
\end{aligned}$$

In the last step we used the fact that $|s_B| \leq 1$ and $\frac{(b-b_B)\chi_B}{|B|\|b\|_{\mathrm{bmo},2}}$ is a $(1, 2)$ atom.

For B with radius $r \geq 1$, we again put $a(x) = \frac{\mathrm{sgn}(b(x))\chi_B(x)}{|B|}$ and use Proposition 4.8:

$$\||b|\frac{\chi_B}{|B|}\|_{h^1} = \|a(b - c_B)\|_{h^1} \lesssim \|\mathcal{M}_b(a)\|_{L^1} + \|b\|_{\mathrm{bmo}}.$$

□

One interesting corollary from Proposition 4.7 and 4.12 is the following.

COROLLARY 4.13. *Let $b \in \mathrm{bmo}(\mathbb{R}^n)$. Then the following are equivalent.*

(1) *The function b is in $\mathrm{lmo}(\mathbb{R}^n)$.*
(2) *The function b satisfies*

$$A_b := \sup_{r(B)<1} \frac{1}{|B|} \||b - b_B|\chi_B\|_{h^1} + \sup_{r(B)\geq 1} \frac{1}{|B|} \||b|\chi_B\|_{h^1} < \infty.$$

PROOF. Suppose $b \in \mathrm{lmo}(\mathbb{R}^n)$. Then from the proof of Theorem 4.10, for an h^1 atom a, γa is an h^1_b atom, where $\gamma = \frac{\log 2\|b\|_{\mathrm{BMO}_{\mathrm{loc},2}}}{\|b\|_{\mathrm{LMO}_{\mathrm{loc},2}}}$. Hence by Proposition 4.7

$$\|\mathcal{M}_b a\|_{L^1} \lesssim \gamma^{-1}\|b\|_{\mathrm{bmo}} \approx \kappa_b := \frac{\|b\|_{\mathrm{LMO}_{\mathrm{loc}}}}{\|b\|_{\mathrm{BMO}_{\mathrm{loc}}}}\|b\|_{\mathrm{bmo}},$$

which is exactly the hypothesis of Proposition 4.12.

Conversely, suppose $A_b < \infty$. Observe that from Proposition 2.6, we can estimate

$$\sup_{r(B)<1} \frac{\log(1 + [r(B)]^{-1})}{|B|} \int_B |b(x) - b_B|dx \lesssim \sup_{r(B)<1} \frac{1}{|B|} \||b - b_B|\chi_B\|_{h^1}.$$

Meanwhile, using the fact that $\|f\|_{L^1(\mathbb{R}^n)} \leq \|f\|_{h^1(\mathbb{R}^n)}$, we have

$$\sup_{r(B)\geq 1} \frac{1}{|B|} \int_B |b(x)|dx \leq \sup_{r(B)\geq 1} \frac{1}{|B|} \||b|\chi_B\|_{h^1}.$$

Thus $\|b\|_{\mathrm{lmo}} \lesssim A_b < \infty$. □

5. Boundedness of the commutator of inhomogeneous singular integral operators with bmo functions

At the end of Section 3, we introduced a class of Calderón-Zygmund operators where the kernel has better decay away from the diagonal (see Definition 3.6). We will now show that the commutators of these operators with bmo functions take h^1_b atoms to L^1.

THEOREM 5.1. *Suppose $b \in \mathrm{bmo}(\mathbb{R}^n)$ and T is a inhomogeneous singular integral operator. Then $[b,T](a) \in L^1(\mathbb{R}^n)$ and*

$$\|[b,T]a\|_{L^1} \lesssim \|b\|_{\mathrm{bmo}}$$

for all h^1_b atoms a. Thus $[b,T] : h^1_{\mathrm{finite},b}(\mathbb{R}^n) \to L^1(\mathbb{R}^n)$, where $h^1_{\mathrm{finite},b}(\mathbb{R}^n)$ denotes the space of finite linear combinations of h^1_b atoms.

PROOF. Let a be an h^1_b atom with $\mathrm{supp}(a) \subset B = B(x_0, r)$. Note that we can write the commutator acting on a as $[b,T](a) = (b - c_B)T(a) - T(a(b - c_B))$.

We first show that $T(a(b - c_B)) \in L^1(\mathbb{R}^n)$. By the L^p boundedness of T for $p = \frac{3}{2}$, we have

$$\begin{aligned}
\|T(a(b-c_B))\|_{L^1(2B)} &\le |2B|^{\frac{1}{3}} \|T(a(b-c_B))\|_{L^{\frac{3}{2}}(2B)} \\
&\lesssim |B|^{\frac{1}{3}} \|T\|_{L^{\frac{3}{2}} \to L^{\frac{3}{2}}} \|a(b-c_B)\|_{L^{\frac{3}{2}}(\mathbb{R}^n)} \\
&\lesssim \|T\|_{L^{\frac{3}{2}} \to L^{\frac{3}{2}}} |B|^{\frac{1}{3}} \|b - c_B\|_{L^6(B)} \|a\|_{L^2(B)} \\
&\lesssim \|T\|_{L^{\frac{3}{2}} \to L^{\frac{3}{2}}} |B|^{-\frac{1}{6}} \|b - c_B\|_{L^6(B)} \lesssim \|T\|_{L^{\frac{3}{2}} \to L^{\frac{3}{2}}} \|b\|_{\mathrm{bmo},6}.
\end{aligned}$$

Looking at $x \in (2B)^c$, first consider the case $r < 1$. From Proposition 4.6, we may assume $\int a = 0$. Since $x \notin \mathrm{supp}(a(b - b_B))$, we can use the kernel representation to write

$$\begin{aligned}
&\|T(a(b-b_B))\|_{L^1((2B)^c)} \\
&\quad \le \int_{(2B)^c} \left| \int_B K(x,y) a(y)(b(y) - b_B) dy \right| dx \\
&\quad \le \int_{(2B)^c} \left| \int_B [K(x,y) - K(x,x_0)] a(y)(b(y) - b_B) dy \right| dx + \int_{(2B)^c} |K(x,x_0)| \left| \int ab \right| dx \\
&\quad =: I + II.
\end{aligned}$$

As $2|y - x_0| < 2r \le |x - x_0|$, we can use the smoothness of K (condition (2) in Definition 3.6) to estimate

$$\begin{aligned}
I &\lesssim \int_{(2B)^c} \int_B \frac{|y - x_0|^\delta}{|x - x_0|^{n+\delta}} |a(y)||b(y) - b_B| dy dx \\
&\lesssim \int_{(2B)^c} \frac{r^\delta}{|x - x_0|^{n+\delta}} \int_B |a(y)||b(y) - b_B| dy dx \\
&\le \int_{(2B)^c} \frac{C r^\delta}{|x - x_0|^{n+\delta}} dx \|b\|_{\mathrm{bmo},2} \\
&\lesssim \|b\|_{\mathrm{bmo}}.
\end{aligned}$$

Using the decay property of K (condition (2) in Definition 3.6) and the cancellation of a against b, recalling that we chose $C_b = \|b\|_{\mathrm{bmo},2}$, we have

$$\begin{aligned}
II &\leq \int_{(2B)^c} |K(x,x_0)| \left| \int ab \right| dx \\
&\lesssim \left(\int_{2r \leq |x-x_0| \leq 1} \frac{1}{|x-x_0|^n} dx + \int_{|x-x_0| \geq 1} \frac{1}{|x-x_0|^{n+\varepsilon}} dx \right) \frac{C_b}{\ln(1+r^{-1})} \\
&\lesssim \|b\|_{\mathrm{bmo},2} \frac{\ln(1+r^{-1})+1}{\ln(1+r^{-1})} \lesssim \|b\|_{\mathrm{bmo}}.
\end{aligned}$$

When $r \geq 1$, using the fact that $2|y-x_0| \leq |x-x_0|$ implies $|x-y| \geq \frac{1}{2}|x-x_0|$, the decay of K again gives

$$\begin{aligned}
\|T(a(b-c_B))\|_{L^1((2B)^c)} &\leq \int_{(2B)^c} \left| \int_B K(x,y)a(y)b(y)dy \right| dx \\
&\lesssim \int_{|x-x_0| \geq 2r} \frac{1}{|x-x_0|^{n+\varepsilon}} \|b\|_{\mathrm{bmo},2} \\
&\lesssim \int_{|x-x_0| \geq 2} \frac{1}{|x-x_0|^{n+\varepsilon}} \|b\|_{\mathrm{bmo},2} \lesssim \|b\|_{\mathrm{bmo}}.
\end{aligned}$$

Now we can handle the term $(b-c_B)T(a)$. Using the boundedness of T on L^2, we have

$$\begin{aligned}
\|(b-c_B)T(a)\|_{L^1(2B)} &\leq \|b-c_B\|_{L^2(2B)} \|T(a)\|_{L^2(\mathbb{R}^n)} \\
&\lesssim \|b\|_{\mathrm{bmo},2} |B|^{\frac{1}{2}} \|T\|_{L^2 \to L^2} \|a\|_{L^2(\mathbb{R}^n)} \\
&\lesssim \|b\|_{\mathrm{bmo}} \|T\|_{L^2 \to L^2}.
\end{aligned}$$

When $r < 1$, we apply the cancellation of a, the smoothness of K and (2.3) to get

$$\begin{aligned}
\|(b-b_B)T(a)\|_{L^1((2B)^c)} &\leq \int_{(2B)^c} |b(x)-b_B| \left| \int_B [K(x,y)-K(x,x_0)]a(y)dy \right| dx \\
&\leq \int_{(2B)^c} |b(x)-b_B| \frac{r^\delta}{|x-x_0|^{n+\delta}} dx \lesssim \|b\|_{\mathrm{bmo}}. \qquad (5.1)
\end{aligned}$$

When $r \geq 1$, similar to the estimate of $\|T(a(b-c_B))\|_{L^1((2B)^c)}$, we have, again by (2.3),

$$\begin{aligned}
\|(b-c_B)T(a)\|_{L^1((2B)^c)} &\leq \int_{(2B)^c} |b(x)| \left| \int_B K(x,y)a(y)dy \right| dx \\
&\leq \int_{(2B)^c} \frac{|b(x)|}{|x-x_0|^{n+\varepsilon}} dx \\
&\leq \int_{(2B)^c} r^\varepsilon \frac{|b(x)-c_B|}{|x-x_0|^{n+\varepsilon}} dx \lesssim \|b\|_{\mathrm{bmo}}. \qquad (5.2)
\end{aligned}$$

We have thus shown $\|[b,T]a\|_{L^1} \lesssim \|b\|_{\mathrm{bmo}}$ for all h^1_b atoms a. Finally, $f \in h^1_{\mathrm{finite},b}(\mathbb{R}^n)$ means it can be written as a finite linear combination $\sum_k \gamma_k a_k$ of h^1_b atoms, so by linearity and taking the infimum over all such decompositions, we have

$$\|[b,T]f\| \lesssim \|b\|_{\mathrm{bmo}} \inf \sum_k |\gamma_k|.$$

□

We cannot extend the conclusion to all of $h^1_{\mathrm{atom},b}(\mathbb{R}^n)$ since we do not know whether for $f \in h^1_{\mathrm{finite},b}(\mathbb{R}^n)$, the infimum $\inf \sum_k |\gamma_k|$ taken over all finite linear combinations of atoms is comparable to $\|f\|_{h^1_{\mathrm{atom},b}}$, the infimum taken over all possible atomic decomposition.

When $h^1_{\mathrm{atom},b}(\mathbb{R}^n) = h^1(\mathbb{R}^n)$, it is true that the infimum over the finite linear combinations of atoms is comparable to the h^1 norm, and it suffices to verify the boundedness on one h^1 atom in order to get a bounded extension to the whole space (see [**36**, Theorem 3.1, Corollary 3.4] for the case of H^1 and [**27**, Proposition 7.1] for the case of h^1 on a space of homogeneous type). This gives us the following corollary.

COROLLARY 5.2. *Let $b \in \mathrm{lmo}(\mathbb{R}^n)$, and T be a inhomogeneous singular integral operator. Then there exists a unique extension L_b of $[b,T]$ that is bounded from $h^1(\mathbb{R}^n)$ to $L^1(\mathbb{R}^n)$.*

In particular, for $b \in \mathrm{lmo}(\mathbb{R}^n)$, we can apply this corollary to show the boundedness of the commutators $[b, \mathcal{R}_{j,\eta}]$ and $[b, \mathcal{R}_j^{\psi}]$ defined in Corollary 3.5, provided the resulting localized convolution kernels have extra decay away from the origin. This requires some extra assumptions on η and ψ in addition to those in Theorem 3.4.

We compare the preceding theorem and its corollary with the results of [**28**]. While in Theorem 5.1 we only assume the h^1_b atoms have approximate cancellation, in [**28**, Theorem 4.2] it was shown that $[b,T]$ maps Pérez h^1_b atoms to L^1 under the stronger assumption of exact cancellation, as in Definition 4.4. Their assumption on b, that $b \in \mathrm{BMO}_\infty$, is weaker than our assumption $b \in \mathrm{bmo}$, in the sense that for large balls it allows the oscillation to grow with the size of the ball. On the other hand, the operators T they consider, which are pseudo-differential operators, have kernels which have much better decay and smoothness—see [**28**, Proposition 2.1]. In fact, it is this decay of the kernel to arbitrarily large order that gives [**28**, Lemma 3.2], which is the key to the proof of the h^1 to h^1 boundedness of the commutator in the case $b \in \mathrm{LMO}_\infty$, part (i) of [**28**, Theorem 1.1].

In our case, since our assumptions on the kernel are much weaker, we need to impose some approximate cancellation conditions in order to get h^1 to h^1 boundedness of $[b,T]$. We follow the ideas in [**14**, Definition 5.1].

DEFINITION 5.3. Suppose T is an inhomogeneous singular integral operator. Given a ball B, denote by $L^2_0(\mathbb{R}^n)$ those $g \in L^2(B)$ with $\int g = 0$. For $b \in \mathrm{bmo}$, define $T^*_B(b)$, relative to this ball B, in the distributional sense, by

$$\langle T^*_B(b), g\rangle := \int_{\mathbb{R}^n} (b - b_B)Tg(x)dx, \qquad \forall\, g \in L^2_0(B). \tag{5.3}$$

As in (5.1), the conditions on the kernel and the vanishing integral of g guarantee that the integral on the right-hand-side of (5.3) converges absolutely and $T^*_B(b)$ is a bounded linear functional on $L^2_0(B)$ with norm bounded by a constant multiple of $\|b\|_{\mathrm{bmo}}|B|^{1/2}$. Thus we can identify $T^*_B(b)$ with an equivalence class of functions in $L^2(B)$ modulo constants (similarly to the case $p = 1$ in [**14**, Remark 5.4]). In particular, we can impose the following condition, denoting $f := T^*_B(b)$, without ambiguity:

$$\left(\fint_B |f - f_B|^2\right)^{1/2} \le [\log(1 + 1/r(B))]^{-1}. \tag{5.4}$$

Similarly to what was pointed out in [**14**, Remark 5.4] for the case of $T^*((\cdot - x_B)^\alpha)$, this condition is not the same as requiring $T^*(b) \in \mathrm{LMO}_{\mathrm{loc}}(\mathbb{R}^n)$ since $T^*_B(b)$ is defined relative to a specific ball B, using the L^2 pairing, not the pairing of bmo and h^1, and we are imposing condition (5.4) only for that ball.

While the local Campanato-type conditions used in [**14**, Theorem 5.3] to guarantee that inhomogeneous singular integral operators are bounded on $h^p(\mathbb{R}^n)$ are also expressed in terms of individual balls, in the case $p = 1$ we do not need to subtract any constant as the power of the monomial is $\alpha = 0$. Thus the condition becomes $T^*1 \in \mathrm{LMO}_{\mathrm{loc}}(\mathbb{R}^n)$, where T^* can be taken in the sense of duality of bmo and h^1. In general, for b bounded, the pairing between bmo and h^1 can be given by integration, but for arbitrary $b \in \mathrm{bmo}$, we need to subtract b_B in the integral in (5.3), unless we have $\int Tg = 0$. This is the case when T is a convolution operator, as Tg inherits the cancellation from g (see also the remarks following Corollary 3.5).

THEOREM 5.4. *Let $b \in \mathrm{bmo}(\mathbb{R}^n)$. Suppose T is an inhomogeneous singular integral operator. If $T^*(1) \in \mathrm{LMO}_{loc}(\mathbb{R}^n)$, and, for any ball B with $r(B) < 1$, $f := T^*_B(b)$ satisfies (5.4), then*

$$\|[b,T](a)\|_{h^1} \lesssim \|b\|_{\mathrm{bmo}}$$

for all h^1_b atoms a.

PROOF. Let a be an h^1_b atom supported in a ball $B = B(x_0, r)$; as above, we may assume that when $r < 1$, $\int a = 0$. Again write $[b,T](a) = (b - c_B)T(a) - T(a(b - c_B))$.

As explained before the statement of the theorem, the assumption $T^*(1) \in \mathrm{LMO}_{\mathrm{loc}}(\mathbb{R}^n)$ allows us to apply [**14**, Theorem 5.3] get the boundedness of T on h^1; thus from (4.4) we get

$$\|T(a(b - c_B))\|_{h^1} \lesssim \|a(b - c_B)\|_{h^1} \lesssim \|b\|_{\mathrm{bmo}}.$$

It remains to estimate the term $\|(b - c_B)T(a)\|_{h^1}$. We will show that $M = (b - c_B)T(a)$ is a multiple of an (s, λ) molecule (see Definition 2.4) associated to the ball $2B$, where we choose

$$s = \frac{3}{2}, \quad \lambda = \frac{n}{2} + \mu, \quad 0 < \mu < \frac{3}{2}\min(\delta, \varepsilon).$$

We first verify condition (M1) of Definition 2.4. Proceeding as in (4.3) and using the boundedness of T on L^2, we have

$$\begin{aligned}\|(b - c_B)T(a)\|_{L^{\frac{3}{2}}(2B)} &\lesssim \|(b - c_B)\|_{L^6(2B)}\|T(a)\|_{L^2}\\ &\lesssim \|b - c_B\|_{L^6(2B)}\|a\|_{L^2}\\ &\lesssim \|b\|_{\mathrm{bmo},6}r^{-\frac{n}{3}}.\end{aligned}$$

Next, we need to show condition (M2) of Definition 2.4, which for our choices of parameters amounts to $\int_{(2B)^c} |M(x)|^{\frac{3}{2}}|x - x_0|^\lambda dx \lesssim r^\mu$.

Suppose that $r < 1$. As in (5.1), using the cancellation on a and the smoothness of the kernel to bound $T(a)$, and (2.3), we have

$$\begin{aligned}&\int_{(2B)^c} |(b(x) - b_B)T(a)(x)|^{\frac{3}{2}}|x - x_0|^\lambda dx\\&\lesssim \int_{(2B)^c} |b(x) - b_B|^{\frac{3}{2}} \frac{r^{\frac{3}{2}\delta}}{|x - x_0|^{\frac{3n}{2}+\frac{3\delta}{2}}}|x - x_0|^\lambda dx = r^{\frac{3}{2}\delta} \int_{(2B)^c} \frac{|b(x) - b_B|^{\frac{3}{2}}}{|x - x_0|^{n+\frac{3}{2}\delta-\mu}} dx\\&\lesssim \|b\|_{\mathrm{bmo}}^{\frac{3}{2}} r^\mu.\end{aligned}$$

Finally, condition (5.4) on $f = T_B^*(b)$ gives us the approximate cancellation for the molecule, condition (M3) of Definition 2.4:

$$\begin{aligned}\left|\int_{\mathbb{R}^n} [b(x) - b_B]T(a)(x)dx\right| = \left|\langle T_B^*(b), a\rangle\right| &\leq \left(\int_B |f - f_B|^2\right)^{1/2} \|a\|_{L^2(B)}\\&\leq [\log(1 + r(B)^{-1})]^{-1}.\end{aligned}$$

Now looking at the case $r \geq 1$, we can proceed as in (5.2) to write

$$\begin{aligned}&\int_{(2B)^c} |b(x)T(a)(x)|^{\frac{3}{2}}|x - x_0|^\lambda dx\\&\lesssim \int_{(2B)^c} \frac{|b(x)|^{\frac{3}{2}}}{|x - x_0|^{\frac{3}{2}n+\frac{3}{2}\varepsilon}}|x - x_0|^\lambda dx\\&\leq r^{\frac{3}{2}\varepsilon} \int_{(2B)^c} \frac{|b(x)|^{\frac{3}{2}}}{|x - x_0|^{n+\frac{3}{2}\varepsilon-\mu}} dx\\&\lesssim \|b\|_{\mathrm{bmo}}^{\frac{3}{2}} r^\mu.\end{aligned}$$

Here we have used (2.3) as well as the fact that $r \geq 1$, $\mu < \frac{3}{2}\varepsilon$. □

Similarly to Corollary 5.2, when $h_b^1 = h^1$, the boundedness extends from one atom to all functions in the space.

COROLLARY 5.5. *Assuming the hypotheses of Theorem 5.4, if in addition* $b \in \mathrm{lmo}(\mathbb{R}^n)$, *then* $[b, T]$ *is bounded from* $h^1(\mathbb{R}^n)$ *to* $h^1(\mathbb{R}^n)$.

Acknowledgments

We are grateful to the volume editors and the referee for their support and helpful comments.

References

[1] M. Agcayazi, A. Gogatishvili, K. Koca, and R. Mustafayev, *A note on maximal commutators and commutators of maximal functions*, J. Math. Soc. Japan **67** (2015), no. 2, 581–593, DOI 10.2969/jmsj/06720581. MR3340187

[2] J. Bastero, M. Milman, and F. J. Ruiz, *Commutators for the maximal and sharp functions*, Proc. Amer. Math. Soc. **128** (2000), no. 11, 3329–3334, DOI 10.1090/S0002-9939-00-05763-4. MR1777580

[3] F. Bernicot, T. Elgindi, and S. Keraani, *On the inviscid limit of the 2D Navier-Stokes equations with vorticity belonging to BMO-type spaces*, Ann. Inst. H. Poincaré C Anal. Non Linéaire **33** (2016), no. 2, 597–619, DOI 10.1016/j.anihpc.2014.12.001. MR3465387

[4] F. Bernicot and T. Hmidi, *On the global well-posedness for Euler equations with unbounded vorticity*, Dyn. Partial Differ. Equ. **12** (2015), no. 2, 127–155, DOI 10.4310/DPDE.2015.v12.n2.a3. MR3361245

[5] A. Bonami and J. Feuto, *Products of functions in Hardy and Lipschitz or BMO spaces*, Recent developments in real and harmonic analysis, Appl. Numer. Harmon. Anal., Birkhäuser Boston, Boston, MA, 2010, pp. 57–71, DOI 10.1007/978-0-8176-4588-5_4. MR2603175

[6] G. Bourdaud, *Remarques sur certains sous-espaces de* $\mathrm{BMO}(\mathbb{R}^n)$ *et de* $\mathrm{bmo}(\mathbb{R}^n)$ (French, with English and French summaries), Ann. Inst. Fourier (Grenoble) **52** (2002), no. 4, 1187–1218. MR1927078

[7] M. Bownik, *Boundedness of operators on Hardy spaces via atomic decompositions*, Proc. Amer. Math. Soc. **133** (2005), no. 12, 3535–3542, DOI 10.1090/S0002-9939-05-07892-5. MR2163588

[8] Bui Huy Qui, *Weighted Hardy spaces*, Math. Nachr. **103** (1981), 45–62, DOI 10.1002/mana.19811030106. MR0653914

[9] D.-C. Chang and S.-Y. Li, *On the boundedness of multipliers, commutators and the second derivatives of Green's operators on* H^1 *and BMO*, Ann. Scuola Norm. Sup. Pisa Cl. Sci. (4) **28** (1999), no. 2, 341–356. MR1736232

[10] R. R. Coifman, R. Rochberg, and G. Weiss, *Factorization theorems for Hardy spaces in several variables*, Ann. of Math. (2) **103** (1976), no. 3, 611–635, DOI 10.2307/1970954. MR412721

[11] R. R. Coifman and G. Weiss, *Extensions of Hardy spaces and their use in analysis*, Bull. Amer. Math. Soc. **83** (1977), no. 4, 569–645, DOI 10.1090/S0002-9904-1977-14325-5. MR0447954

[12] G. D. Dafni, *Hardy spaces on strongly pseudoconvex domains in C('n) and domains of finite type in C('2)*, ProQuest LLC, Ann Arbor, MI, 1993. Thesis (Ph.D.)–Princeton University. MR2689452

[13] G. Dafni, *Local VMO and weak convergence in* h^1, Canad. Math. Bull. **45** (2002), no. 1, 46–59, DOI 10.4153/CMB-2002-005-2. MR1884133

[14] G. Dafni, C. H. Lau, T. Picon, and C. Vasconcelos, *Inhomogeneous cancellation conditions and Calderón-Zygmund type operators on* h^p, Nonlinear Anal. **225** (2022), Paper No. 113110, 22, DOI 10.1016/j.na.2022.113110. MR4475438

[15] G. Dafni, C. H. Lau, T. Picon, and C. Vasconcelos, *Necessary cancellation conditions for the boundedness of operators on local Hardy spaces*, Potential Anal. (2023), `https://doi.org/10.1007/s11118-023-10100-w`.

[16] G. Dafni and E. Liflyand, *A local Hilbert transform, Hardy's inequality and molecular characterization of Goldberg's local Hardy space*, Complex Anal. Synerg. **5** (2019), no. 1, Paper No. 10, 9, DOI 10.1007/s40627-019-0036-2. MR3936310

[17] G. Dafni, E. Mombourquette, and H. Yue, *Some characterizations of local bmo and* h^1 *on spaces of homogeneous type*, Harmonic analysis and nonlinear partial differential equations, RIMS Kôkyûroku Bessatsu, B56, Res. Inst. Math. Sci. (RIMS), Kyoto, 2016, pp. 187–215. MR3617720

[18] G. Dafni and H. Yue, *Some characterizations of local bmo and* h^1 *on metric measure spaces*, Anal. Math. Phys. **2** (2012), no. 3, 285–318, DOI 10.1007/s13324-012-0034-5. MR2958361

[19] W. Ding, Y. Han, and Y. Zhu, *Boundedness of singular integral operators on local Hardy spaces and dual spaces*, Potential Anal. **55** (2021), no. 3, 419–441, DOI 10.1007/s11118-020-09863-3. MR4313085

[20] C. Fefferman and E. M. Stein, H^p *spaces of several variables*, Acta Math. **129** (1972), no. 3-4, 137–193, DOI 10.1007/BF02392215. MR0447953

[21] X. Fu, D. Yang, and S. Yang, *Endpoint boundedness of linear commutators on local Hardy spaces over metric measure spaces of homogeneous type*, J. Geom. Anal. **31** (2021), no. 4, 4092–4164, DOI 10.1007/s12220-020-00429-8. MR4236554

[22] J. García-Cuerva, E. Harboure, C. Segovia, and J. L. Torrea, *Weighted norm inequalities for commutators of strongly singular integrals*, Indiana Univ. Math. J. **40** (1991), no. 4, 1397–1420, DOI 10.1512/iumj.1991.40.40063. MR1142721

[23] D. Goldberg, *Local Hardy spaces*, Harmonic analysis in Euclidean spaces (Proc. Sympos. Pure Math., Williams Coll., Williamstown, Mass., 1978), Proc. Sympos. Pure Math., XXXV, Part 1, Amer. Math. Soc., Providence, RI, 1979, pp. 245–248. MR545262

[24] D. Goldberg, *A local version of real Hardy spaces*, Duke Math. J. **46** (1979), no. 1, 27–42. MR523600

[25] L. Grafakos, *Modern Fourier analysis*, 3rd ed., Graduate Texts in Mathematics, vol. 250, Springer, New York, 2014, DOI 10.1007/978-1-4939-1230-8. MR3243741

[26] E. Harboure, C. Segovia, and J. L. Torrea, *Boundedness of commutators of fractional and singular integrals for the extreme values of p*, Illinois J. Math. **41** (1997), no. 4, 676–700. MR1468874

[27] Z. He, D. Yang, and W. Yuan, *Real-variable characterizations of local Hardy spaces on spaces of homogeneous type*, Math. Nachr. **294** (2021), no. 5, 900–955, DOI 10.1002/mana.201900320. MR4288480

[28] H. D. Hung and L. D. Ky, *An Hardy estimate for commutators of pseudo-differential operators*, Taiwanese J. Math. **19** (2015), no. 4, 1097–1109, DOI 10.11650/tjm.19.2015.5003. MR3384681

[29] D. Quoc Huy and L. Dang Ky, *Weighted Hardy space estimates for commutators of Calderón-Zygmund operators*, Vietnam J. Math. **49** (2021), no. 4, 1065–1077, DOI 10.1007/s10013-020-00406-2. MR4319538

[30] T. P. Hytönen, *The L^p-to-L^q boundedness of commutators with applications to the Jacobian operator* (English, with English and French summaries), J. Math. Pures Appl. (9) **156** (2021), 351–391, DOI 10.1016/j.matpur.2021.10.007. MR4338459

[31] S. Janson, *Mean oscillation and commutators of singular integral operators*, Ark. Mat. **16** (1978), no. 2, 263–270, DOI 10.1007/BF02386000. MR524754

[32] F. John and L. Nirenberg, *On functions of bounded mean oscillation*, Comm. Pure Appl. Math. **14** (1961), 415–426, DOI 10.1002/cpa.3160140317. MR0131498

[33] Y. Komori, *Calderón-Zygmund operators on $H^p(\mathbf{R}^n)$*, Sci. Math. Jpn. **53** (2001), no. 1, 65–73. MR1821600

[34] Luong Dang Ky, *Bilinear decompositions and commutators of singular integral operators*, Trans. Amer. Math. Soc. **365** (2013), no. 6, 2931–2958, DOI 10.1090/S0002-9947-2012-05727-8. MR3034454

[35] S.-Y. Li, *Toeplitz operators on Hardy space $H^p(S)$ with $0 < p \leq 1$*, Integral Equations Operator Theory **15** (1992), no. 5, 807–824, DOI 10.1007/BF01200702. MR1177324

[36] S. Meda, P. Sjögren, and M. Vallarino, *On the H^1-L^1 boundedness of operators*, Proc. Amer. Math. Soc. **136** (2008), no. 8, 2921–2931, DOI 10.1090/S0002-9939-08-09365-9. MR2399059

[37] M. Milman and T. Schonbek, *Second order estimates in interpolation theory and applications*, Proc. Amer. Math. Soc. **110** (1990), no. 4, 961–969, DOI 10.2307/2047743. MR1075187

[38] M. M. Peloso and S. Secco, *Local Riesz transforms characterization of local Hardy spaces*, Collect. Math. **59** (2008), no. 3, 299–320, DOI 10.1007/BF03191189. MR2452309

[39] C. Pérez, *Endpoint estimates for commutators of singular integral operators*, J. Funct. Anal. **128** (1995), no. 1, 163–185, DOI 10.1006/jfan.1995.1027. MR1317714

[40] D. Sarason, *Functions of vanishing mean oscillation*, Trans. Amer. Math. Soc. **207** (1975), 391–405, DOI 10.2307/1997184. MR0377518

[41] D. Sarason, *Function theory on the unit circle*, Virginia Polytechnic Institute and State University, Department of Mathematics, Blacksburg, VA, 1978. Notes for lectures given at a Conference at Virginia Polytechnic Institute and State University, Blacksburg, Va., June 19–23, 1978. MR521811

[42] C. Segovia and J. L. Torrea, *Weighted inequalities for commutators of fractional and singular integrals*, Publ. Mat. **35** (1991), no. 1, 209–235, DOI 10.5565/PUBLMAT_35191_09. Conference on Mathematical Analysis (El Escorial, 1989). MR1103616

[43] X. L. Shi and A. Torchinsky, *Functions of vanishing mean oscillation*, Math. Nachr. **133** (1987), 289–296, DOI 10.1002/mana.19871330119. MR0912434

[44] D. A. Stegenga, *Bounded Toeplitz operators on H^1 and applications of the duality between H^1 and the functions of bounded mean oscillation*, Amer. J. Math. **98** (1976), no. 3, 573–589, DOI 10.2307/2373807. MR0420326

[45] E. M. Stein, *Singular integrals and differentiability properties of functions*, Princeton Mathematical Series, vol. No. 30, Princeton University Press, Princeton, NJ, 1970. MR0290095

[46] E. M. Stein, *Harmonic analysis: real-variable methods, orthogonality, and oscillatory integrals*, Princeton Mathematical Series, vol. 43, Princeton University Press, Princeton, NJ, 1993. With the assistance of Timothy S. Murphy; Monographs in Harmonic Analysis, III. MR1232192

[47] L. Tang, *Interior h^1 estimates for parabolic equations with LMO coefficients*, Canad. J. Math. **62** (2010), no. 1, 202–217, DOI 10.4153/CJM-2010-011-1. MR2597030

[48] L. Tang, *Weighted local Hardy spaces and their applications*, Illinois J. Math. **56** (2012), no. 2, 453–495. MR3161335

[49] A. Uchiyama, *On the compactness of operators of Hankel type*, Tohoku Math. J. (2) **30** (1978), no. 1, 163–171, DOI 10.2748/tmj/1178230105. MR467384

Department of Mathematics and Statistics, Concordia University, Montréal, Québec, H3G-1M8, Canada

Email address: galia.dafni@concordia.ca

Department of Mathematical Sciences, University of Cincinnati, Cincinnati, Ohio 45221-0025

Email address: lauco@ucmail.uc.edu

Contemporary Mathematics
Volume **799**, 2024
https://doi.org/10.1090/conm/799/16022

On interpolation by functions in ℓ^p_A

Raymond Cheng and Christopher Felder

ABSTRACT. This work explores several aspects of interpolating sequences for ℓ^p_A, the space of analytic functions on the unit disk with p-summable Maclaurin coefficients. Much of this work is communicated through a Carlesonian lens. We investigate various analogues of Gramian matrices, for which we show boundedness conditions are necessary and sufficient for interpolation, including a characterization of universal interpolating sequences in terms of Riesz systems. We also discuss weak separation, giving a characterization of such sequences using a generalization of the pseudohyperbolic metric. Lastly, we consider Carleson measures and embeddings.

1. Introduction

Let $\{z_k\}_{k=0}^{\infty}$ be a sequence of distinct points in the open unit disk $\mathbb{D}$ of the complex plane, and let $\{w_k\}_{k=0}^{\infty}$ be a sequence of complex numbers. An interpolation problem, broadly speaking, is to find an analytic function f on $\mathbb{D}$ such that

$$f(z_k) = w_k$$

for all $k = 0, 1, 2, \ldots$. Furthermore, it is natural to seek a characterization of those sequences $\{z_k\}$ and $\{w_k\}$ for which such a function f always exists, and belongs to a certain class of analytic functions on $\mathbb{D}$. As explained by Duren [**16**, p. 148], intuition suggests that the points $\{z_k\}$ must not lie too "close together," lest a highly oscillatory choice of targets $\{w_k\}$ fail to be interpolated by a function of the prescribed class.

This notion is illustrated by a theorem of Carleson [**5**], which characterizes interpolation by functions in H^∞, the space of bounded analytic functions on $\mathbb{D}$. A sequence $\{z_k\}$ in $\mathbb{D}$ is said to be *uniformly separated* if there exists $\delta > 0$ such that

$$\prod_{j=0,\, j\neq k}^{\infty} \left| \frac{z_k - z_j}{1 - \bar{z}_j z_k} \right| \geqslant \delta$$

for all $k = 1, 2, 3, \ldots$.

THEOREM 1.1 (Carleson). *Let $\{z_k\}$ be a sequence of points in $\mathbb{D}$. Then $\{z_k\}$ has the property that for any bounded sequence $\{w_k\}$ of complex numbers, there exists $f \in H^\infty$ such that*

$$f(z_k) = w_k$$

for all $k = 1, 2, 3, \ldots$, if and only if $\{z_k\}$ is uniformly separated.

2020 *Mathematics Subject Classification.* Primary 46E15; Secondary 30J99, 30H99.

This was extended to H^p by Shapiro and Shields [**23**]. See the books by Seip [**22**], and by Agler and McCarthy [**1**], for a modern exposition of interpolation in numerous other spaces of functions.

The present paper is concerned with interpolation by functions in ℓ^p_A, the space of analytic functions in $\mathbb{D}$ whose Maclaurin coefficients are p-summable, with $1 < p < \infty$. Vinogradov [**24,25**] (for English translations, with Khavin, see [**26,27**]) derived exact conditions for interpolation in ℓ^p_A provided that the sequence $\{z_k\}$ lies in a Stoltz domain, or that it tends rapidly enough to the boundary. Our approach connects interpolation in ℓ^p_A with associated sequences of functionals, infinite matrices, a nonlinear functional equation, notions of separation, and Carleson measures. In particular, after providing some background information on the ℓ^p_A spaces in Section 2, we move to Section 3, which is concerned with various upper and lower bounds that characterize universal interpolating sequences. Section 4 involves studying limits of truncated interpolation problems to deduce an interpolation result based on the geometry of the Banach space in terms of a minimality condition. Criteria for interpolation are expressed in terms of matrix conditions in Section 5. There arises a pair of nonlinear operators that extend the notion of a Gramian matrix to the case $p \neq 2$. Section 6 characterizes sequences which are weakly separated by the multiplier algebra of ℓ^p_A. We close with Section 7 on Carleson measures for ℓ^p_A, and a handful of open questions.

2. Preliminaries

For $0 < p < \infty$, the space ℓ^p_A is defined to be space of analytic functions on the open unit disk $\mathbb{D}$ for which the Maclaurin coefficients are pth order summable, i.e.,

$$\ell^p_A := \left\{ f(z) = \sum_{n\geq 0} a_n z^n \in \mathrm{Hol}(\mathbb{D}) : \sum_{n\geq 0} |a_n|^p < \infty \right\}.$$

This function space is endowed with the norm (or quasinorm, if $0 < p \leqslant 1$) that it inherits from the sequence space ℓ^p. Thus let us write

$$\|f\|_p = \|\{a_k\}_{k=0}^\infty\|_{\ell^p}$$

for any $f(z) = \sum_{k=0}^\infty a_k z^k$ belonging to ℓ^p_A. When $p = 2$, we recover the classical Hardy space H^2 on the disk, however, we emphasize that $\|\cdot\|_p$ refers to the norm on ℓ^p_A, and not the norm on the Hardy space H^p, or some other function space parametrized by p.

We limit our attention to the range $1 < p < \infty$. For then ℓ^p_A is reflexive, smooth and uniformly convex. In particular, it enjoys the unique nearest point property, and each nonzero vector has a unique norming functional.

Throughout this paper, q will be the Hölder conjugate to p, that is, $1/p+1/q = 1$ holds. We recall that for $1 < p < \infty$, $p \neq 2$, the dual space of ℓ^p_A can be identified with ℓ^q_A, under the pairing

$$\langle f, g\rangle = \sum_{k=0}^\infty f_k g_k,$$

where $f(z) = \sum_{k=0}^\infty f_k z^k \in \ell^p_A$ and $g(z) = \sum_{k=0}^\infty g_k z^k \in \ell^q_A$.

Point evaluation at any $w \in \mathbb{D}$ is a bounded linear functional on ℓ^p_A. It is implemented by the kernel function $\Lambda_w \in \ell^q_A$ given by

$$\Lambda_w(z) := \sum_{k=0}^{\infty} w^k z^k, \qquad z \in \mathbb{D}.$$

Indeed, for any $f(z) = \sum_{k=0}^{\infty} a_k z^k \in \ell^p_A$ we have

$$f(w) = \langle f, \Lambda_w \rangle = \sum_{k=0}^{\infty} a_k w^k.$$

The norm of Λ_w (as either a vector or a functional) is

$$\|\Lambda_w\|_q = \frac{1}{(1 - |w|^q)^{1/q}}.$$

There is a sensible way to define "inner function" in the context of ℓ^p_A, that employs a notion of orthogonality in general normed linear spaces.

Let $\mathbf{x}$ and $\mathbf{y}$ be vectors belonging to a normed linear space $\mathscr{X}$. We say that $\mathbf{x}$ is *orthogonal* to $\mathbf{y}$ in the Birkhoff–James sense [**3, 19**] if

$$\|\mathbf{x} + \beta \mathbf{y}\|_{\mathscr{X}} \geqslant \|\mathbf{x}\|_{\mathscr{X}} \tag{2.1}$$

for all scalars β, and in this case we write $\mathbf{x} \perp_{\mathscr{X}} \mathbf{y}$. Birkhoff–James orthogonality is one way to extend the concept of orthogonality from an inner product space to normed spaces. There are other ways to generalize orthogonality, but this approach is particularly useful since it is relates directly to an extremal condition, namely (2.1).

If $\mathscr{X}$ is a Hilbert space, then the usual orthogonality relation $\mathbf{x} \perp \mathbf{y}$ is equivalent to $\mathbf{x} \perp_{\mathscr{X}} \mathbf{y}$. More generally, however, the relation $\perp_{\mathscr{X}}$ is neither symmetric nor linear. When $\mathscr{X} = \ell^p_A$, let us write $\perp_p$ instead of $\perp_{\ell^p_A}$. There is a practical criterion for the relation $\perp_p$ when $1 < p < \infty$.

THEOREM 2.1 (James, [**19**]). *Suppose that $1 < p < \infty$. Then for $f(z) = \sum_{k=0}^{\infty} f_k z^k$ and $g(z) = \sum_{k=0}^{\infty} g_k z^k$ belonging to ℓ^p_A we have*

$$f \perp_p g \iff \sum_{k=0}^{\infty} |f_k|^{p-2} \overline{f}_k g_k = 0, \tag{2.2}$$

where any occurrence of "$|0|^{p-2}0$" in the right side is interpreted as zero.

In light of (2.2) we define, for a complex number $\alpha = re^{i\theta}$, and any $s > 0$, the quantity

$$\alpha^{\langle s \rangle} = (re^{i\theta})^{\langle s \rangle} := r^s e^{-i\theta}. \tag{2.3}$$

It is readily seen that for any complex numbers α and β, exponent $s > 0$, and integer $n \geqslant 0$, we have

$$\begin{aligned} (\alpha\beta)^{\langle s \rangle} &= \alpha^{\langle s \rangle} \beta^{\langle s \rangle} \\ |\alpha^{\langle s \rangle}| &= |\alpha|^s \\ \alpha^{\langle s \rangle} \alpha &= |\alpha|^{s+1} \\ (\alpha^{\langle s \rangle})^n &= (\alpha^n)^{\langle s \rangle} \\ (\alpha^{\langle p-1 \rangle})^{\langle q-1 \rangle} &= \alpha. \end{aligned}$$

Further to the notation (2.3), for $f(z) = \sum_{k=0}^{\infty} f_k z^k$, let us write

$$f^{\langle s \rangle}(z) := \sum_{k=0}^{\infty} f_k^{\langle s \rangle} z^k \tag{2.4}$$

for any $s > 0$. Similarly, for any matrix or vector with complex-valued entries $A = [a_{j,k}]$, we take $A^{\langle s \rangle}$ to mean the matrix $[a_{j,k}^{\langle s \rangle}]$. By comparing with the case $p = 2$, we can think of taking the $\langle s \rangle$ power as generalizing complex conjugation.

If $f \in \ell^p_A$, it is easy to verify that $f^{\langle p-1 \rangle} \in \ell^q_A$. Thus from (2.2) we get

$$f \perp_p g \iff \langle g, f^{\langle p-1 \rangle} \rangle = 0. \tag{2.5}$$

Consequently the relation $\perp_p$ is linear in its second argument, when $1 < p < \infty$, and it then makes sense to speak of a vector being orthogonal to a subspace of ℓ^p_A. In particular, if $f \perp_p g$ for all g belonging to a subspace $\mathscr{M}$ of ℓ^p_A, then

$$\|f + g\|_p \geqslant \|f\|_p$$

for all $g \in \mathscr{M}$. That is, f solves a nearest-point problem in relation to the subspace $\mathscr{M}$. Direct calculation will also confirm that

$$\langle f, f^{\langle p-1 \rangle} \rangle = \|f\|_p^p,$$

and hence $f^{\langle p-1 \rangle}/\|f\|_p^{p-1} \in \ell^q_A$ is the norming functional of $f \in \ell^p_A \backslash \{0\}$ (smoothness ensures that the norming functional is unique). With this concept of orthogonality established, we may now introduce a definition of inner function that is particular to ℓ^p_A.

DEFINITION 2.2. Let $1 < p < \infty$. A function $f \in \ell^p_A$ is said to be p-inner if it is not identically zero and it satisfies

$$f(z) \ \perp_p \ z^k f(z)$$

for all positive integers k.

That is, f is p-inner if it is nontrivially orthogonal to all of its forward shifts. Apart from a multiplicative constant, this definition, originating from [**11**], is equivalent to the traditional meaning of "inner" when $p = 2$. Furthermore, this approach to defining an inner property is consistent with that taken in other function spaces [**2, 4, 13, 15, 17, 18, 20, 21**].

Birkhoff–James Orthogonality also plays a role when we utilize a version of the Pythagorean theorem for ℓ^p_A. It takes the form of a family of inequalities relating the lengths of orthogonal vectors with that of their sum [**8**, Corollary 3.4].

THEOREM 2.3. *Let $1 < r < \infty$, and $1 < p < \infty$. If $1 < p \leqslant 2 \leqslant r < \infty$ or $2 \leqslant p \leqslant r < \infty$, then there exists $K > 0$ such that*

$$\|f\|_p^r + K\|g\|_p^r \leqslant \|f + g\|_p^r \tag{2.6}$$

whenever $f \perp_p g$ in ℓ^p_A.

If $1 < r \leqslant p \leqslant 2$ or $1 < r \leqslant 2 \leqslant p < \infty$, then there exists $K > 0$ such that

$$\|f\|_p^r + K\|g\|_p^r \geqslant \|f + g\|_p^r \tag{2.7}$$

whenever $f \perp_p g$ in ℓ^p_A.

Actually, this theorem holds with any Lebesgue space L^p in place of ℓ^p_A. When $p = 2$, the parameters are $K = 1$ and $r = 2$, and the Pythagorean inequalities reduce to the familiar Pythagorean theorem for a Hilbert space. More generally, these Pythagorean inequalities enable the application of some Hilbert space methods and techniques to smooth Banach spaces satisfying the weak parallelogram laws; see, for example, [**12**, Propositions 4.8.1 and 4.8.3 and Theorem 8.8.1].

For further background on the function space ℓ^p_A, we refer to the book [**12**].

3. Boundedness conditions

Much of the modern treatment of interpolating sequences has been connected to the boundedness of Gramian matrices. For example, in H^∞, Carleson's Theorem (Theorem 1.1) tells us that a sequence $\{z_k\} \subseteq \mathbb{D}$ is interpolating if and only if the corresponding Gramian G, given by

$$G_{i,j} = \langle s_i, s_j \rangle_2,$$

where s_i is the normalized Szegő kernel at z_i, is both bounded and bounded below (see [**1**, Chapter 9] for a more general view). However, the boundedness of G is equivalent to the existence of a positive constant M such that

$$\Big\| \sum_i a_i s_i \Big\|_2^2 \leq M \sum_i |a_i|^2,$$

for any sequence $\{a_i\}$ of scalars. When $p \neq 2$, the conditions in Propositions 3.2 and 3.3 and Corollary 3.4, using Hölder duality, should be read as replacements for this phenomenon. We focus on a direct matrix interpretation of this in Section 5.

Let us begin by interpolating a fixed finite target when $1 < p < \infty$, $p \neq 2$, and letting $1/p + 1/q = 1$. Fixing a positive integer N, distinct points $z_0, z_1, z_2, \ldots, z_N \in \mathbb{D}$ and points $w_0, w_1, w_2, \ldots, w_N \in \overline{\mathbb{D}}$, we should first convince ourselves that there exists $h \in \ell^p_A$ such that $h(z_k) = w_k$ for all k, $0 \leqslant k \leqslant N$. If we let B_k be the finite Blaschke product with zeros at $\{z_0, z_1, z_2, \ldots, z_N\} \setminus \{z_k\}$, then the function

$$h_*(z) = w_0 B_0(z)/B_0(z_0) + w_1 B_1(z)/B_1(z_1) + \cdots + w_N B_N(z)/B_N(z_N) \tag{3.1}$$

does the job. Of course we would also want norm control over such a function. To obtain it, the following observation is useful.

LEMMA 3.1. *Let $1 < p < \infty$ and $1/p + 1/q = 1$. Let $\{z_0, z_1, \ldots, z_N\}$ be distinct points in $\mathbb{D}$, and let $\mathscr{J}$ be the subspace of ℓ^p_A consisting of functions vanishing on $\{z_0, z_1, \ldots, z_N\}$. Then*

$$\mathscr{J}^\perp = \bigvee_{\ell^q_A} \{\Lambda_{z_0}, \Lambda_{z_1}, \ldots, \Lambda_{z_N}\}.$$

PROOF. Trivially it holds that

$$\mathscr{J}^\perp \supseteq \bigvee_{\ell^q_A} \{\Lambda_{z_0}, \Lambda_{z_1}, \ldots, \Lambda_{z_N}\}.$$

Suppose that g belongs to $\mathscr{J}^\perp$, but fails to belong to the subspace $\bigvee_{\ell^q_A} \{\Lambda_{z_0}, \Lambda_{z_1}, \ldots, \Lambda_{z_N}\}$; in fact, we may assume that

$$g \perp_q \bigvee_{\ell^q_A} \{\Lambda_{z_0}, \Lambda_{z_1}, \ldots, \Lambda_{z_N}\}.$$

Then, according to (2.5), $g^{\langle q-1\rangle} \in \ell^p_A$ satisfies

$$\langle \Lambda_{z_k}, g^{\langle q-1\rangle}\rangle = 0$$

for each k, $0 \leqslant k \leqslant N$, which is to say that $g^{\langle q-1\rangle}$ resides in $\mathscr{J}$. Finally, this forces

$$\langle g, g^{\langle q-1\rangle}\rangle = \|g\|_q^q = 0,$$

or $g = 0$. □

With that, we have the following interpolation condition for a finite set.

PROPOSITION 3.2. *Let $1 < p < \infty$ and $1/p + 1/q = 1$. Fix a positive integer N, distinct points $z_0, z_1, \ldots, z_N \in \mathbb{D}$ and points $w_0, w_1, \ldots, w_N \in \mathbb{C}$. There exists $h \in \ell^p_A$ with $\|h\|_p \leqslant 1$ such that $h(z_k) = w_k$ for all k, $0 \leqslant k \leqslant N$, if and only if*

$$\sup\left\{\frac{|\beta_0 w_0 + \beta_1 w_1 + \cdots + \beta_N w_N|}{\|\beta_0\Lambda_{z_0} + \beta_1\Lambda_{z_1} + \cdots + \beta_N\Lambda_{z_N}\|_q} : (\beta_0, \beta_1, \ldots, \beta_N) \in \mathbb{C}^{N+1} \setminus \{(0,\ldots,0)\}\right\} \leqslant 1.$$

PROOF. Let $\mathscr{J}$ be the subspace of functions in ℓ^p_A that vanish at the points $z_0, z_1, \ldots, z_N$ (and possibly elsewhere). Any member f of ℓ^p_A satisfying the conditions $f(z_k) = w_k$, $0 \leqslant k \leqslant N$, must be of the form $f = h_* + J$, where $J \in \mathscr{J}$ and h_* is as given in (3.1). The interpolation problem is equivalent to asking whether the coset $h_* + \mathscr{J}$ contains an element of norm less than or equal to unity.

The distance from h_* to $\mathscr{J}$ is equal to the norm of h_*, viewed as a bounded linear functional on the annihilator of $\mathscr{J}$. But Lemma 3.1 gives

$$\mathscr{J}^\perp = \bigvee_{\ell^q_A}\{\Lambda_{z_0}, \Lambda_{z_1}, \ldots, \Lambda_{z_N}\}.$$

Consequently,

$$\begin{aligned} I &= \inf\left\{\|h\|_p : \ h(z_k) = w_k, \ 0 \leqslant k \leqslant N\right\} \\ &= \inf\left\{\|h_* + J\|_p : \ J \in \mathscr{J}\right\} \\ &= \sup\left\{\frac{|\langle h_*, \psi\rangle|}{\|\psi\|_q} : \ \psi \in \mathscr{J}^\perp \setminus \{0\}\right\} \\ &= \sup\left\{\frac{|\beta_0 w_0 + \beta_1 w_1 + \cdots + \beta_N w_N|}{\|\beta_0\Lambda_{z_0} + \beta_1\Lambda_{z_1} + \cdots + \beta_N\Lambda_{z_N}\|_q} : \right. \\ &\qquad\qquad \left. (\beta_0, \beta_1, \ldots, \beta_N) \in \mathbb{C}^{N+1} \setminus \{(0, 0, \ldots, 0)\}\right\}. \end{aligned}$$

□

This extends to interpolation on an infinite sequence.

PROPOSITION 3.3. *Let $1 < p < \infty$ and $1/p + 1/q = 1$. Fix distinct points $z_0, z_1, \ldots \in \mathbb{D}$ and points $w_0, w_1, \ldots \in \mathbb{C}$. There exists $h \in \ell^p_A$ such that $h(z_k) = w_k$ for all k, $k \geqslant 0$ if and only if*

$$\lim_{N\to\infty} \sup\left\{\frac{|\beta_0 w_0 + \beta_1 w_1 + \cdots + \beta_N w_N|}{\|\beta_0\Lambda_{z_0} + \beta_1\Lambda_{z_1} + \cdots + \beta_N\Lambda_{z_N}\|_q} : (\beta_0, \beta_1, \ldots, \beta_N) \in \mathbb{C}^{N+1} \setminus \{0\}\right\} < \infty. \tag{3.2}$$

PROOF. Suppose an interpolating function h exists. By Proposition 3.2 for every positive integer N,

$$\|h\|_p \geqslant \inf \left\{ \|H\|_p : \ H(z_k) = w_k, \ 0 \leqslant k \leqslant N \right\}$$
$$= \sup \left\{ \frac{|\beta_0 w_0 + \beta_1 w_1 + \cdots + \beta_N w_N|}{\|\beta_0 \Lambda_{z_0} + \beta_1 \Lambda_{z_1} + \cdots + \beta_N \Lambda_{z_N}\|_q} : \ (\beta_0, \beta_1, \ldots, \beta_N) \in \mathbb{C}^{N+1} \setminus \{0\} \right\}.$$

Then take $N \longrightarrow \infty$.

Conversely, suppose that the bound (3.2) holds. For each N, there is the corresponding interpolation function $h_{*,N}$ from (3.1), and a function $J = J_N \in \ell^p_A$ for which the extreme value in

$$\inf \left\{ \|h_{*,N} + J\|_p : \ J(z_k) = 0, \ 0 \leqslant k \leqslant N \right\}$$

is attained (uniform convexity of ℓ^p_A makes this possible).

By the Banach-Alaoglu theorem, there is a weakly convergent subsequence $\{h_{*,N_k} + J_{N_k}\}_{k=0}^{\infty}$. Let $h \in \ell^p_A$ be the weak limit (the weak and weak-$*$ topologies coincide here since $1 < p < \infty$). By weak convergence, applied to the point evaluations at z_k, we see that $h(z_k) = w_k$ for all k. □

Our next objective is to interpolate to a target sequence belonging to a particular space, namely ℓ^p. Specifically, given $W = (w_k)_{k=0}^{\infty} \in \ell^p$ we wish to find a function $f \in \ell^p_A$ such that

$$(1 - |z_k|^q)^{1/q} f(z_k) = w_k$$

for all k. The presence of the weight $(1 - |z_k|^q)^{1/q}$ is justified in part by analogous results from other spaces; we shall see that it is a natural choice in the present situation.

COROLLARY 3.4. *Let $1 < p < \infty$, and $1/p + 1/q = 1$. Suppose that $\{z_0, z_1, z_2, \ldots\}$ is a sequence of distinct points in $\mathbb{D}$ satisfying the condition*

$$\text{(3.3)} \qquad \sup \left\{ \frac{\left(|c_0|^q/(1 - |z_0|^q) + |c_1|^q/(1 - |z_1|^q) + \cdots\right)^{1/q}}{\|c_0 \Lambda_{z_0} + c_1 \Lambda_{z_1} + \cdots\|_q} : \ c_k \in \mathbb{C}, \text{ not all zero} \right\} < \infty.$$

Then for every sequence $(w_0, w_1, w_2 \ldots) \in \ell^p$ there exists a function $f \in \ell^p_A$ such that

$$(1 - |z_k|^q)^{1/q} f(z_k) = w_k$$

for all k.

PROOF. Replace w_k with $w_k/(1 - |z_k|^q)^{1/q}$ in Proposition 3.3, and apply Hölder's inequality to get

$$\frac{|c_0 w_0/(1 - |z_0|^q)^{1/q} + c_1 w_1/(1 - |z_1|^q)^{1/q} + \cdots + c_N w_N/(1 - |z_N|^q)^{1/q}|}{\|c_0 \Lambda_{z_0} + c_1 \Lambda_{z_1} + \cdots + c_N \Lambda_{z_N}\|_q}$$
$$\leqslant \frac{\left(|w_0|^p + \cdots + |w_N|^p\right)^{1/p} \left(|c_0|^q/(1 - |z_0|^q) + \cdots + |c_N|^q/(1 - |z_N|^q) + \cdots\right)^{1/q}}{\|c_0 \Lambda_{z_0} + c_1 \Lambda_{z_1} + \cdots + c_N \Lambda_{z_N}\|_q}.$$

Take $N \longrightarrow \infty$. □

As noted before, $1/(1-|\zeta|^q)^{1/q} = \|\Lambda_\zeta\|_q$ for any $\zeta \in \mathbb{D}$; hence the condition (3.3) could also be expressed as

$$\sup\left\{\frac{\left(|c_0|^q\|\Lambda_{z_0}\|_q^q + |c_1|^q\|\Lambda_{z_1}\|_q^q + \cdots\right)^{1/q}}{\|c_0\Lambda_{z_0} + c_1\Lambda_{z_1} + \cdots\|_q} : c_k \in \mathbb{C},\ (c_0, c_1, \ldots) \neq 0\right\} < \infty. \tag{3.4}$$

The boundedness conditions on quantities of this type will be important in results to come; let us now give them labels.

DEFINITION 3.5. Say that a sequence of distinct points $Z = \{z_0, z_1, z_2, \ldots\}$ in $\mathbb{D}$ satisfies the condition (LB) if there exists $C_1 > 0$ such that

$$C_1 \leqslant \frac{\left\|c_0\Lambda_{z_0} + c_1\Lambda_{z_1} + \cdots\right\|_q}{\left(|c_0|^q\|\Lambda_{z_0}\|_q^q + |c_1|^q\|\Lambda_{z_1}\|_q^q + \ldots\right)^{1/q}} \tag{LB}$$

for all sequences (c_k), not all entries being zero.

Similarly we say that Z satisfies (UB) if there exists $C_2 > 0$ such that

$$\frac{\left\|c_0\Lambda_{z_0} + c_1\Lambda_{z_1} + \cdots\right\|_q}{\left(|c_0|^q\|\Lambda_{z_0}\|_q^q + |c_1|^q\|\Lambda_{z_1}\|_q^q + \ldots\right)^{1/q}} \leqslant C_2 \tag{UB}$$

for all sequences (c_k), not all entries being zero.

Notice that both (LB) and (UB) depend on the parameter q, the conjugate exponent to p, and that we previously encountered (LB) in the form of condition (3.4).

Of course, these conditions on a sequence $\{z_0, z_1, z_2, \ldots\}$ could also be viewed as conditions on the corresponding point evaluation kernel functions. There is existing terminology for this situation.

DEFINITION 3.6. Given a sequence of distinct points $Z = \{z_k\}_{k=0}^\infty$ of $\mathbb{D}$ we call the set of normalized kernels $\{\Lambda_{z_k}/\|\Lambda_{z_k}\|_q\}_{k=0}^\infty$ a *Riesz system* in ℓ_A^q if both (LB) and (UB) hold for some positive constants C_1 and C_2.

Let us distinguish those sequences on which we are able to interpolate to any target in ℓ^p.

DEFINITION 3.7. We call a sequence $Z = \{z_0, z_1, z_2 \ldots\}$ of distinct points in $\mathbb{D}$ a *universal interpolation sequence* for ℓ_A^p if for every $\{w_k\}_{k=0}^\infty \in \ell^p$ there exists a function $f \in \ell_A^p$ such that $f(z_k)(1-|z_k|^q)^{1/q} = w_k$ for all k.

The following two theorems presented in [**27**, p. 1044] furnish numerous examples of universal interpolation sequences for ℓ_A^p.

THEOREM 3.8 (Vinogradov). *Suppose that $1 < p < \infty$. Let the sequence $Z = \{z_0, z_1, z_2, \ldots\}$ lie in a Stoltz domain. Then Z is a universal interpolation sequence for ℓ_A^p if and only if Z is uniformly separated.*

THEOREM 3.9 (Vinogradov). *Suppose that $1 < p < \infty$. If the sequence $Z = \{z_0, z_1, z_2, \ldots\}$ satisfies the conditions*

$$|z_k| \leqslant |z_{k+1}|, \quad k = 0, 1, 2, \ldots; \quad \textit{and}$$

$$\sup_n (1-|z_n|)^{-1} \sum_{k \geqslant n} (1-|z_k|) < \infty,$$

and there exists $\gamma > 0$ such that

$$\inf_{j \neq k} \left| \frac{z_j - z_k}{1 - \overline{z}_k z_j} \right| \geqslant \gamma,$$

then Z is a universal interpolation sequence for ℓ^p_A.

Previously we showed (in Corollary 3.4) that a certain Riesz system for ℓ^q_A gives rise to a universal interpolation sequence for ℓ^p_A. It turns out that the converse holds, and so we have the following theorem.

THEOREM 3.10. *Let $1 < p < \infty$ and $1/p + 1/q = 1$. A sequence $Z = \{z_0, z_1, z_2, \dots\}$ of distinct points in $\mathbb{D}$ is a universal interpolation sequence for ℓ^p_A if and only if $\{\Lambda_{z_k}/\|\Lambda_{z_k}\|_q\}_{k=0}^{\infty}$ is a Riesz system in ℓ^q_A.*

PROOF. It suffices to prove the converse of Corollary 3.4. Define the linear mapping T on ℓ^p_A by

$$Tf := \big(f(z_k)/\|\Lambda_{z_k}\|_q\big)_{k=0}^{\infty}.$$

The kernel $\mathscr{K}$ of T is the subspace of functions of ℓ^p_A vanishing on Z. To say that Z is a universal interpolating sequence is equivalent to T being surjective from ℓ^p_A to ℓ^p; this is equivalent to the induced mapping

$$\tilde{T} : \ell^p_A/\mathscr{K} \longmapsto \ell^p$$

given by

$$\tilde{T}(f + \mathscr{K}) := Tf$$

being invertible.

But the adjoint of $\tilde{T}$ is given by $(a_k) \longmapsto (a_k \Lambda_{z_k}/\|\Lambda_{z_k}\|_q)$, as can be seen by

$$\begin{aligned}\langle \tilde{T}(f + \mathscr{K}), (a_k)\rangle &= \langle (f(z_k)\|\Lambda_{z_k}\|_q), (a_k)\rangle \\ &= \langle f, (a_k \Lambda_{z_k}/\|\Lambda_{z_k}\|_q)\rangle.\end{aligned}$$

So boundedness of $\tilde{T}$ and its inverse imply that

$$C_1 \leqslant \frac{\big\|a_0 \Lambda_{z_0}/\|\Lambda_{z_0}\|_q + a_1 \Lambda_{z_1}/\|\Lambda_{z_1}\|_q + \cdots\big\|_q}{\big(|a_0|^q + |a_1|^q + \cdots\big)^{1/q}} \leqslant C_2$$

for some positive constants C_1 and C_2. Now it is a trivial matter to substitute $a_k = c_k \|\Lambda_{z_k}\|_q$ to see that this condition is equivalent to $\{z_k\}$ being a Riesz system in ℓ^q_A. □

There is a connection between zero sets of ℓ^p_A and interpolation. We say that a sequence of distinct points Z in $\mathbb{D}$ is a *pre-zero set* of ℓ^p_A if there is a nontrivial function in ℓ^p_A that vanishes on Z, but possibly elsewhere, however. For $1 < p < \infty$, a characterization of the pre-zero sets of ℓ^p_A, expressed in terms of associated p-inner functions, was obtained in [**10**]. (A precise description of the zero sets is an ongoing challenge.)

PROPOSITION 3.11. *Let $1 < p < \infty$ and $1/p + 1/q = 1$. Suppose that $Z = \{z_k\}$ is a sequence of distinct nonzero points in $\mathbb{D}$. If Z is a universal interpolation sequence for ℓ^p_A, then Z is a pre-zero set for ℓ^p_A.*

PROOF. By hypothesis there exists $f \in \ell^p_A$ such that $f(z_0) = 1$ and $f(z_k) = 0$ for all $k > 0$. Obviously, this tells us that $\{z_k\}_{k=1}^{\infty}$ is a pre-zero set for ℓ^p_A. But then, all of Z is a pre-zero set as well, since the nontrivial function $(z - z_0)f(z)$ belongs to ℓ^p_A and vanishes on Z. □

4. Continuity

We turn to the problem of interpolating on an infinite sequence in relation to interpolating on a finite truncation of the sequence. The limiting process is well behaved under certain conditions.

Fix $1 < p < \infty$ and let $1/p + 1/q = 1$. Let $Z = \{z_k\}$ be a sequence of distinct nonzero points in $\mathbb{D}$. For each $N \in \mathbb{N}$, and each j, $0 \leqslant j \leqslant N$, let $f_{j,N}$ be the function in ℓ^p_A of minimum norm such that

$$(1 - |z_j|^q)^{1/q} f_{j,N}(z_j) = 1 \tag{4.1}$$

$$(1 - |z_k|^q)^{1/q} f_{j,N}(z_k) = 0, \;\; 0 \leqslant k \leqslant N, \;\; k \neq j. \tag{4.2}$$

Such $f_{j,N}$ exists uniquely, since it is the minimum norm element of a nonempty convex set in a uniformly convex space.

Similarly define f_j be the function in ℓ^p_A of minimum norm such that

$$(1 - |z_j|^q)^{1/q} f_j(z_j) = 1 \tag{4.3}$$

$$(1 - |z_k|^q)^{1/q} f_j(z_k) = 0, \;\; k \neq j. \tag{4.4}$$

Such f_j need not exist in general, but they do exist of course if Z is a universal interpolation sequence.

We need the following tool to establish the main result of this section.

LEMMA 4.1. *Let $1 < p < \infty$, and $1/p + 1/q = 1$. Suppose that $v_1, v_2, \ldots$ are unit vectors in ℓ^p_A, converging in norm to v. Let λ_n be the norming functional of v_n, and let λ be that of v. Then λ_n converges in norm to λ in ℓ^q_A.*

PROOF. By definition, all of the norming functionals have unit norm. Thus by the Banach-Alaoglu theorem, there is a subsequence $\{v_{n_k}\}$ that converges weakly to some $\tilde{\lambda} \in \ell^q_A$ (by reflexivity, the weak and weak* topologies coincide).

If $\tilde{\lambda}$ is normed within ϵ by the unit vector $\tilde{v}$, then

$$\|\tilde{\lambda}\|_q \leqslant \epsilon + \langle \tilde{v}, \tilde{\lambda} \rangle \leqslant \epsilon + \lim_{k \to \infty} \left| \langle \tilde{v}, \lambda_{n_k} \rangle \right| \leqslant \epsilon + \|\tilde{v}\|_p \|\lambda_{n_k}\|_q = \epsilon + 1.$$

Consequently, the norm of $\tilde{\lambda}$ is at most one.

Next, for each k we have

$$\begin{aligned} \langle v, \tilde{\lambda} \rangle &= \langle v, \tilde{\lambda} - \lambda_{n_k} \rangle + \langle v, \lambda_{n_k} \rangle \\ &= \langle v, \tilde{\lambda} - \lambda_{n_k} \rangle + \langle v - v_{n_k}, \lambda_{n_k} \rangle + \langle v_{n_k}, \lambda_{n_k} \rangle \\ &= \langle v, \tilde{\lambda} - \lambda_{n_k} \rangle + \langle v - v_{n_k}, \lambda_{n_k} \rangle + 1. \end{aligned}$$

The first two terms can be made arbitrarily small by choosing k sufficiently large. This shows that $\langle v, \tilde{\lambda} \rangle = 1$.

But in fact from

$$1 = \langle v, \tilde{\lambda} \rangle \leqslant \|v\|_p \|\tilde{\lambda}\|_q = \|\tilde{\lambda}\|_q \leqslant 1,$$

we see that $\tilde{\lambda}$ must have unit norm. This forces $\tilde{\lambda}$ to be a norming functional for v. Since ℓ^p_A is smooth, it can only be that $\tilde{\lambda} = \lambda$.

Finally, from the weak convergence of λ_{n_k} to λ, along with $\|\lambda_{n_k}\|_q \to \|\lambda\|_q$, we conclude that λ_{n_k} converges to λ in norm.

This argument applies to any subsequence of $\{\lambda_n\}$, with the same limit λ, and so convergence in ℓ^q_A of the sequence itself is assured. $\square$

We are now in a position to show that under broad circumstances, interpolation is well behaved with respect to finite approximation.

THEOREM 4.2. *Let the interpolating functions $f_{j,N}$ be defined as in (4.1) and (4.2), and let f_j be defined as in (4.3) and (4.4). For any $j \geqslant 0$, f_j exists if and only if $\lim_{N\to\infty} \|f_{j,N}\|_p < \infty$. In this case, $f_{j,N}$ converges in norm to f_j in ℓ^p_A, as N increases without bound.*

PROOF. To prove this, and establish when the functions $\{f_j\}$ exist, we need the following apparatus from duality.

Fix $j = 0$ and $N \geqslant 1$, and let $\mathscr{M}$ be the subspace of ℓ^p_A consisting of functions vanishing on $z_0, z_1, \ldots, z_N$ (and possibly elsewhere). Then by a basic duality theorem (see, for example, [**14**, Theorem 10.2, p. 91]),

$$
\begin{aligned}
\|f_{0,N}\|_p &= \operatorname{dist}(f_{0,N}, \mathscr{M}) \\
&= \sup\left\{ \frac{|\langle f_{0,N}, \psi\rangle|}{\|\psi\|_q} : \psi \in \mathscr{M}^\perp \right\} \\
&= \sup\left\{ \frac{|\langle f_{0,N}, b_0\Lambda_{z_0} + b_1\Lambda_{z_1} + \cdots + b_N\Lambda_{z_N}\rangle|}{\|b_0\Lambda_{z_0} + b_1\Lambda_{z_1} + \cdots + b_N\Lambda_{z_N}\|_q} : (b_0, b_1, \ldots, b_N) \in \mathbb{C}^{N+1} \setminus \{0\} \right\} \\
&= \sup\left\{ \frac{|b_0 f_{0,N}(z_0) + b_1 f_{0,N}(z_1) + \cdots + b_N f_{0,N}(z_N)|}{\|b_0\Lambda_{z_0} + b_1\Lambda_{z_1} + \cdots + b_N\Lambda_{z_N}\|_q} : (b_0, b_1, \ldots, b_N) \in \mathbb{C}^{N+1} \setminus \{0\} \right\} \\
&= \sup\left\{ \frac{|b_0|/(1-|z_0|^q)^{1/q}}{\|b_0\Lambda_{z_0} + b_1\Lambda_{z_1} + \cdots + b_N\Lambda_{z_N}\|_q} : (b_0, b_1, \ldots, b_N) \in \mathbb{C}^{N+1} \setminus \{0\} \right\} \\
&= \sup\left\{ \frac{1/(1-|z_0|^q)^{1/q}}{\|\Lambda_{z_0} + b_1\Lambda_{z_1} + \cdots + b_N\Lambda_{z_N}\|_q} : (b_0, b_1, \ldots, b_N) \in \mathbb{C}^{N+1} \setminus \{0\} \right\} \\
&= \left[\operatorname{dist}(\Lambda_{z_0}/\|\Lambda_{z_0}\|_q, \mathscr{N})\right]^{-1},
\end{aligned}
$$

where $\mathscr{N}$ is the span in ℓ^q_A of $\{\Lambda_{z_1}, \Lambda_{z_2}, \ldots, \Lambda_{z_N}\}$.

Accordingly, let us define $g_{0,N}$ to be the unique metric projection of $\Lambda_{z_0}/\|\Lambda_{z_0}\|_q$ onto the span in ℓ^q_A of $\Lambda_{z_1}, \Lambda_{z_2}, \ldots, \Lambda_{z_N}$. We can carry out the analogous exercise to obtain $f_{j,N}$ and $g_{j,N}$ for all N and $0 \leqslant j \leqslant N$.

By direct calculation we see that $f_{j,N}/\|f_{j,N}\|_p$ is a norming functional for $g_{j,N}$, and likewise $g_{j,N}/\|g_{j,N}\|_q$ is a norming functional for $f_{j,N}$. This further implies that

$$
\frac{f_{j,N}^{\langle p-1\rangle}}{\|f_{j,N}\|_p^{p-1}} = \frac{g_{j,N}}{\|g_{j,N}\|_q}. \tag{4.5}
$$

Moreover, provided only that f_j exists, we see that g_j is analogously defined, with $f_j/\|f_j\|_p$ and $g_j/\|g_j\|_q$ being mutually norming, and

$$
\frac{f_j^{\langle p-1\rangle}}{\|f_j\|_p^{p-1}} = \frac{g_j}{\|g_j\|_q} \tag{4.6}
$$

as well.

Notice that as N increases $g_{j,N}$ is a metric projection onto successively larger subspaces of ℓ^q_A. By [**12**, Proposition 4.8.3], the metric projection is well behaved in the limit, and we have $g_{j,N}$ convergent in norm to g_j in ℓ^q.

Apply Lemma 4.1, in conjunction with the norm relations (4.5) and (4.6). The only way f_j can fail to exist is if $g_{j,N}$ tends to zero. That implies $\lim_{N\to\infty} \|f_{j,N}\|_p = \infty$. □

COROLLARY 4.3. *Let the functions f_j be defined as in (4.3) and (4.4). For f_j to exist for all $j \geqslant 0$ it is necessary and sufficient that*

$$\operatorname{dist}_{\ell^q_A}\left(\Lambda_{z_j}, \operatorname{span}\{\Lambda_{z_k}\}_{k\neq j}\right) > 0$$

for all j.

The condition in the Corollary is called "topologically free" in [**1**]. It goes by the name "minimality" in other works, for example [**7**].

The next assertion connects two conditions encountered in our analysis.

PROPOSITION 4.4. *Let $1 < p < \infty$, $1/p + 1/q = 1$, and let $\{z_k\}$ be a sequence of distinct points in the disk $\mathbb{D}$. If the condition* (LB) *holds, then so does the separation condition*

$$\inf\left\{\left\|\frac{\Lambda_{z_j}}{\|\Lambda_{z_j}\|_q} - \sum_{k\neq j} c_k \frac{\Lambda_{z_k}}{\|\Lambda_{z_k}\|_q}\right\|_q : j \geqslant 0,\ c_k \in \mathbb{C}\right\} > 0.$$

PROOF.

$$\begin{aligned}
&\inf\left\{\left\|\frac{\Lambda_{z_j}}{\|\Lambda_{z_j}\|_q} - \sum_{k\neq j} c_k \frac{\Lambda_{z_k}}{\|\Lambda_{z_k}\|_q}\right\|_q : j \geqslant 0,\ c_k \in \mathbb{C}\right\} \\
&= \inf\left\{\frac{\left\|c_j \frac{\Lambda_{z_j}}{\|\Lambda_{z_j}\|_q} - \sum_{k\neq j} c_k \frac{\Lambda_{z_k}}{\|\Lambda_{z_k}\|_q}\right\|_q}{|c_j|} : j \geqslant 0,\ c_k \in \mathbb{C},\ c_j \neq 0\right\} \\
&\geqslant \inf\left\{\frac{\left\|c_j \frac{\Lambda_{z_j}}{\|\Lambda_{z_j}\|_q} - \sum_{k\neq j} c_k \frac{\Lambda_{z_k}}{\|\Lambda_{z_k}\|_q}\right\|_q}{\left(\sum_k |c_k|^q\right)^{1/q}} : c \neq 0\right\} \\
&> 0,
\end{aligned}$$

the last inequality being equivalent to (LB). □

Next, we will discuss the convergence of the coefficients of the dual extremal functions.

For $N \in \mathbb{N}$ and $0 \leqslant j \leqslant N$, and with $g_{j,N}$ defined as in the proof of Theorem 4.2, let us write

$$g_{j,N} = \frac{\Lambda_j}{\|\Lambda_j\|_q} - \sum_{k\neq j} \gamma_k^{(j,N)} \frac{\Lambda_k}{\|\Lambda_k\|_q},$$

with the understanding that $\gamma_j^{(j,N)} = 1$ and $\gamma_k^{(j,N)} = 0$ whenever $k > N$.

PROPOSITION 4.5. *Suppose that* $1 < p < \infty$, *and* $1/p + 1/q = 1$. *Let a sequence of distinct points* $Z = \{z_0, z_1, z_2, \ldots\}$ *be given. If the condition*

$$M_j := \operatorname{dist}(\Lambda_{z_j}/\|\Lambda_{z_j}\|_q, \operatorname{span}\{\Lambda_{z_k}\}_{k\neq j}) > 0$$

holds for all j, *then* $\lim_{N\to\infty} \gamma_k^{(j,N)}$ *exists for all* j *and* k.

PROOF. The claim is trivial when $j = k$, so we assume otherwise. Let r and K be suitable Pythagorean parameters for ℓ^q_A from Theorem 2.3, line (2.6) (see also [**12**, p. 58]). Suppose that $1 \leqslant M < N$. Then $g_{j,M} - g_{j,N}$ lies in the span of $\{\Lambda_k : k \neq j, \ 0 \leqslant k \leqslant N\}$. Thus by the extremal property of $g_{j,N}$,

$$g_{j,N} \perp_q (g_{j,N} - g_{j,M}).$$

Consequently, inequality (2.6) gives

$$\begin{aligned}
\|g_{j,M}\|_q^r &\geqslant \|g_{j,N}\|_q^r + K\|g_{j,M} - g_{j,N}\|_q^r \\
&\geqslant \|g_{j,N}\|_q^r + K\Big\| \sum_{m\neq j} (\gamma_m^{(j,M)} - \gamma_m^{(j,N)})\Lambda_{z_m}/\|\Lambda_{z_m}\|_q \Big\|_q^r \\
&\geqslant \|g_{j,N}\|_q^r + K|\gamma_k^{(j,M)} - \gamma_k^{(j,N)}|^r \cdot M_k^r.
\end{aligned}$$

We already know that $g_{j,N} \to g_j$ in norm as $N \to \infty$. This forces $\{\gamma_k^{(j,N)}\}_{N=1}^{\infty}$ to be a Cauchy sequence. □

Let us give the name $\gamma_k^{(j)}$ to $\lim_{N\to\infty} \gamma_k^{(j,N)}$. We have thus shown that g_j has the representation

$$g_j \sim \frac{\Lambda_{z_j}}{\|\Lambda_{z_j}\|_q} - \sum_{k\neq j} \gamma_k^{(j)} \frac{\Lambda_{z_k}}{\|\Lambda_{z_j}\|_q},$$

with convergence in the sense that the representations for $g_{j,N}$ converge to g_j in ℓ^q_A.

5. Matrix analysis

Next we represent the information obtained thus far about the extremal functions $\{f_j\}$ and their respective dual extremal functions $\{g_j\}$ in terms of matrices, and uncover another criterion for a universal interpolation sequence for ℓ^p_A. We also obtain interesting nonlinear operators relating the associated extremal vectors when $p \neq 2$, extending the notion of a Gramian matrix.

With the extremal functions f_j as previously defined in (4.3) and (4.4), let their coefficients by expressed via

$$f_j(z) := \sum_{k=0}^{\infty} \varphi_{j,k} z^k.$$

Define the matrix

$$\Phi := \begin{bmatrix} \varphi_{0,0} & \varphi_{0,1} & \varphi_{0,2} & \cdots \\ \varphi_{1,0} & \varphi_{1,1} & \varphi_{1,2} & \cdots \\ \varphi_{2,0} & \varphi_{2,1} & \varphi_{2,2} & \cdots \\ \vdots & \vdots & \vdots & \ddots \end{bmatrix}$$

and the matrix Ψ by

$$\big[\Psi\big]_{j,k} := z_k^j (1 - |z_k|^q)^{1/q},$$

where the indices j and k vary from 0 to ∞.

Thus the jth row of Φ consists of the coefficients of f_j, and the kth column of Ψ is populated by the coefficients of the normalized kernel function $\Lambda_{z_k}/\|\Lambda_{z_k}\|_q$. Notice also that the matrix $\Psi^*\Psi$ is the familiar Gramian matrix in the $p = 2$ case, where Ψ^* stands for the conjugate transpose of Ψ. In what follows, we write Ψ^T for the transpose of the matrix Ψ, without conjugation.

THEOREM 5.1. *Let $1 < p < \infty$, and $1/p + 1/q = 1$. Suppose that $Z = \{z_j\}$ is a sequence of distinct points of $\mathbb{D}$. If the condition*

$$\operatorname{dist}(\Lambda_{z_j}/\|\Lambda_{z_j}\|_q, \operatorname{span}\{\Lambda_{z_k}\}_{k\neq j}) > 0$$

holds for all j, then the following are equivalent.

(1) *Z is a universal interpolating sequence.*
(2) *The matrix Ψ represents a bounded operator from ℓ^q into ℓ^q that is bounded below.*
(3) *The matrix Φ represents a bounded invertible operator from ℓ^q onto $\ell^q/\ker\Psi^T$.*

In this case, $I = \Phi\Psi$, and for any $W = \{w_j\} \in \ell^p$ a function $f \in \ell^p_A$ that interpolates to the target W is given by $f = \Phi^T W$.

PROOF. The separation condition ensures that each f_j exists, and hence the matrix Φ exists. Previously, in Theorem 3.10, it was established that (1) is equivalent to $\{\Lambda_{z_j}/\|\Lambda_{z_j}\|_q\}_{j=0}^{\infty}$ being a Riesz system; this, in turn, says that Φ is a bounded, invertible operator from ℓ^q to its range in ℓ^q; that is equivalent to (2). If (3) holds, then for any $W = \{w_j\} \in \ell^p$ a function $f \in \ell^p_A$ that interpolates to the target W is given by $f = \Phi^T W$, thus implying (1). Finally, in any case the matrix equation $I = \Phi\Psi$ holds algebraically. If (2) holds as well, then $\Phi^T = (\Psi^T)^{-1}$ (viewed as operating on the range of Ψ), and we have (3). □

This enables us to solve the interpolation problem in terms of $\{f_j\}$, ensuring that the interpolating function can be represented as

$$f(z) := \sum_{k=0}^{\infty} \frac{f_k(z)w_k}{(1 - |z_k|^q)^{1/q}},$$

in the sense that the matrix operation supplies the coefficients of $f \in \ell^p_A$. Vinogradov [**25**] obtained similar results under a different set of assumptions on Z.

Next, we also obtain interesting nonlinear conditions for the associated extremal vectors.

The remainder of this section comprises a set of results that in principle enable us to test an infinite sequence $\{z_0, z_1, z_2, \ldots\}$ for the condition (UB) based on the finite fragments $\{z_0, z_1, \ldots, z_n\}$.

Let distinct points $\{z_0, z_1, \ldots, z_n\}$ of the disk $\mathbb{D}$ be chosen. In this situation the matrix Ψ is given by

$$\Psi = \begin{bmatrix} (1-|z_0|^q)^{1/q} & (1-|z_1|^q)^{1/q} & \cdots & (1-|z_n|^q)^{1/q} \\ z_0(1-|z_0|^q)^{1/q} & z_1(1-|z_1|^q)^{1/q} & \cdots & z_n(1-|z_n|^q)^{1/q} \\ z_0^2(1-|z_0|^q)^{1/q} & z_1^2(1-|z_1|^q)^{1/q} & \cdots & z_n^2(1-|z_n|^q)^{1/q} \\ \vdots & \vdots & \ddots & \vdots \end{bmatrix}. \tag{5.1}$$

We denote its operator norm from ℓ^q to $\ell^q(\{0, 1, \ldots, n\})$ by the usual $\|\Psi\|$.

LEMMA 5.2. *Let distinct points $\{z_0, z_1, \ldots, z_n\}$ of the disk $\mathbb{D}$ be chosen. With the matrix Ψ defined as in (5.1), there exist unit vectors $\tilde{a} \in \ell^q(\{0, 1, \ldots, n\})$ and $\tilde{b} \in \ell^q_A$ such that*

$$\|\Psi\tilde{a}\|_q = \|\Psi\|$$
$$\|\Psi^T\tilde{b}\|_q = \|\Psi\|$$
$$\langle\tilde{b}, \Psi\tilde{a}\rangle = \|\Psi\|.$$

PROOF. First, let us observe that the condition (UB) always holds for a finite collection of points $\{z_0, z_1, \ldots, z_n\}$. This is because

$$\begin{aligned}\big\|a_0\Lambda_{z_0}/\|\Lambda_{z_0}\|_q + \cdots + a_n\Lambda_{z_n}/\|\Lambda_{z_n}\|_q\big\|_q &\leqslant \|a\|_q\left[\Big(\frac{\|\Lambda_{z_0}\|_q}{\|\Lambda_{z_0}\|_q}\Big)^p + \cdots + \Big(\frac{\|\Lambda_{z_n}\|_q}{\|\Lambda_{z_n}\|_q}\Big)^p\right]^{1/p}\\ &= (n+1)^{1/p}\|a\|_q.\end{aligned}$$

Next, for $a = [a_0 \ \ a_1 \ \ \cdots \ \ a_n]^T$ we can express the condition (UB) as

$$\sup_{a\neq 0}\frac{\|\Psi a\|_q}{\|a\|_q} < \infty.$$

The supremum in the condition (UB) is attained by some choice of unit vector $\tilde{a}$. This is because $\|\Psi a\|_q$ is a continuous function of a as a varies varies over the unit sphere of $\mathbb{C}^{n+1}$, a compact set.

With some selection of $\tilde{a}$ fixed, we obtain $\tilde{b} \in \ell^p$ as follows. We know that

$$\|\Psi\| = \|\Psi\tilde{a}\|_q = \sup\left\{\frac{|\langle b, \Psi\tilde{a}\rangle|}{\|b\|_p} : \ b \in \ell^p_A, \ b \neq 0\right\}.$$

We can therefore choose unit vectors $b^{(n)} \in \ell^p$ so that each $\langle b^{(n)}, \Psi\tilde{a}\rangle \geqslant 0$, and

$$\|\Psi\| = \|\Psi\tilde{a}\|_q = \lim_{n\to\infty}\langle b^{(n)}, \Psi\tilde{a}\rangle.$$

By passing to a subsequence, we can assume that the $b^{(n)}$ converge weakly to some $b \in \ell^p$. If $\lambda \in \ell^q$ is the norming functional for b, then from

$$1 = \|b^{(n)}\|_p = \|b^{(n)}\|_p\|\lambda\|_q \geqslant \big|\langle b^{(n)}, \lambda\rangle\big| \to \langle b, \lambda\rangle = \|b\|,$$

we see that

$$\|\Psi\| = \lim_{n\to\infty}\langle b^{(n)}, \Psi\tilde{a}\rangle \leqslant \frac{\langle b, \Psi\tilde{a}\rangle}{\|b\|_p} \leqslant \|\Psi\|.$$

Equality must hold throughout the previous line. Put $\tilde{b} = b/\|b\|_p$, and note

$$\|\Psi\| = \langle\tilde{b}, \Psi\tilde{a}\rangle = \langle\Psi^T\tilde{b}, \tilde{a}\rangle \leqslant \|\Psi^T\tilde{b}\|_p\|\tilde{a}\|_q = \frac{\|\Psi^T\tilde{b}\|_p}{\|\tilde{b}\|_p} \leqslant \|\Psi\|.$$

Again, equality prevails, showing that $\|\Psi^T\tilde{b}\|_p = \|\Psi\|$. □

Next we show that the extremal vectors $\tilde{a}$ and $\tilde{b}$ are related by a nonlinear equation when $p \neq 2$. Recall that we use the notation

$$a^{\langle p-1\rangle} = \big[a_0^{\langle p-1\rangle} \ \ a_1^{\langle p-1\rangle} \ \ \cdots \ \ a_n^{\langle p-1\rangle}\big]^T$$

when

$$a = \big[a_0 \ \ a_1 \ \ \cdots \ \ a_n\big]^T.$$

THEOREM 5.3. *Let $\{z_0, z_1, \ldots, z_n\}$ be distinct points of $\mathbb{D}$. If Ψ is the matrix given by* (5.1), *and $\tilde{a}$ and $\tilde{b}$ are associated extremal unit vectors from Lemma* 5.2, *then*

$$\frac{\Psi\tilde{a}}{\|\Psi\|} = \tilde{b}^{\langle p-1\rangle} \quad \textit{and} \quad \frac{\Psi^T\tilde{b}}{\|\Psi\|} = \tilde{a}^{\langle q-1\rangle}.$$

PROOF. In Lemma (5.2), it was established that

$$\langle \tilde{b}, \Psi\tilde{a}\rangle = \|\Psi\|. \tag{5.2}$$

We check by inspection that

$$\big\|\Psi\tilde{a}/\|\Psi\|\big\|_q = 1 \quad \text{and} \quad \big\|\Psi^T\tilde{b}/\|\Psi\|\big\|_p = 1.$$

We can therefore read (5.2) as saying that $\Psi\tilde{a}/\|\Psi\|$ is a norming functional for $\tilde{b}$, and $\Psi^T\tilde{b}/\|\Psi\|$ is norming for $\tilde{a}$.

Routine calculation confirms that $\|\tilde{a}^{\langle q-1\rangle}\|_p = \|\tilde{a}\|_q = 1$, and $\|\tilde{b}^{\langle p-1\rangle}\|_q = \|\tilde{b}\|_p = 1$. In addition, we have

$$\langle \tilde{a}, \tilde{a}^{\langle q-1\rangle}\rangle = \|\tilde{a}\|_q^q = 1$$
$$\langle \tilde{b}, \tilde{b}^{\langle p-1\rangle}\rangle = \|\tilde{b}\|_p^p = 1.$$

Thus we also find that $\tilde{a}$ is normed by $\tilde{a}^{\langle q-1\rangle}$, and $\tilde{b}$ is normed by $\tilde{b}^{\langle p-1\rangle}$.

But the spaces ℓ^p and ℓ^q are smooth, being that $1 < p < \infty$, which is to say that norming functionals for these spaces are unique. Consequently we obtain

$$\frac{\Psi\tilde{a}}{\|\Psi\|} = \tilde{b}^{\langle p-1\rangle} \quad \text{and} \quad \frac{\Psi^T\tilde{b}}{\|\Psi\|} = \tilde{a}^{\langle q-1\rangle}.$$

□

By using the property

$$(a^{\langle p-1\rangle})^{\langle q-1\rangle} = a$$

for all $a \in \mathbb{C}$, we immediately derive a fixed-point property for $\tilde{a}$ and $\tilde{b}$.

COROLLARY 5.4. *Let $\{z_0, z_1, \ldots, z_n\}$ be distinct points of $\mathbb{D}$. If Ψ is the matrix given by* (5.1), *and $\tilde{a}$ and $\tilde{b}$ are associated extremal unit vectors from Lemma* 5.2, *then*

$$\tilde{a} = \frac{\left[\Psi^T(\Psi\tilde{a})^{\langle q-1\rangle}\right]^{\langle p-1\rangle}}{\|\Psi\|^p} \tag{5.3}$$

$$\tilde{b} = \frac{\left[\Psi(\Psi^T\tilde{b})^{\langle p-1\rangle}\right]^{\langle q-1\rangle}}{\|\Psi\|^q}. \tag{5.4}$$

REMARK 5.5. When $p = 2$, the mapping

$$x \longmapsto \frac{\left[\Psi^T(\Psi x)^{\langle q-1\rangle}\right]^{\langle p-1\rangle}}{\|\Psi\|^p}$$

simplifies to

$$x \longmapsto \frac{\Psi^*(\Psi x)}{\|\Psi\|^2},$$

which is to say that the nonlinear operator appearing in (5.3) generalizes the Gramian matrix when $p \neq 2$, apart from a multiplicative constant.

There are converse statements to Corollary 5.4, enabling us to retrieve both extremal vectors, if either fixed-point condition (5.3) or (5.4) holds. Here is the precise statement when a unit vector $\tilde{a}$ satisfies (5.3).

THEOREM 5.6. *Let $\{z_0, z_1, \ldots, z_n\}$ be distinct points of $\mathbb{D}$ and let Ψ be the matrix from (5.1). If there exists a unit vector $\tilde{a} \in \ell^q(\{0, 1, \ldots, n\})$ such that*

$$\tilde{a} = \frac{\left[\Psi^T(\Psi\tilde{a})^{\langle q-1\rangle}\right]^{\langle p-1\rangle}}{\|\Psi\|^p},$$

and

$$\tilde{b} := \left(\frac{\Psi\tilde{a}}{\|\Psi\|}\right)^{\langle q-1\rangle}$$

is a unit vector in ℓ^p, then $\|\Psi\tilde{a}\|_q = \|\Psi\|$, $\|\Psi^T\tilde{b}\|_p = \|\Psi\|$, $\langle\tilde{b}, \Psi\tilde{a}\rangle = \|\Psi\|$,

$$\frac{\Psi\tilde{a}}{\|\Psi\|} = \tilde{b}^{\langle p-1\rangle} \quad \textit{and} \quad \frac{\Psi^T\tilde{b}}{\|\Psi\|} = \tilde{a}^{\langle q-1\rangle},$$

and $\tilde{b}$ satisfies the fixed-point condition (5.4).

PROOF. With $\tilde{b}$ as defined, we immediately have

$$\frac{\Psi\tilde{a}}{\|\Psi\|} = \tilde{b}^{\langle p-1\rangle}$$

and

$$\begin{aligned}
\tilde{a} &= \frac{\left[\Psi^T(\Psi\tilde{a})^{\langle q-1\rangle}\right]^{\langle p-1\rangle}}{\|\Psi\|^p} \\
&= \frac{\left[\Psi^T(\Psi\tilde{a}/\|\Psi\|)^{\langle q-1\rangle}\right]^{\langle p-1\rangle}}{\|\Psi\|^p/\|\Psi\|^{\langle q-1\rangle\langle p-1\rangle}} \\
&= \frac{[\Psi^T\tilde{b}]^{\langle p-1\rangle}}{\|\Psi\|^{p-1}} \\
\tilde{a}^{\langle q-1\rangle} &= \frac{\Psi^T\tilde{b}}{\|\Psi\|},
\end{aligned}$$

as claimed.

Next, we obtain the fixed-point condition (5.4) from

$$\begin{aligned}
\tilde{a} &= \frac{[\Psi^T\tilde{b}]^{\langle p-1\rangle}}{\|\Psi\|^{p-1}} \\
\frac{\Psi\tilde{a}}{\|\Psi\|} &= \frac{\Psi[\Psi^T\tilde{b}]^{\langle p-1\rangle}}{\|\Psi\|^p} \\
\left(\frac{\Psi\tilde{a}}{\|\Psi\|}\right)^{\langle q-1\rangle} &= \frac{(\Psi[\Psi^T\tilde{b}]^{\langle p-1\rangle})^{\langle q-1\rangle}}{\|\Psi\|^{p(q-1)}} \\
\tilde{b} &= \frac{\left[\Psi(\Psi^T\tilde{b})^{\langle p-1\rangle}\right]^{\langle q-1\rangle}}{\|\Psi\|^q}.
\end{aligned}$$

Finally,

$$1 = \langle\tilde{a}, (\tilde{a})^{\langle q-1\rangle}\rangle = \frac{\langle\tilde{a}, \Psi^T\tilde{b}\rangle}{\|\Psi\|}$$

yields $\langle \tilde{a}, \Psi^T \tilde{b} \rangle = \|\Psi\|$, and the inequalities

$$\|\Psi\| = \langle \tilde{a}, \Psi^T \tilde{b} \rangle \leqslant \|\Psi \tilde{a}\|_q \|\tilde{b}\|_p \leqslant \|\Psi\| \|\tilde{a}\|_q \|\tilde{b}\|_p = \|\Psi\|$$
$$\|\Psi\| = \langle \tilde{a}, \Psi^T \tilde{b} \rangle \leqslant \|\tilde{a}\|_q \|\Psi^T \tilde{b}\|_p \leqslant \|\Psi\| \|\tilde{a}\|_q \|\tilde{b}\|_p = \|\Psi\|$$

force $\|\Psi \tilde{a}\|_q = \|\Psi\|$ and $\|\Psi^T \tilde{b}\|_p = \|\Psi\|$. □

If instead (5.4) holds, the statement and proof are analogous.

Finally let us confirm that the extremal property of $\tilde{a}$ is well behaved in the limit.

PROPOSITION 5.7. *Let $1 < p < \infty$ and $1/p + 1/q = 1$. Suppose that $\{z_0, z_1, z_2, \ldots\}$ is a sequence of distinct points in the disk $\mathbb{D}$. If Ψ is the infinite matrix with (j,k)th entry $z_k^j (1 - |z_k|^q)^{1/q}$, then either Ψ is a bounded operator on ℓ^q, and*

$$\lim_{n\to\infty} \left(\sup_{(a_1,\ldots,a_n)\neq 0} \frac{\|\Psi(a_1,\ldots,a_n,0,0,\ldots)^T\|_q}{\|(a_1,\ldots,a_n)\|_q} \right) = \sup_{a\neq 0} \frac{\|\Psi a\|_q}{\|a\|_q}, \tag{5.5}$$

or Ψ is unbounded and both sides of (5.5) *are infinite.*

PROOF. First, let us suppose that the infinite matrix Ψ is a bounded operator from ℓ^q into ℓ^q. For each $N = 1, 2, 3, \ldots$ let $\Psi^{(N)}$ be the $\infty \times (N+1)$ matrix consisting of the 0th through Nth columns of Ψ, and let $a^{(N)}$ be an $(N+1)$-dimensional unit vector for which the supremum

$$\sup_{a\neq 0} \frac{\|\Psi^{(N)} a\|_q}{\|a\|_q}$$

is attained. It is obvious that

$$\lim_{n\to\infty} \frac{\|\Psi^{(N)} a^{(N)}\|_q}{\|a^{(N)}\|_q} \leqslant \sup_{a\neq 0} \frac{\|\Psi a\|_q}{\|a\|_q}. \tag{5.6}$$

On the other hand, there exist unit vectors $a^{[n]}$ in ℓ^q, supported on the indices $\{0, 1, 2, \ldots, n\}$, such that $\|\Psi a^{[n]}\|_q$ converges to $\|\Psi\|$.

Let $\epsilon > 0$. For some N, we have

$$\|\Psi\| - \|\Psi a^{[n]}\|_q < \epsilon$$

whenever $n \geqslant N$.

Then

$$\|\Psi\| - \epsilon \leqslant \frac{\|\Psi a^{[N]}\|_q}{\|a^{[N]}\|_q} \leqslant \frac{\|\Psi^{(N)} a^{(N)}\|_q}{\|a^{(N)}\|_q}.$$

This shows that equality holds in (5.6).

If Ψ is not a bounded operator on ℓ^q, then both sides of (5.5) diverge to infinity. □

These results say that the condition (UB) can be approached by approximating with finite dimensional cases. Furthermore, they identify the nonlinear relationships between the extremal vectors that apply when $p \neq 2$.

6. Weak separation

Another facet of Carleson's Theorem (Theorem 1.1) equates interpolating sequences with a geometric condition and a measure-theoretic condition. Namely, a sequence $Z = \{z_j\}$ is interpolating for H^∞ if and only if both of the following conditions hold:

(1) The points of Z are weakly separated in the pseudohyperbolic metric; that is,
$$\inf_{j\neq k}\left|\frac{z_j - z_k}{1-\overline{z_k}z_j}\right| > 0.$$

(2) There exists a constant $C > 0$ so that the atomic measure $\mu = \sum_{j\geq 0}(1 - |z_j|^2)\delta_{z_j}$ satisfies
$$\int_{\mathbb{D}} |f|^2\, d\mu \leq C\|f\|^2_{H^2}$$
for all $f \in H^2$.

In the present section and the one to follow, we will explore both corresponding conditions for ℓ^p_A. We begin here by providing a definition for weak separation in the multiplier algebra of ℓ^p_A, which we will denote by
$$\mathscr{M}_p := \{\varphi \in \ell^p_A : \varphi f \in \ell^p_A,\ \forall f \in \ell^p_A\}.$$
We endow this space with the norm
$$\|\varphi\|_{\mathscr{M}_p} = \sup\{\|\varphi f\|_p : f \in \ell^p_A,\ \|f\|_p = 1\}.$$

We point the reader to [**9, 12**] for further background and properties of these spaces.

DEFINITION 6.1. Let $1 < p < \infty$. A sequence $Z = \{z_0, z_1, \ldots\} \subseteq \mathbb{D}$ is *weakly separated* by $\mathscr{M}_p$ if there exists $\epsilon > 0$ such that, whenever $i \neq j$, there exists $\varphi_{ij} \in \mathscr{M}_p$, with $\|\varphi_{ij}\|_{\mathscr{M}_p} \leqslant 1$, that satisfies
$$\varphi_{ij}(z_i) = \epsilon, \quad \varphi_{ij}(z_j) = 0.$$

It is well known that when $p = 2$, this definition is equivalent to separation in terms of the pseudohyperbolic metric (see [**5**] or [**1**]). We would like to uncover a similar *geometric* condition for $p \neq 2$.

Fix $1 < p < \infty$, and let $1/p + 1/q = 1$. Let us define
$$\rho_p(z_1, z_2) := \left|\frac{z - z_2}{1 - z_2^{\langle q-1\rangle} z_1}\right| = \left|\frac{z_1 - z_2}{1 - |z_2|^{q-2}\overline{z_2}z_1}\right|.$$
We see that when $p = 2$, we recover the pseudohyperbolic metric. We also note that the function
$$\varphi(z) = \frac{z-\alpha}{1-\alpha^{\langle q-1\rangle} z}$$
is extremal in the sense that it solves the problem
$$\sup\{|f(0)| : \|f\|_p = 1, f(\alpha) = 0\}.$$
Again, notice that when $p = 2$, we recover the familiar Blaschke factor. More generally, $\varphi(z)$ is p-inner, as discussed in Section 2. However, products of such functions fail to be p-inner when $p \neq 2$; this is one of the ways that the function theory of ℓ^p_A is a tough nut to crack. See [**10**] for details.

It is immediate upon inspection that ρ_p will not be symmetric, i.e., in general $\rho_p(z_1, z_2) \neq \rho_p(z_2, z_1)$. In turn, the best we can hope for is that ρ_p be a *quasimetric* (i.e., a metric lacking symmetry), which turns out to be the case.

PROPOSITION 6.2. *Let $1 < p < \infty$. The function $\rho_p \colon \mathbb{D} \times \mathbb{D} \to \mathbb{R}$ given by*

$$\rho_p(z_1, z_2) := \left| \frac{z_1 - z_2}{1 - z_2^{\langle q-1 \rangle} z_1} \right|$$

is a quasimetric on the unit disk.

PROOF. The function ρ_p is clearly positive and positive definite. We need now only to prove a triangle inequality. Since the Blaschke factor φ_w, with zero at $w \in \mathbb{D}$, is an automorphism of $\mathbb{D}$, we have

$$\rho_p(z_1, z_2) \leqslant \rho_p(z_1, w) + \rho_p(w, z_2) \iff \rho_p(\varphi_w(z_1), \varphi_w(z_2)) \leqslant \rho_p(\varphi_w(z_1), 0) + \rho_p(0, \varphi_w(z_2)).$$

With $\gamma_1 = \varphi_w(z_1)$ and $\gamma_2 = \varphi_w(z_2)$, it suffices to show that

$$\rho_p(\gamma_1, \gamma_2) \leqslant \rho_p(\gamma_1, 0) + \rho_p(0, \gamma_2),$$

or equivalently that

$$\left| \frac{\gamma_1 - \gamma_2}{1 - \gamma_2^{\langle q-1 \rangle} \gamma_1} \right| \leqslant |\gamma_1| + |\gamma_2|.$$

What we will actually show is that

$$\rho_p(\gamma_1, \gamma_2) \leqslant \frac{|\gamma_1| + |\gamma_2|}{1 + |\gamma_2|^{\langle q-1 \rangle} |\gamma_1|},$$

which gives the result, as $1 + |\gamma_2|^{\langle q-1 \rangle} |\gamma_1| \geqslant 1$. Notice that

$$\begin{aligned} \rho_p(-|\gamma_1|, |\gamma_2|) &= \left| \frac{-|\gamma_1| - |\gamma_2|}{1 - |\gamma_2|^{\langle q-1 \rangle}(-|\gamma_1|)} \right| \\ &= \frac{|\gamma_1| + |\gamma_2|}{1 + |\gamma_2|^{\langle q-1 \rangle} |\gamma_1|} \end{aligned}$$

which is the upper bound we would like to establish for $\rho_p(\gamma_1, \gamma_2)$. In order to do this, we show that

$$1 - \rho_p(\gamma_1, \gamma_2)^2 \geqslant 1 - \rho_p(-|\gamma_1|, |\gamma_2|)^2.$$

Suppose γ_1 and γ_2 are nonzero (otherwise the inequality is trivial). Notice that $\Re\{\gamma_1 \overline{\gamma_2}\} \geqslant -1$ and $|\gamma_1 \gamma_2| \leqslant 1$, which gives

$$\frac{\Re\{\gamma_1 \overline{\gamma_2}\}}{|\gamma_1 \gamma_2|} \geqslant \Re\{\gamma_1 \overline{\gamma_2}\} \geqslant -1,$$

and yields that

$$\Re\{\gamma_1 \overline{\gamma_2}\} \geqslant -|\gamma_1 \gamma_2|.$$

Let us work backwards from this inequality to get what we want:

$$\begin{aligned}
&\Re\{\gamma_1\overline{\gamma_2}\} \geqslant -|\gamma_1\gamma_2| \\
\implies &\Re\{\gamma_1\overline{\gamma_2}\} \geqslant \left(\frac{|\gamma_2|^{q-2}-1}{1-|\gamma_2|^{q-2}}\right)|\gamma_1\gamma_2| \\
\implies &\Re\{\gamma_1\overline{\gamma_2}\} - \Re\left\{\gamma_1\gamma_2^{\langle q-1\rangle}\right\} \geqslant \left(|\gamma_2|^{q-2}-1\right)|\gamma_1\gamma_2| \\
\implies &\Re\{\gamma_1\overline{\gamma_2}\} - \Re\left\{\gamma_1\gamma_2^{\langle q-1\rangle}\right\} \geqslant |\gamma_2|^{\langle q-1\rangle>}|\gamma_1| - |\gamma_1\gamma_2|
\end{aligned}$$

We now multiply each side by 2 and add $1+|\gamma_2^{\langle q-1\rangle}\gamma_1|^2 - |\gamma_1|^2 - |\gamma_2|^2$ to the last inequality above:

$$\begin{aligned}
&2\Re\{\gamma_1\overline{\gamma_2}\} - 2\Re\left\{\gamma_1\gamma_2^{\langle q-1\rangle}\right\} + 1 + |\gamma_2^{\langle q-1\rangle}\gamma_1|^2 - |\gamma_1|^2 - |\gamma_2|^2 \\
&\qquad \geqslant 2|\gamma_2|^{\langle q-1\rangle}|\gamma_1| - 2|\gamma_1\gamma_2| + 1 + |\gamma_2^{\langle q-1\rangle}\gamma_1|^2 - |\gamma_1|^2 - |\gamma_2|^2 \\
\implies &\underbrace{1 - 2\Re\left\{\gamma_2^{\langle q-1\rangle}\gamma_1\right\} + \left|\gamma_2^{\langle q-1\rangle}\gamma_1\right|^2}_{|1-\gamma_2^{\langle q-1\rangle}\gamma_1|^2} - \underbrace{\left(|\gamma_1|^2 - 2\Re\{\gamma_1\overline{\gamma_2}\} + |\gamma_2|^2\right)}_{|\gamma_1-\gamma_2|^2} \\
&\qquad \geqslant \underbrace{1 + 2|\gamma_2|^{\langle q-1\rangle}|\gamma_1| + |\gamma_2^{\langle q-1\rangle}\gamma_1|^2}_{(1+|\gamma_2|^{\langle q-1\rangle}|\gamma_1|)^2} - \underbrace{\left(|\gamma_1|^2 - 2|\gamma_1\gamma_2| + |\gamma_2|^2\right)}_{(|\gamma_1|+|\gamma_1|)^2}.
\end{aligned}$$

Hence, we have

$$|1-\gamma_2^{\langle q-1\rangle}\gamma_1|^2 + |\gamma_1-\gamma_2|^2 \geqslant (1+|\gamma_2|^{\langle q-1\rangle}|\gamma_1|)^2 + (|\gamma_1|+|\gamma_1|)^2.$$

This was all to say, using the triangle inequality in the second line below, that

$$\begin{aligned}
1-\rho_p(\gamma_1,\gamma_2)^2 &= \frac{|1-\gamma_2^{\langle q-1\rangle}\gamma_1|^2 - |\gamma_1-\gamma_2|^2}{|1-\gamma_2^{\langle q-1\rangle}\gamma_1|^2} \\
&\geqslant \frac{|1-\gamma_2^{\langle q-1\rangle}\gamma_1|^2 - |\gamma_1-\gamma_2|^2}{1+|\gamma_2^{\langle q-1\rangle}\gamma_1|^2} \\
&\geqslant \frac{(1+|\gamma_2|^{\langle q-1\rangle}|\gamma_1|)^2 + (|\gamma_1|+|\gamma_1|)^2}{1+|\gamma_2^{\langle q-1\rangle}\gamma_1|^2} \\
&= 1-\rho_p(-|\gamma_1|,|\gamma_2|)^2.
\end{aligned}$$

Thus, we have

$$\begin{aligned}
\rho_p(\gamma_1,\gamma_2) &\leqslant \rho_p(-|\gamma_1|,|\gamma_2|) \\
&\leqslant |\gamma_1| + |\gamma_2| \\
&\leqslant \rho_p(\gamma_1,0) + \rho_p(0,\gamma_2),
\end{aligned}$$

which establishes the result. □

In turn, there is some hope that ρ_p may do the job of describing weak separation. That is, for the sequence $Z=\{z_j\}$, we hope to show the separation condition

$$\inf_{n\neq k} \rho_p(z_n,z_k) > 0 \tag{6.1}$$

is equivalent to Z being weakly separated by $\mathscr{M}_p$. Before doing this, we need a couple of observations, which are variants of the classical Schwarz and Schwarz–Pick lemmas.

LEMMA 6.3. *Let $1 < p < \infty$ and suppose $\varphi \in \mathscr{M}_p$, with $\|\varphi\|_{\mathscr{M}_p} \leqslant M$ and $\varphi(0) = 0$. Then, for all $z \in \mathbb{D}$,*

$$|\varphi(z)| \leqslant M|z|.$$

PROOF. The hypotheses ensure, by [**12**, Proposition 12.2.4], that $\varphi \in H^\infty$ with $\|\varphi\|_{H^\infty} \leqslant M$. This now becomes just a straightforward modification of the classical Schwarz lemma. Let

$$g(z) = \begin{cases} \frac{\varphi(z)}{z}, & z \neq 0 \\ \varphi'(0), & z = 0 \end{cases}$$

which is holomorphic on $\mathbb{D}$. Let D_r be the disk of radius r centered at the origin $(0 < r < 1)$. Now, let $z \in D_r$ and invoke the maximum modulus principle to find $z_r \in \partial D_r$ such that

$$|g(z)| \leqslant |g(z_r)| = \left|\frac{\varphi(z_r)}{z_r}\right| \leqslant \frac{M}{r}.$$

As $r \to 1$, we see that $|g(z)| \leqslant M$, which gives $|\varphi(z)| \leqslant M|z|$. □

We can now prove a replacement of the classical Schwarz–Pick Lemma.

LEMMA 6.4 (p-Schwarz–Pick Lemma). *Let $1 < p < \infty$. If $f \in \mathscr{M}_p$ with $\|f\|_{\mathscr{M}_p} \leq 1$, then, for each $z, w \in \mathbb{D}$, there is a constant M such that*

$$\rho_p(f(z), f(w)) \leqslant M\rho_p(z, w)$$

Note that the constant M depends on f, p, and w.

PROOF. Let $w \in \mathbb{D}$ and define

$$\varphi_{p,w}(z) := \frac{z - w}{1 - w^{\langle q-1\rangle} z}.$$

Note that

$$\varphi_{p,w}^{-1}(z) = \frac{z + w}{1 + w^{\langle q-1\rangle} z} = \varphi_{p,-w}(z).$$

Consider the function

$$\varphi_{p,f(w)} \circ f \circ \varphi_{p,w}^{-1},$$

which is defined and bounded on $\mathbb{D}$, and fixes the origin. Thus, we can apply Lemma 6.3 to get

$$\left|\left(\varphi_{p,f(w)} \circ f \circ \varphi_{p,w}^{-1}\right)(z)\right| \leqslant M|z|$$

for all $z \in \mathbb{D}$, with $M = \|\varphi_{p,f(w)} \circ f \circ \varphi_{p,w}^{-1}\|_\infty$. As $\varphi_{p,w}^{-1}$ is injective, we have

$$\left|\left(\varphi_{p,f(w)} \circ f\right)(z)\right| \leqslant M\left|\varphi_{p,w}(z)\right|,$$

which is precisely

$$\rho_p(f(z), f(w)) \leqslant M\rho_p(z, w).$$

□

We now characterize weakly separated sequences with Theorem 6.5.

THEOREM 6.5. *Let $1 < p < \infty$ and let $Z = \{z_k\}$ be a sequence in $\mathbb{D}$. Then Z is weakly separated by $\mathscr{M}_p$ if and only if*

$$\inf_{j \neq k} \rho_p(z_j, z_k) > 0.$$

PROOF. For the forward direction, suppose there exists $\epsilon > 0$ such that, whenever $i \neq j$, there exists $\psi_{ij} \in \mathscr{M}_p$, with $\|\psi_{ij}\|_{\mathscr{M}_p} \leqslant 1$, that satisfies

$$\psi_{ij}(z_i) = \epsilon, \quad \psi_{ij}(z_j) = 0.$$

It suffices to show that there exists $\delta > 0$ such that $\rho_p(z_j, z_k) \geqslant \delta$ for all $j \neq k$. Now, by the p-Schwarz–Pick Lemma, we know that for each $z_i, z_j \in Z$, there is a constant M_{ij} such that

$$\epsilon = \rho_p(\psi_{ij}(z_i), \psi_{ij}(z_j)) \leqslant M_{ij}\rho_p(z_i, z_j).$$

Here, $M_{ij} = \|\varphi_{p,\psi_{ij}(z_j)} \circ \psi_{ij} \circ \varphi^{-1}_{p,z_j}\|_{H^\infty} = \|\psi_{ij} \circ \varphi^{-1}_{p,z_j}\|_{H^\infty}$. If $C = \sup_{i \neq j} M_{ij}$ is finite, then it suffices to take $\delta = \epsilon/C$. But notice that $\varphi^{-1}_{p,z_j}(\mathbb{D}) \subseteq \mathbb{D}$, so $\|\varphi^{-1}_{p,z_j}\|_{H^\infty} \leqslant 1$. And since the same bound is true for ψ_{ij} by hypothesis, we have $M_{ij} \leqslant 1$, and thus, $C \leqslant 1$.

For the backward direction, let

$$\psi_k(w) = \frac{w - z_k}{1 - z_k^{\langle q-1 \rangle} w}$$

and let

$$\epsilon = \left(\inf_{j \neq k} \rho_p(z_j, z_k)\right)\left(\sup_{k \geqslant 1} \|\psi_k\|_1\right)^{-1}$$

which is positive and finite by assumption.

Further, let $\hat{\psi}_k = \psi_k/\|\psi_k\|_1$ and let $\hat{\psi}_k(z_j) = \epsilon_{jk}$. Note that

$$|\epsilon_{jk}| = \left|\hat{\psi}_k(z_j)\right| = \left|\frac{z_j - z_k}{1 - z_k^{\langle q-1 \rangle} z_j}\right| \frac{1}{\|\psi_k\|_1} \geqslant \epsilon.$$

Now take $\varphi_{jk} = \frac{\epsilon}{\epsilon_{jk}}\hat{\psi}_k$ and notice that

$$\|\varphi_{jk}\|_1 = \left|\frac{\epsilon}{\epsilon_{jk}}\right| \|\hat{\psi}_k\|_1 \leqslant 1,$$

which means φ_{jk} is in the closed unit ball of $\mathscr{M}_p$ since $\|\varphi_{jk}\|_{\mathscr{M}_p} \leqslant \|\varphi_{jk}\|_1$. Further, whenever $j \neq k$, we have

$$\varphi_{jk}(z_j) = \frac{\epsilon}{\epsilon_{jk}}\hat{\psi}_k(z_j) = \epsilon$$

and

$$\varphi_{jk}(z_k) = \frac{\epsilon}{\epsilon_{jk}}\hat{\psi}_k(z_k) = 0.$$

Thus, Z is weakly separated. □

7. Carleson measures

In this final section we relate the condition (UB) to another class of inequalities, as well as to Carleson measures.

Recall that a finite measure μ on $\mathbb{D}$ is said to be a *Carleson measure* if for some $C > 0$ we have

$$\mu(S) \leqslant Ch$$

for every set S of the form

$$S = \{re^{i\theta} \in \mathbb{D} : \ 1 - h \leqslant r < 1, \ \theta_0 \leqslant \theta \leqslant \theta_0 + h\},$$

where $0 \leqslant \theta_0 \leqslant 2\pi$ and $0 < h < 1$. That is, μ is a Carleson measure in terms of "Carleson windows."

Theorem 7.1 characterizes the condition (UB). It is the dual statement.

THEOREM 7.1. *Assume $1 < p < \infty$ and $1/p + 1/q = 1$, and $\{z_0, z_1, z_2 \ldots\}$ is a sequence of distinct points in $\mathbb{D}$. There exists $K > 0$ such that*

$$\sum_{k=0}^{\infty}(1-|z_k|^q)^{p-1}|f(z_k)|^p \leqslant K\|f\|_p^p \tag{7.1}$$

for all $f \in \ell_A^p$, if and only if (UB) *holds, that is, there exists $C > 0$ such that*

$$\Big\|\sum_{k=0}^{\infty} c_k\Lambda_{z_k}\Big\|_q \leqslant C\Big(\sum_{k=0}^{\infty}|c_k|^q\|\Lambda_{z_k}\|_q^q\Big)^{1/q} \tag{7.2}$$

for all constants $\{c_k\}_{k=0}^{\infty}$.

PROOF. If (7.1) holds, then for any constants $\{c_k\}_{k=0}^{\infty}$,

$$\begin{aligned}
\Big\|\sum_{k=0}^{\infty} c_k\Lambda_{z_k}\Big\|_q &= \sup_{f\neq 0}\frac{\big|\sum_{k=0}^{\infty} c_k\Lambda_{z_k}(f)\big|}{\|f\|_p} \\
&\leqslant \sup_{f\neq 0}\frac{\big|\sum_{k=0}^{\infty} c_k f(z_k)\big|}{\|f\|_p} \\
&\leqslant \sup_{f\neq 0}\frac{\sum_{k=0}^{\infty}|c_k||f(z_k)|}{\|f\|_p} \\
&= \sup_{f\neq 0}\frac{\sum_{k=0}^{\infty}|c_k|\|\Lambda_{z_k}\|_q(1-|z_k|^q)^{1/q}|f(z_k)|}{\|f\|_p} \\
&\leqslant \sup_{f\neq 0}\frac{\big(\sum_{k=0}^{\infty}|c_k|^q\|\Lambda_{z_k}\|_q^q\big)^{1/q}\big(\sum_{k=0}^{\infty}(1-|z_k|^q)^{p/q}|f(z_k)|^p\big)^{1/p}}{\|f\|_p} \\
&\leqslant \sup_{f\neq 0}\frac{\big(\sum_{k=0}^{\infty}|c_k|^q\|\Lambda_{z_k}\|_q^q\big)^{1/q}(K\|f\|_p^p)^{1/p}}{\|f\|_p} \\
&= K^{1/p}\Big(\sum_{k=0}^{\infty}|c_k|^q\|\Lambda_{z_k}\|_q^q\Big)^{1/q},
\end{aligned}$$

where we have used $p/q = p(1-1/p) = p-1$. This shows that (7.2) holds with $C = K^{1/p}$.

Conversely, suppose we select points $z_0, z_1, \ldots, z_N$ such that

$$\|c_0\Lambda_{z_0} + c_1\Lambda_{z_1} + \cdots + c_N\Lambda_{z_N}\|_q \leqslant K\Big(|c_0|^q\|\Lambda_{z_0}\|_q^q + \cdots + |c_N|^q\|\Lambda_{z_N}\|_q^q\Big)^{1/q} \tag{7.3}$$

for all N and all constants $\{c_k\}_{k=0}^{N}$.

Then

$$\begin{aligned}\|f\|_p &\geqslant \frac{|c_0 f(z_0) + c_1 f(z_1) + \cdots + c_N f(z_N)|}{\|c_0\Lambda_{z_0} + c_1\Lambda_{z_1} + \cdots + c_N\Lambda_{z_N}\|_q} \\ &\geqslant \frac{|c_0 f(z_0) + c_1 f(z_1) + \cdots + c_N f(z_N)|}{K\Big(|c_0|^q\|\Lambda_{z_0}\|_q^q + \cdots + |c_N|^q\|\Lambda_{z_N}\|_q^q\Big)^{1/q}} \\ &= \frac{\big|\beta_0(1-|z_0|^q)^{1/q}|f(z_0)| + \beta_1(1-|z_1|^q)^{1/q}|f(z_1)| + \cdots + \beta_N(1-|z_N|^q)^{1/q}|f(z_N)|\big|}{K\Big(|\beta_0|^q + |\beta_1|^q + \cdots + |\beta_N|^q\Big)^{1/q}},\end{aligned}$$

where we have chosen

$$\beta_k = \begin{cases} \frac{c_k f(z_k)}{(1-|z_k|^q)^{1/q}|f(z_k)|} & , \ f(z_k) \neq 0 \\ 0 & , \ f(z_k) = 0. \end{cases}$$

That is, for every sequence $\beta = \{\beta_k\}_{k=0}^{\infty} \in \ell^q$, we have

$$\sum_{k=0}^{\infty} \beta_k (1-|z_k|^q)^{1/q}|f(z_k)| \leqslant K\|\beta\|_q\|f\|_p.$$

Finally, divide by $\|\beta\|_q$ then take the supremum over $\beta \in \ell^q\setminus\{0\}$ to obtain (7.1). $\square$

REMARK 7.2. In the course of solving the corona problem, Carleson showed [**5, 6**] that μ is a Carleson measure if and only if for some $C > 0$,

$$\int_{\mathbb{D}} |f(z)|^p \, d\mu \leqslant C\|f\|_{H^p}^p \tag{7.4}$$

for all f belonging to the Hardy space H^p, where $0 < p < \infty$. For this reason, in some texts the condition (7.4) is used as the definition of Carleson measure. However, we will reserve the name "Carleson measure" to those measures satisfying the definition given in the beginning of this section (in terms of "Carleson windows"), as we shall see that Carleson's result does not carry over from H^p to the spaces ℓ^p_A when $2 < p < \infty$.

A necessary condition for the points Z to satisfy the condition (UB) follows immediately.

COROLLARY 7.3. *Assume $1 < p < \infty$ and $1/p + 1/q = 1$. Suppose that $\{z_k\}_{k=0}^{\infty}$ is a sequence of distinct points in $\mathbb{D}$. If there exists $K > 0$ such that*

$$\Big\|\sum_{k=0}^{\infty} \beta_k \Lambda_{z_k}\Big\|_q \leqslant K\Big(\sum_{k=0}^{\infty} |\beta_k|^q \|\Lambda_{z_k}\|_q^q\Big)^{1/q}$$

for all sequences $\{\beta_k\}_{k=0}^{\infty}$, then

$$\sum_{k=0}^{\infty} (1-|z_k|^q)^{p-1} < \infty.$$

To prove this, use $f(z) = 1$ in the theorem.

Other choices of f in Theorem 7.1 provide additional necessary conditions on a measure μ for the norm inequality (7.1) to hold. This idea is illustrated in the next proposition.

PROPOSITION 7.4. *Let $1 < p < \infty$ and $1/p + 1/q = 1$. Suppose that the finite measure μ on $\mathbb{D}$ satisfies*

$$\int_{\mathbb{D}} |f(z)|^p \, d\mu \leqslant C\|f\|_p^p$$

for all $f \in \ell_A^p$. Then there exist constants A_1 and A_2 such that

$$A_1\Big(|a|^p + \frac{|1-a^2|^p}{1-|a|^p}\Big) \geqslant c^p \mu\Big(\Big|\frac{a-z}{1-\bar{a}z}\Big| \geqslant c\Big)$$

whenever $a \in \mathbb{D}$ and $0 < c < 1$, and

$$A_2\Big(\frac{1}{1-|a|^p}\Big) \geqslant c^p \mu\Big(\frac{1}{|1-az|} \geqslant c\Big)$$

whenever $a \in \mathbb{D}$ and $c > 1$.

PROOF. Suppose that μ satisfies

$$\int_{\mathbb{D}} |f(z)|^p \, d\mu \leqslant C\|f\|_p^p$$

for all $f \in \ell_A^p$.

We use

$$\int_{\mathbb{D}} |f(z)|^p \, d\mu = \|f\|_{L^p(\mu)}^p = \Big[\sup\Big\{\Big|\int_{\mathbb{D}} f(z)g(z)\, d\mu\Big| : \ g(z) \in L^q(\mu), \ \|g\|_{L^q(\mu)} = 1\Big\}\Big]^p,$$

along with the choice $g(z) = \overline{f(z)}\chi_E(z)/|f(z)|$, where $E = \{z \in \mathbb{D} : \ |f(z)| \geqslant c\}$ for $c > 0$. The result is

$$\begin{aligned} C\|f\|_p^p &\geqslant \Big[\Big|\int_{\mathbb{D}} f(z)g(z)\, d\mu\Big| \Big/ \Big(\int_{\mathbb{D}} |g(z)|^q \, d\mu\Big)^{1/q}\Big]^p \\ &\geqslant \big[c\mu(E)/\mu(E)^{1/q}\big]^p \\ &= c^p \mu(E). \end{aligned}$$

By choosing $f(z) = \varphi_a(z) = (a-z)/(1-\bar{a}z)$, a Möbius transformation, we get bounds on the measures of certain circles and anti-circles in $\mathbb{D}$.

Using

$$\|\varphi\|_p^p = |a|^p + \frac{|1-a^2|^p}{1-|a|^p},$$

we have

$$C\Big(|a|^p + \frac{|1-a^2|^p}{1-|a|^p}\Big) \geqslant c^p \mu\Big(\Big|\frac{a-z}{1-\bar{a}z}\Big| \geqslant c\Big), \tag{7.5}$$

for all $0 < c < 1$ and $a \in \mathbb{D}$.

We can apply the same idea with $f = \Lambda_a$, with the resulting condition

$$C\Big(\frac{1}{1-|a|^p}\Big) \geqslant c^p \mu\Big(\frac{1}{|1-az|} \geqslant c\Big). \tag{7.6}$$

□

We now connect the inequality (7.1) to Carleson measures.

Suppose that $1 < p < 2$ and $h > 0$ is small. Choose $c = 1 - h$ and $a = h^{1/p}c$. The region

$$\Big\{z : \ \Big|\frac{a-z}{1-\bar{a}z}\Big| \geqslant c\Big\}$$

appearing in (7.5) is the exterior of a disk contained inside $\mathbb{D}$. Its boundary circle intersects the x-axis at the solutions of a quadratic equation. The region also encloses a Carleson window S_δ of maximal width δ. From elementary geometry, and taking into account the continuity of the square root function, etc., we can estimate the value of δ by

$$\begin{aligned}\delta &= 1 - \frac{-2a(1-c^2) + \sqrt{4a^2(1-c^2)^2 + (1-a^2c^2)(c^2-a^2)}}{-2(1-a^2c^2)} \\ &= 1 - \frac{-(1-h)h^{1/p}(1-[1-h]^2) + \kappa}{-(1-h^{2/p}[1-h]^4)}, \\ &\qquad \text{where } \kappa = \sqrt{h^{2/p}(1-h)^2(1-[1-h]^2)^2 + (1-h^{2/p}[1-h]^4)(1-h)^2(1-h^{2/p})}, \\ &\approx 1 - \frac{-2h^{1+1/p} + (1-h)}{-(1-h^{2/p})} \\ &\approx h.\end{aligned}$$

(Since $1 < p < 2$, we have that h dominates $h^{2/p}$ in the above estimates.) The notation $\approx$ means that the expressions are equal apart from terms of higher than first order in h.

And now (7.5) gives

$$\begin{aligned}\mu(S_\delta) &\leqslant \mu\left(\left\{z : \left|\frac{a-z}{1-\bar{a}z}\right| \geqslant c\right\}\right) \\ &\leqslant \frac{C_1}{c^p}\left(|a|^p + \frac{|1-a^2|^p}{1-|a|^p}\right) \\ &\leqslant 2C_1 h\end{aligned}$$

for h sufficiently small.

In Figure 1, we plot the above inequalities; the most heavily shaded region depicts the previously discussed Carleson window.

Suppose that $2 \leqslant p < \infty$, and $h > 0$ is small. Choose $c = 1/h$ and $a = (1-h^{p-1})^{1/p}$.

The region appearing in (7.6) can be described by

$$\left|\frac{1}{a} - z\right| \leqslant \frac{1}{ac},$$

which is the intersection of $\mathbb{D}$ with the disk of radius $1/ac$ centered at $1/a$. The width of the intersection region is equal to

$$\frac{1}{ac} - \left(\frac{1}{a} - 1\right) = \frac{1-(1-a)c}{ac} \approx \frac{h - h^{p-1}/p}{(1-h^{p-1})^{1/p}} \approx h.$$

(It is in this step that $p \geqslant 2$ was needed.)

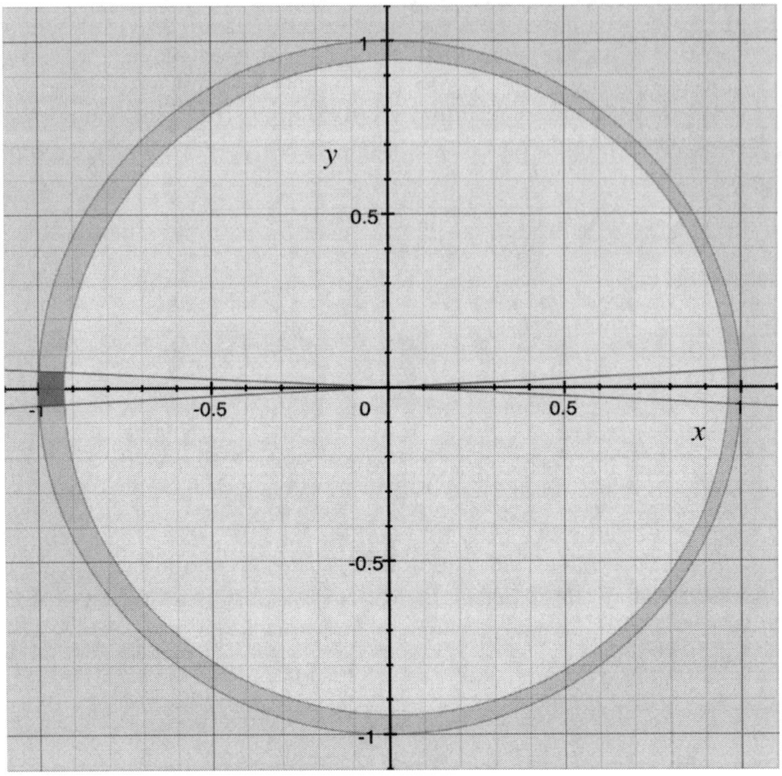

FIGURE 1. The intersection of the unit disk and the region described in (7.5), for $p = \frac{3}{2}$ and $h = \frac{1}{10}$, along with a Carleson window.

Thus the region contains a Carleson window

$$S_\delta := \{re^{i\theta} : \quad 1 - \delta < r < 1; \ |\theta| < \delta/2\},$$

where $\delta \approx h$.

Figure 2 shows the Carleson window in this case.

Now (7.6) leads to

$$\begin{aligned} \mu(S_\delta) &\approx \mu\Big(\frac{1}{|1 - az|} \geqslant c\Big) \\ &\leqslant \frac{C_2}{c^p}\Big(\frac{1}{1 - |a|^p}\Big) \\ &= \frac{C_2 h^p}{h^{p-1}} \\ &= C_2 h. \end{aligned}$$

The argument is rotationally symmetric, and so the following proposition has been proved. (Of course, the $p = 2$ case was already known.)

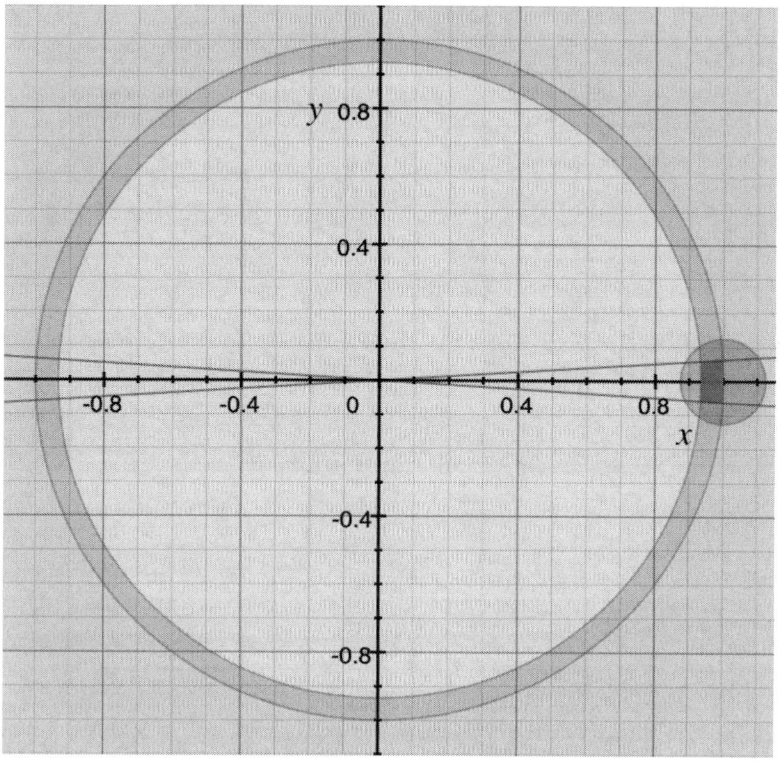

FIGURE 2. The intersection of the unit disk and the region described in (7.6), for $p = 4$ and $h = \frac{1}{8}$, along with a Carleson window (depicted with the heaviest shading).

PROPOSITION 7.5. *Let $1 < p < \infty$. Suppose that the measure μ on $\mathbb{D}$ satisfies*

$$\int_{\mathbb{D}} |f(z)|^p \, d\mu \leqslant C\|f\|_p^p \tag{7.7}$$

for all $f \in \ell^p_A$. Then μ is a Carleson measure.

Here's part of the converse.

PROPOSITION 7.6. *Let $1 < p \leqslant 2$. If μ is a Carleson measure, then there exists $C > 0$ such that*

$$\int_{\mathbb{D}} |f|^p \, d\mu \leqslant C\|f\|^p_{\ell^p_A}$$

for all $f \in \ell^p_A$.

PROOF. First, by Hölder's inequality, we have

$$\int_{\mathbb{D}} F \, d\mu \leqslant \Big(\int_{\mathbb{D}} F^r \, d\mu\Big)^{1/r} \mu(\mathbb{D})^{1/r'}$$

for any nonnegative integrable F, with $1 < r < \infty$ and $1/r + 1/r' = 1$. Take $1 < p < 2$ and $r = q/p > 1$, where $1/p + 1/q = 1$. Then for $F = |f|^p$, we have

$$\Big(\int_{\mathbb{D}} |f|^p \, d\mu\Big)^{1/p} \leqslant \Big(\int_{\mathbb{D}} |f|^q \, d\mu\Big)^{1/q} \mu(\mathbb{D})^{1/(r'p)}.$$

Thus if μ is a Carleson measure, and $1 < p < 2$, then for some $C > 0$

$$\begin{aligned}\Big(\int_{\mathbb{D}} |f|^p \, d\mu\Big)^{1/p} &\leqslant \Big(\int_{\mathbb{D}} |f|^q \, d\mu\Big)^{1/q} \mu(\mathbb{D})^{1/(r'p)} \\ &\leqslant C\mu(\mathbb{D})^{1/(r'p)} \|f\|_{H^q} \\ &\leqslant C\mu(\mathbb{D})^{1/(r'p)} \|f\|_{\ell^p_A}\end{aligned}$$

for all $f \in \ell^p_A$, with Hausdorff–Young inequality being used in the last step. □

The converse to Proposition 7.5 fails when $2 < p < \infty$, as can be seen from the following proposition.

PROPOSITION 7.7. *If $2 < p < \infty$, there exists a Carleson measure μ such that for no $C > 0$ does*

$$\int_{\mathbb{D}} |f(z)|^p \, d\mu(z) \leqslant C\|f\|_p^p$$

hold for all $f \in \ell^p_A$.

PROOF. For $n = 1, 2, 3, \ldots$ let $r_n = 1 - 1/n$. Then of course $\lim_{n\to\infty} r_n = 1$ monotonically, and

$$\sum_{n=1}^{\infty} (r_{n+1} - r_n) = \sum_{n=1}^{\infty} \frac{1}{n(n+1)} = 1.$$

Let μ be the measure defined on $\mathbb{D}$ consisting of point masses at the points $\{r_n\}$ with the respective weights

$$\mu(\{r_n\}) = \frac{1}{n(n+1)}.$$

Then μ is a Carleson measure, since the amount of mass in a Carleson window of width $1 - r_m$ is at most

$$\sum_{n=m}^{\infty} (r_{n+1} - r_n) = 1 - r_m.$$

Select $\epsilon > 0$, and let $2 < p < \infty$, and consider the function

$$f(z) = \sum_{k=0}^{\infty} \frac{z^k}{(k+1)^{\epsilon+1/p}}.$$

Then $f \in \ell^p_A$. On the other hand,

$$\begin{aligned}\int_{\mathbb{D}} |f(z)|^p \, d\mu(z) &= \sum_{n=1}^{\infty} \frac{1}{n(n+1)} \left| \sum_{k=0}^{\infty} \frac{(1-1/n)^k}{(k+1)^{\epsilon+1/p}} \right|^p \\ &\geqslant \sum_{n=1}^{\infty} \frac{1}{n(n+1)} \left| \sum_{k=0}^{n} \frac{(1-1/n)^n}{(k+1)^{\epsilon+1/p}} \right|^p \\ &\geqslant C \sum_{n=1}^{\infty} \frac{1}{n(n+1)} \left| \frac{1}{e} (n+1)^{1-\epsilon-1/p} \right|^p \\ &\geqslant \frac{C}{e^p} \sum_{n=1}^{\infty} (n+1)^{p-2-p\epsilon-1},\end{aligned}$$

where the constant C comes from estimating the sum in k using an integral.

Since $p > 2$, we can choose ϵ sufficiently small that $p - 2 - \epsilon \geqslant 0$, so that the final expression diverges to $+\infty$. This proves the claim. □

It remains, therefore, to fully characterize the norm inequality (7.1) when $2 < p < \infty$. Looking back to earlier sections, we would like to uncover any connection between weak separation and the condition (LB), and find other potential uses for the nonlinear operator extending the Gramian. These matters are the subject of ongoing work.

References

[1] J. Agler and J. E. McCarthy, *Pick interpolation and Hilbert function spaces*, Graduate Studies in Mathematics, vol. 44, American Mathematical Society, Providence, RI, 2002, DOI 10.1090/gsm/044. MR1882259

[2] A. Aleman, S. Richter, and C. Sundberg, *Beurling's theorem for the Bergman space*, Acta Math. **177** (1996), no. 2, 275–310, DOI 10.1007/BF02392623. MR1440934

[3] J. Alonso, H. Martini, and S. Wu, *On Birkhoff orthogonality and isosceles orthogonality in normed linear spaces*, Aequationes Math. **83** (2012), no. 1-2, 153–189, DOI 10.1007/s00010-011-0092-z. MR2885507

[4] C. Bénéteau, M. C. Fleeman, D. S. Khavinson, D. Seco, and A. A. Sola, *Remarks on inner functions and optimal approximants*, Canad. Math. Bull. **61** (2018), no. 4, 704–716, DOI 10.4153/CMB-2017-058-4. MR3846742

[5] L. Carleson, *An interpolation problem for bounded analytic functions*, Amer. J. Math. **80** (1958), 921–930, DOI 10.2307/2372840. MR117349

[6] L. Carleson, *Interpolations by bounded analytic functions and the corona problem*, Ann. of Math. (2) **76** (1962), 547–559, DOI 10.2307/1970375. MR141789

[7] R. Cheng, A. G. Miamee, and M. Pourahmadi, *Regularity and minimality of infinite variance processes*, J. Theoret. Probab. **13** (2000), no. 4, 1115–1122, DOI 10.1023/A:1007874226636. MR1820505

[8] R. Cheng and W. T. Ross, *Weak parallelogram laws on Banach spaces and applications to prediction*, Period. Math. Hungar. **71** (2015), no. 1, 45–58, DOI 10.1007/s10998-014-0078-4. MR3374700

[9] R. Cheng, J. Mashreghi, and W. T. Ross, *Multipliers of sequence spaces*, Concr. Oper. **4** (2017), no. 1, 76–108, DOI 10.1515/conop-2017-0007. MR3714456

[10] R. Cheng, J. Mashreghi, and W. T. Ross, *Inner functions and zero sets for ℓ^p_A*, Trans. Amer. Math. Soc. **372** (2019), no. 3, 2045–2072, DOI 10.1090/tran/7675. MR3976584

[11] R. Cheng, J. Mashreghi, and W. T. Ross, *Inner functions in reproducing kernel spaces*, Analysis of operators on function spaces, Trends Math., Birkhäuser/Springer, Cham, 2019, pp. 167–211, DOI 10.1007/978-3-030-14640-5_6. MR4019469

[12] R. Cheng, J. Mashreghi, and W. T. Ross, *Function theory and ℓ^p spaces*, University Lecture Series, vol. 75, American Mathematical Society, Providence, RI, [2020] ©2020. MR4249569

[13] Raymond Cheng, Javad Mashreghi, and William T. Ross, *Inner vectors for Toeplitz operators*, Complex Analysis and Spectral Theory (G. Dales, D. Khavinson, and Mashreghi, eds.), Contemporary Math., American Mathematical Society, (2020).

[14] J. B. Conway, *A course in functional analysis*, Graduate Texts in Mathematics, vol. 96, Springer-Verlag, New York, 1985, DOI 10.1007/978-1-4757-3828-5. MR768926

[15] P. Duren, D. Khavinson, H. S. Shapiro, and C. Sundberg, *Invariant subspaces in Bergman spaces and the biharmonic equation*, Michigan Math. J. **41** (1994), no. 2, 247–259, DOI 10.1307/mmj/1029004992. MR1278431

[16] P. L. Duren, *Theory of H^p spaces*, Pure and Applied Mathematics, Vol. 38, Academic Press, New York-London, 1970. MR268655

[17] P. Duren, D. Khavinson, and H. S. Shapiro, *Extremal functions in invariant subspaces of Bergman spaces*, Illinois J. Math. **40** (1996), no. 2, 202–210. MR1398090

[18] P. Duren, D. Khavinson, H. S. Shapiro, and C. Sundberg, *Contractive zero-divisors in Bergman spaces*, Pacific J. Math. **157** (1993), no. 1, 37–56. MR1197044

[19] R. C. James, *Orthogonality and linear functionals in normed linear spaces*, Trans. Amer. Math. Soc. **61** (1947), 265–292, DOI 10.2307/1990220. MR21241

[20] S. Richter, *Invariant subspaces of the Dirichlet shift*, J. Reine Angew. Math. **386** (1988), 205–220, DOI 10.1515/crll.1988.386.205. MR936999

[21] D. Seco, *A characterization of Dirichlet-inner functions*, Complex Anal. Oper. Theory **13** (2019), no. 4, 1653–1659, DOI 10.1007/s11785-018-0786-5. MR3957006

[22] K. Seip, *Interpolation and sampling in spaces of analytic functions*, University Lecture Series, vol. 33, American Mathematical Society, Providence, RI, 2004, DOI 10.1090/ulect/033. MR2040080

[23] H. S. Shapiro and A. L. Shields, *On some interpolation problems for analytic functions*, Amer. J. Math. **83** (1961), 513–532, DOI 10.2307/2372892. MR133446

[24] S. A. Vinogradov, *On the interpolation and zeros of power series with a sequence of coefficients in l^p* (Russian), Dokl. Akad. Nauk SSSR **160** (1965), 263–266. MR176041

[25] S. A. Vinogradov, *The interpolation of power series whose sequence of coefficients is from l^p* (Russian), Funkcional. Anal. i Priložen. **1** (1967), no. 3, 83–85. MR219730

[26] S. A. Vinogradov and V. P. Khavin, *Free interpolation in H^∞ and in some other classes of functions. I.*, J. Soviet Math. **9** (1978), 137–171.

[27] S. A. Vinodgradov and V. P. Khavin, *Free interpolation in H^∞ and in some other classes of functions. II.*, J. Soviet Math. **14** (1980), 1027–1065.

Department of Mathematics and Statistics, Old Dominion University, Norfolk, Virginia 23529

Email address: `rcheng@odu.edu`

Department of Mathematics, Indiana University, Bloomington, Indiana 47405

Email address: `cfelder@iu.edu`

Contemporary Mathematics
Volume **799**, 2024
https://doi.org/10.1090/conm/799/16023

Kernel-summability methods and the Silverman–Toeplitz Theorem

Pierre-Olivier Parisé

ABSTRACT. We introduce kernel-summability methods in Banach spaces using vector-valued integrals and prove an analogue of the Silverman–Toeplitz Theorem. We also show that inclusion of kernel-summability methods for scalar-valued functions implies inclusion for vector-valued functions. This extends a previous result from Mashreghi, Ransford and the author. We then apply these abstract results to the summability of Taylor series of functions in a Banach space of holomorphic functions on the unit disk.

1. Introduction

A summability method, as defined by Boos [**1**], is a triplet $(A, c_A, \lim_A)$, where A is a linear function, c_A is called the summability domain, and $\lim_A$ is the new operation of limit induced by the mapping A. For instance, in the case of a matrix-summability method, the mapping A would be determined by an infinite matrix $(a_{m,n})_{m,n\geq 0}$ of complex numbers such that a sequence of complex numbers $(v_n)_{n\geq 0}$ is transformed into the new sequence

$$A(v_n)_{n\geq 0} := \Big(\sum_{n\geq 0} a_{m,n} v_n \Big)_{m\geq 0}.$$

The space c_A is the space of all sequences such that $A(v_n)_{n\geq 0}$ converges and $\lim_A$ is the new operation of limit defined by $\lim_A (v_n)_{n\geq 0} = \lim_{m\to 0} \sum_{n\geq 0} a_{m,n} v_n$. The Cesàro summability method is an example of a matrix-summability method. In Section 3, we introduce in more detail the terminology attached to this more recent point of view of the theory of summability.

An important and classical result in the theory of summability is the famous Silverman–Toeplitz Theorem (see [**5**, Theorem 1 and Theorem 2]). This result gives necessary and sufficient conditions for a matrix-summability method to preserve limits of convergent sequences of complex numbers. The matrix-summability methods that preserve limits of convergent sequences are called *regular*. The precise statement of the theorem is presented as Theorem 1.1.

2020 *Mathematics Subject Classification.* Primary 40C10, 40D05; Secondary 41A10, 46E20, 40J05.

The auther is supported by NSERC and FRQNT postdoctoral scholarships.

THEOREM 1.1 (Silverman–Toeplitz Theorem). *A matrix-summability method* $(A, c_A, \lim_A)$, *with* $A = (a_{m,n})_{m,n\geq 0}$ *and* $a_{m,n} \in \mathbb{C}$, *is regular if and only if the following conditions are satisfied:*

(1) $\sup_{m\geq 0} \sum_{n\geq 0} |a_{m,n}| < \infty$;
(2) $\lim_{m\to\infty} a_{m,n} = 0$, *for every integer* $n \geq 0$; *and*
(3) $\lim_{m\to\infty} \sum_{n\geq 0} a_{m,n} = 1$.

The aim of this paper is to explore generalizations and variants of the Silverman–Toeplitz Theorem.

In Section 4, we adapt this theorem to kernel-summability methods applied to vector-valued functions. Kernel-summability methods are summability methods given by a (Borel) measurable function $a : F \times E \to \mathbb{C}$ from two topological spaces E and F into $\mathbb{C}$. These methods are applied to functions $v : E \to X$, where X is a Banach space, in the following way:

$$\int_E a(r,t)v(t)\, d\mu(t), \tag{1.1}$$

where μ is a Borel measure on E. In Section 2, we give an introduction to vector-valued integrals, which is based on [**11**], to make sense of (1.1). Therefore, the idea behind the kernel-summability methods is to transform possibly divergent integrals into convergent integrals. Notice when $E = F = \mathbb{N}$, the set of natural numbers including 0, we recover classical matrix-summability methods.

The Silverman–Toeplitz Theorem is a result on inclusion of summability methods. Indeed, let $\mathcal{I}$ denote the summability method of strong convergence. This means that the operator in $\mathcal{I}$ is given by the identity $I(v) = v$, where $v : E \to X$ and $\lim_I$ is the standard convergence of sequences. Then the Silverman–Toeplitz Theorem says that the summability method of strong convergence is included in any regular kernel-summability method, meaning that if a function $v : E \to X$ converges to some vector $x \in X$, then the integral in (1.1) also converges to x. By adopting this point of view, we give a result of the same flavor as in Section 5 by replacing the summability of strong convergence by an arbitrary kernel-summability method. Loosely speaking (see Theorem 5.4 for the precise statement), we show that if one kernel-summability method $\mathcal{A}$ is included in another kernel-summability method $\mathcal{B}$ only when restricted to complex-valued functions, then $\mathcal{A}$ is also included in $\mathcal{B}$ when restricted to vector-valued functions. This extends [**15**, Theorem 5.1] to kernel-summability methods. The statement of Theorem 5.4 involves operators from a Banach space X into another Banach space Y. These operators enable the use of the Banach–Steinhauss Theorem. Is it possible to remove these operators in the statement of this theorem and still obtain the same conclusion? We partially answer this question in the context of reflexive Banach spaces by introducing the concept of *weak inclusion* of summability methods.

Originally, the inspiration behind the statement of Theorem 5.4 comes from the applications of summability theory to approximation in Banach spaces of holomorphic functions that were studied in [**15**, Section 6]. With this original motivation in mind, we present, in Section 6, general consequences of the results from Section 5 on the summability of Taylor series in Banach spaces of holomorphic functions in the unit disk.

2. Vector-valued integrals

Let $(E, \mathcal{T}_E)$ be a topological space. Throughout the paper, we assume that E is locally compact and Hausdorff. We will also always assume that E is not compact. Let X be a Banach space. The space of functions from E into X is denoted by $\mathcal{F}(E, X)$, which is a complex vector space with the usual addition and scalar multiplication of functions.

We also equip E with a Radon measure μ defined on the family of Borel sets, meaning that $\mu(K) < \infty$ for any compact set $K \subseteq E$. We will also suppose that E is σ-compact. As a consequence, we can find a sequence of compact sets[1] $(K_n)_{n\geq 0}$ such that $K_n \subset K_{n+1}$ and $\cup_n K_n = E$. In particular, this means that E is σ-finite. We denote by $L^1_{\mathbb{C}}(E)$ the space of integrable complex-valued functions on E and by $L^\infty_{\mathbb{C}}(E)$ the space of essentially bounded complex-valued functions on E. This section is based on work by Lewis [**11**].

2.1. Measurability. A function $s \in \mathcal{F}(E, X)$ is a *step function* if there are pairwise disjoint measurable sets E_1, E_2, ..., E_n of finite measure and vectors x_1, x_2, ..., x_n in X such that

$$s = \sum_{j=1}^{n} x_j \chi_{E_j},$$

where χ_{E_j} is the characteristic function of E_j.

DEFINITION 2.1. A function $v \in \mathcal{F}(E, X)$ is *strongly measurable* if there exists a sequence of step functions $(s_n)_{n\geq 0}$ and a measurable set Z of measure zero such that $\|s_n(t) - v(t)\|_X \to 0$ as $n \to \infty$ for any $t \in E\backslash Z$.

An important fact we will use later is the following: if $(v_n)_{n\geq 0}$ is a sequence of strongly measurable functions and if v is a function such that $v_n(t) \to v(t)$ as $n \to \infty$ μ-a.e. on E, then v is strongly measurable.

DEFINITION 2.2. A function $v \in \mathcal{F}(E, X)$ is *weakly measurable* if for any continuous linear functional $\phi \in X^*$, the function $\phi \circ v : E \to \mathbb{C}$ is measurable.

There is an important result connecting weak measurability to strong measurability. To state this connection, we need the notion of separably valued functions. A function $v : E \to X$ is *essentially separably valued* if there exist a measurable set Z of measure zero and a countable set $C \subset X$ such that $f(E\backslash Z) \subset \overline{C}$.

THEOREM 2.3 (Pettis Measurability Theorem). *A function $v : E \to X$ is strongly measurable if and only if the two following conditions are satisfied:*

(1) *v is weakly measurable; and*
(2) *v is essentially separably valued.*

In particular, if v is strongly measurable, then it is weakly measurable. We now prove, as a consequence of the Pettis Measurability Theorem, the strong measurability of continuous functions, which is interesting in its own right. We present it as a proposition for reference later.

[1]From now on, the symbol $\subset$ means strict inclusion of sets and $\subseteq$ means inclusion of sets, with the possibility of equality.

PROPOSITION 2.4. *If $v : E \to X$ is continuous, then it is strongly measurable.*

PROOF. Notice first that v must be weakly measurable. This comes from the fact that $\phi \circ v : E \to \mathbb{C}$ is a continuous function for every $\phi \in X^*$ and therefore measurable. Furthermore, if $K \subset E$ is a compact set, then $v(K)$ is (essentially) separably valued. To see this, for any given m, cover $v(K)$ by a finite union of balls of radius $1/m$ and centered at x_n^m, with $n = 1, 2, \ldots, N_m$. Then, let $C_m := \{x_1^m, \ldots, x_{N_m}^m\}$ and take $C = \cup_{m \geq 1} C_m$. The restriction $v|_K : K \to X$ is strongly measurable by the Pettis Measurability Theorem. Since we assumed that $E = \cup_n K_n$, where $(K_n)_{n \geq 0}$ is an increasing sequence of compact sets, we see that $v|_{K_n} \to v$ pointwise on E. Therefore, the function v is strongly measurable. □

2.2. Bochner integral. The integral of a step function $s = \sum_{j=1}^n x_j \chi_{E_j}$ is defined by

$$\int_E s(t)\, d\mu(t) := \sum_{j=1}^n \mu(E_j) x_j.$$

It does not depend on the representation of s. A strongly measurable function $v \in \mathcal{F}(E, X)$ is *Bochner–integrable* if there is a sequence of step functions $(s_n)_{n \geq 0}$ and a measurable set $Z \subset E$ of measure zero such that

(1) $s_n(t) \to v(t)$ as $n \to \infty$, for any $t \in E \backslash Z$;
(2) $t \mapsto \|v(t) - s_n(t)\|_X \in L^1_{\mathbb{C}}(E)$ for any integer $n \geq 1$ and

$$\lim_{n \to \infty} \int_E \|v(t) - s_n(t)\|_X \, d\mu = 0; \text{ and}$$

(3) the limit $\lim_{n \to \infty} \displaystyle\int_A s_n(t)\, d\mu$ exists for any Borel set A.

The *Bochner integral* of a Bochner–integrable function v is then defined by

$$\int_E v(t)\, d\mu(t) := \lim_{n \to \infty} \int_E s_n(t)\, d\mu(t).$$

We denote by $L^1(E, X)$ the space of Bochner–integrable functions on E. The Bochner integral satisfies the following properties.

PROPERTY 2.5.

(I) It is unique and linear.
(II) A strongly measurable function $v \in \mathcal{F}(E, X)$ is Bochner–integrable if and only if $t \mapsto \|v(t)\|_X \in L^1_{\mathbb{C}}(E)$.
(III) For any $v \in L^1(E, X)$, we have $\left\| \displaystyle\int_E v\, d\mu \right\|_X \leq \displaystyle\int_E \|v\|_X \, d\mu$.
(IV) If $T : X \to Y$ is a bounded linear operator, where Y is a Banach space, and $v \in L^1(E, X)$, then $t \mapsto T(v(t)) \in L^1(E, Y)$ and

$$T\Big(\int_E v(t)\, d\mu(t)\Big) = \int_E T(v(t))\, d\mu(t).$$

(V) If $f : E \to \mathbb{C}$ is in $L^1_{\mathbb{C}}(E)$ and if $v(t) = f(t)x$ with $x \in X$, then v is Bochner–integrable and

$$\int_E v(t)\, d\mu(t) = \Big(\int_E f(t)\, d\mu(t)\Big) x.$$

We will also make use of the space $L^\infty(E, X)$ of essentially bounded vector-valued functions on E.

2.3. Pettis integral. We introduce a weaker notion of integral, called in the literature the *Pettis integral.* For a weakly measurable function $v : E \to X$, we say that v is Pettis integrable if

(1) the function $\phi \circ v \in L^1_{\mathbb{C}}(E)$, for any continuous linear functional $\phi : X \to \mathbb{C}$; and
(2) for any Borel set $E' \subset E$, there exists a vector $I_{v,E'} \in X$ such that
$$\phi(I_{v,E'}) = \int_{E'} \phi(v)\, d\mu,$$
for any continuous linear functional $\phi : X \to \mathbb{C}$.

In this case, the vector $I_v := I_{v,E}$ is the *weak integral* of v on E and we also write, when the context is clear,
$$I_v = \int_E v(t)\, d\mu(t).$$
We also say that $v : E \to X$ is *weakly integrable* when the weak integral I_v exists. The weak integral enjoys the following properties.

PROPERTY 2.6.

(I) It is unique and linear, when it exists.
(II) If $T : X \to Y$ is a bounded linear operator, where Y is a Banach space, and if v is a weakly integrable function, then $t \mapsto T(v(t))$ is weakly integrable and
$$T\Big(\int_E v(t)\, d\mu(t)\Big) = \int_E T(v(t))\, d\mu(t).$$
(III) If v is Bochner–integrable, then v is weakly integrable and the Bochner integral is equal to the weak integral.

3. What is a summability method?

The main goal of summability theory is to transform a divergent sequence into a convergent one. In this section, we will extend this notion to transform functions into other functions which, hopefully, behave better than the original function. The notions and point of view adopted in this section are adapted from Boos' book [**1**]. We let X and Y be Banach spaces with norm $\|\cdot\|_X$ and $\|\cdot\|_Y$, respectively. Let $(E, \mathcal{T}_E)$, $(F, \mathcal{T}_F)$ be two noncompact, locally compact and Hausdorff topological spaces.

3.1. Convergence at infinity. To mimic the notion of convergence of sequences, we introduce the notion of *convergence at infinity.* A function $v \in \mathcal{F}(E, X)$ *converges at infinity* if there is an element $x \in X$ such that for any $\varepsilon > 0$, there is a compact set $K \subset E$ such that
$$\|v(t) - x\|_X < \varepsilon \quad (t \notin K).$$
We denote this by $\lim_{t\to\infty} v(t) = x$. The fact that E is not compact makes this definition nonvoid.

The set of all strongly measurable functions $v : E \to X$ that are locally bounded and converge at infinity is denoted by $c(E, X)$. A function $v \in \mathcal{F}(E, X)$ is *locally bounded* if for any compact set $K \subset E$, $\sup_{t \in K} \|v(t)\|_X < \infty$. We see immediately that $c(E, X) \subset L^\infty(E, X)$.

Here are two examples to illustrate the general theory of summability outlined in Section 3.2.

EXAMPLE 3.1. Let $E = \mathbb{N} := \{0, 1, 2, \ldots\}$ endowed with the topology $\mathcal{T} = \mathcal{P}(\mathbb{N})$, the family of all subsets of $\mathbb{N}$. In this case, since all functions $v : \mathbb{N} \to X$ are measurable, the set $c(E, X)$ is the set of all convergent sequences on $\mathbb{N}$ and the set $L^\infty(E, X)$ is the set of all bounded sequences on $\mathbb{N}$. The usual notation for these sets are $c(X)$ and $\ell^\infty(X)$.

EXAMPLE 3.2. Let $E = [0, R)$, with $0 < R \leq \infty$, endowed with the topology generated by the intervals of the form (a, b), $[0, b)$. In this case, the set $\mathcal{F}(E, X)$ is the set of functions $v : [0, R) \to X$. The space $c(E, X)$ is the set of all strongly measurable functions on E which have a limit at R. The set $L^\infty(E, X)$ is the set of essentially bounded functions on E.

3.2. Definition of a summability method. We define a summability method in the following way.

DEFINITION 3.3. A *summability method* is a triplet $\mathcal{A} = (A, c_A(X), \lim_A)$, where

(1) $A : D_A \to \mathcal{F}(F, Y)$ is a mapping, with $D_A \subseteq \mathcal{F}(E, X)$ a vector subspace called the *domain* of A;
(2) $c_A(X)$ is the *summability domain* defined by
$$c_A(X) := \big\{v \in \mathcal{F}(E, X) \,:\, v \in D_A \text{ and } \lim_{r \to \infty} A(v)(r) \text{ exists}\big\}; \text{ and}$$
(3) $\lim_A : c_A(X) \to X$ is the *summability operator* defined by $\lim_A := \lim \circ A|_{c_A(X)}$, where $\lim : c(E, X) \to X$ is the limit operator at infinity.

A function $v \in \mathcal{F}(E, X)$ is $\mathcal{A}$-*summable* (or summable by the method $\mathcal{A}$) if $v \in c_A(X)$. We say that a function $v \in \mathcal{F}(E, X)$ is $\mathcal{A}$-*summable to* some $x \in X$ if $\lim_A v(t) = x$. When the mapping A is linear on D_A, the summability operator is a linear mapping. We will assume for the rest of the paper that A is linear. Some authors call the mapping $\lim_A$ an *operator of limits* (see, for example, [**6**, **8**]).

We now give a list of classical summability methods to show how they are defined using Definition 3.3. We will also introduce three types of summability methods that will be our main focus in this paper.

3.3. Matrix-summability methods.

EXAMPLE 3.4. Let $E = F = \mathbb{N}$ and $\mathcal{T}_E = \mathcal{T}_F := \mathcal{P}(\mathbb{N})$. Then we denote an element $v \in \mathcal{F}(\mathbb{N}, X)$ as $(v_n)_{n \geq 0}$. We define the mapping $I : \mathcal{F}(\mathbb{N}, X) \to \mathcal{F}(\mathbb{N}, X)$ by
$$I(v) := (v_n)_{n \geq 0} \quad (v := (v_n)_{n \geq 0} \in \mathcal{F}(\mathbb{N}, X)).$$
We have $D_I = \mathcal{F}(\mathbb{N}, X)$ and
$$c_I(X) = \big\{(v_n)_{n \geq 0} \,:\, \lim_{n \to \infty} v_n \text{ exists}\big\} = c(X)$$
is the space of all convergent sequences in X. The triplet $(I, c_I(X), \lim_I)$ is the *strong convergence* in X.

EXAMPLE 3.5. Let E and F be as in the previous example. We define the mapping $C^0 : \mathcal{F}(\mathbb{N}, X) \to \mathcal{F}(\mathbb{N}, X)$ by

$$C^0(v) := \big(v_0 + v_1 + \cdots + v_n\big)_{n\geq 0} \quad \big(v := (v_n)_{n\geq 0} \in \mathcal{F}(\mathbb{N}, X)\big).$$

We have $D_{C^0}(X) = \mathcal{F}(\mathbb{N}, X)$, the subspace $c_{C_0}(X)$ is the space of convergent series and the summability operator $\lim_{C^0}$ is the value of the convergent series $\sum_{n\geq 0} x_n$. The summability method $(C^0, c_{C^0}(X), \lim_{C^0})$ is called the *method of summation of series*. We denote the summability domain $c_{C^0}(X)$ by $cs(X)$:

$$cs(X) := \Big\{(v_n)_{n\geq 0} \, : \, \sum_{n\geq 0} v_n \text{ converges}\Big\}.$$

We usually use the notation $(s_n)_{n\geq 0}$ for the sequence $C^0(v)$ and each s_n is called the n-partial sums of the series $\sum_{n\geq 0} x_n$.

EXAMPLE 3.6. Let E and F be as in the previous examples. We define the mapping $C : \mathcal{F}(\mathbb{N}, X) \to \mathcal{F}(\mathbb{N}, X)$ by

$$C(v) := \Big(\frac{v_0 + v_1 + \ldots + v_n}{n+1}\Big)_{n\geq 0} \quad \big(v := (v_n)_{n\geq 0} \in \mathcal{F}(\mathbb{N}, X)\big).$$

Each member of the sequence $C(v)$ is called the *Cesàro mean* (of order 1) of the sequence v and is denoted by $(\sigma_n)_{n\geq 0}$. We have $D_C = X^{\mathbb{N}}$ and

$$c_C(X) = \big\{(v_n)_{n\geq 0} \, : \, \lim \circ C(v_n)_{n\geq 0} = \lim_{n\to\infty} \sigma_n \text{ exists}\big\}.$$

The triplet $(C, c_C(X), \lim_C)$ is called the *Cesàro summability method of order* 1 or simply the *Cesàro method.*

There is a more general definition of the Cesàro method involving the generalized binomial coefficients. We will however not mention its definition here. We would rather only mention a surprising fact about the summability domain of the generalized Cesàro methods. A complete characterization of the summability domain of the Cesàro summability methods was only achieved recently by Leonetti [**9**]. This shows that, in general, it is hard to describe explicitly the summability domain of the mapping A in a summability method.

The previous examples regroup a special type of summability methods. They can be seen as multiplying an infinite matrix with the infinite sequence (v_n). For example, the mapping C in the Cesàro summability method can be rewritten as

$$C(v) = \begin{pmatrix} 1 & 0 & 0 & \cdots \\ 1/2 & 1/2 & 0 & \cdots \\ 1/3 & 1/3 & 1/3 & \cdots \\ \vdots & \vdots & \vdots & \ddots \end{pmatrix} \begin{pmatrix} v_0 \\ v_1 \\ v_2 \\ \vdots \end{pmatrix}.$$

This inspires Definition 3.7.

DEFINITION 3.7. Let $E = F = \mathbb{N}$. A summability method $(A, c_A(X), \lim_A)$ is called a *matrix-summability method* if there exist scalars $a_{m,n} \in \mathbb{C}$, $m, n \in \mathbb{N}$, such that the mapping A is given by

$$A(v) = \Big(\sum_{n\geq 0} a_{m,n} v_n\Big)_{n\geq 0} \quad \big(v := (v_n)_{n\geq 0} \in D_A\big).$$

In this case, we let $A = (a_{m,n})_{m,n\geq 0}$ be the infinite matrix with coefficients $a_{m,n}$. Hardy's book *Divergent series* [**5**] is the classical reference for matrix-summability methods.

3.4. Sequence-to-function summability methods. Alongside the Cesàro summability method, one of the well-known summability methods is the Abel summability method.

EXAMPLE 3.8. Let $E = \mathbb{N}$ with $\mathcal{T}_E = \mathcal{P}(\mathbb{N})$ and $F = [0, 1)$ with the topology $\mathcal{T}_F$ generated by the sets of the form $[0, b)$ and (a, b) $(0 < a \leq b < 1)$. Notice here that the limit at infinity becomes the limit at 1. For a sequence $v := (v_n)_{n\geq 0}$, we define the mapping $A_v^0 : [0, r) \to X$ by

$$A_v^0(r) := \sum_{n=0}^{\infty} (1-r)r^n v_n \quad (0 \leq r < 1)$$

if the series converges in X. The expression for $A_v^0(r)$ is called the *Abel mean* (of order 0) of the sequence v. We then define the mapping $A : D_{A^0} \subseteq \mathcal{F}(\mathbb{N}, X) \to \mathcal{F}([0, 1), X)$ by

$$A^0(v) := A_v^0 \quad (v \in D_{A^0}).$$

In general, we have $D_{A^0} \subset X^{\mathbb{N}}$ and

$$c_{A^0}(X) = \big\{v := (v_n)_{n\geq 0} \,:\, \lim \circ A^0(v) = \lim_{r\to 1^-} \sum_{n=0}^{\infty} (1-r)r^n v_n \text{ exists}\big\}.$$

The triplet $(A^0, c_{A^0}(X), \lim_{A^0})$ is called the *Abel summability method* (of order 0).

The Abel summability method is an example of a sequence-to-function summability method.

DEFINITION 3.9. Let $E = \mathbb{N}$ and $(F, \mathcal{T}_F)$ be a locally compact Hausdorff topological space which is noncompact. A summability method $(A, c_A(X), \lim_A)$ is called a *sequence-to-function* summability method if there is a sequence $(a_n)_{n\geq 0}$ of functions $a_n : F \to \mathbb{C}$ such that

$$A_v(r) = \sum_{n\geq 0} a_n(r) v_n \quad \big(v := (v_n)_{n\geq 0} \in D_A,\, r \in F\big).$$

3.5. Kernel-summability methods. Sequence-to-function summability methods were recently used in [**15**] to study the summability of Taylor series of holomorphic functions in the unit disk. One of them can be seen as a special type of kernel-summability method.

EXAMPLE 3.10. Let $E = F = [0, 1)$ and $\mathcal{T}_E = \mathcal{T}_F$ be the topology in the last example. We define the mapping $L : D_L \subseteq \mathcal{F}([0, 1), X) \to \mathcal{F}([0, 1), X)$ by

$$L(v)(r) := \frac{-1}{\log(1-r)} \int_0^r \frac{v(t)}{1-t}\, dt,$$

where $v \in D_L$. The subspace D_L is defined as

$$D_L = \big\{v \in \mathcal{F}([0, 1), X) \,:\, L(v)(r) \text{ exists for any } r \in [0, 1)\big\}$$

and the summability domain is

$$c_L(X) = \big\{v \in \mathcal{F}([0, 1), X) \,:\, \lim_{r\to 1^-} L(v)(r) \text{ exists}\big\}.$$

The triplet $(L, c_L(X), \lim_L)$ is called the *logarithmic summability method.* This summability method was introduced by Borwein [**2**].

Inspired by the previous example, we introduce the following type of summability methods.

DEFINITION 3.11. Let $(E, \mathcal{T}_E)$ and $(F, \mathcal{T}_F)$ be two locally compact Hausdorff topological spaces which are noncompact. Let μ be a Borel measure on E. A summability method $(A, c_A(X), \lim_A)$ is called a *kernel-summability method* if there exists a (Borel) measurable function $a : F \times E \to \mathbb{C}$ such that the mapping $A : D_A \subset \mathcal{F}(E, X) \to \mathcal{F}(F, X)$ is given by

$$A(v)(r) := \int_E a(r,t)v(t)\, d\mu(t),$$

where the weak integral exists for any $v \in D_A$ and $r \in F$. The function $a : F \times E \to \mathbb{C}$ is called the *summability kernel.*

We still need to introduce two more concepts to see where the Silverman–Toeplitz Theorem sits in the general theory of summability.

3.6. Conservative summability methods. As the Cesàro summability method has shown, it is hard to describe explicitly the domain of the mapping in a summability method. We can relax the goal and instead search for subspaces that are contained in the domain D_A. We can also try to describe the image $A(M)$ for some subspace $M \subseteq D_A$. To make these last statements more precise, we introduce the following concept.

DEFINITION 3.12. Let $\mathcal{A} := (A, c_A(X), \lim_A)$ be a kernel-summability method. Let $M \subseteq \mathcal{F}(E, X)$ and $N \subseteq \mathcal{F}(F, X)$ be two (vector) subspaces. The method $\mathcal{A}$ is called (M, N)*-conservative* if $M \subseteq D_A$ and $A(M) \subseteq N$. We denote by (M, N) the class of (M, N)-conservative summability method.

The first chapter of Maddox's memoir [**12**] has an exhaustive summary of the problem of describing the family (M, N) of matrix-summability methods for various choices of spaces M and N. The ones that are closely related to the Silverman–Toeplitz Theorem are the classes of bounded summability methods, and of conservative summability methods. In the case $X = \mathbb{C}$, they are usually defined as follows.

(1) $\mathcal{A}$ is *bounded* if $A \in (\ell^\infty(\mathbb{C}), \ell^\infty(\mathbb{C}))$.
(2) $\mathcal{A}$ is *conservative* if $A \in (c(X), c(Y))$.

These classes can be defined without the notion of measurability. For kernel-summability methods, the above definitions must be adapted to incorporate measurable functions.

DEFINITION 3.13. Let $\mathcal{A} = (A, c_A(X), \lim_A)$ be a kernel-summability method.

(1) $\mathcal{A}$ is *bounded* if $A \in (L^\infty(E, X), L^\infty(F, X))$.
(2) $\mathcal{A}$ is *conservative* if $A \in (c(E, X), c(F, X))$.

Since $c(E, X) \subseteq L^\infty(E, X)$, we then have that the class $(c(E, X), c(F, Y))$ is included in the class $(L^\infty(E, X), L^\infty(F, X))$. The strong convergence method, the Cesàro method, the Abel method and the logarithmic method are all conservative. The method of summation of series is however not conservative, nor bounded. Indeed, taking $x_n = x$, with $\|x\|_X = 1$, we have $s_n = nx$ which is not bounded.

We will mainly use these classes of summability methods to define regularity and scalar-regularity.

4. The Silverman–Toeplitz Theorem

From Cesàro's Lemma, we have that

$$c(X) \subset c_C(X) \quad \text{and} \quad \lim_{n\to\infty} \sigma_n = \lim_{n\to\infty} v_n \quad \big((v_n)_{n\geq 0} \in c(X)\big).$$

Also, from Abel's Theorem, we know that

$$c(X) \subset c_{A^0}(X) \quad \text{and} \quad \lim_{r\to 1^-} A_r^0(v_n)_{n\geq 0} = \lim_{n\to\infty} v_n \quad \big((v_n)_{n\geq 0} \in c(X)\big).$$

We say that the Cesàro and Abel methods are *regular.*

DEFINITION 4.1. Let $\mathcal{A} := (A, c_A(X), \lim_A)$ be a kernel-summability method. We say that $\mathcal{A}$ is *regular* if

(1) the method $\mathcal{A}$ is conservative; and
(2) the method $\mathcal{A}$ preserves the value of limits, that is, for any $v \in c(E, X)$, we have

$$\lim_A v(t) = \lim_{r\to\infty} A(v)(r) = \lim_{t\to\infty} v(t).$$

In the introduction, we stated the conditions for a matrix-summability method to be regular in the statement of the Silverman–Toeplitz Theorem. We will now present generalizations of this theorem starting with matrix-summability methods applied to vector-valued sequences.

4.1. Vector-valued matrix-summability methods. The Silverman–Toeplitz Theorem can be extended to matrix-summability methods applied to sequences of vectors in a Banach space X. The fact the conditions are necessary comes from the fact that $\mathbb{C}$ embeds isometrically in X via the transformation $\lambda \mapsto \lambda x$, for a unit vector x. Therefore, the summability method is regular for *scalar*-valued sequences and the Silverman–Toeplitz Theorem can be applied to conclude that the conditions are necessary. On the other hand, the fact that the conditions are sufficient comes from elementary estimates. A more general proof for kernel-summability methods will be given in Section 4.2.

The Silverman–Toeplitz Theorem was also extended by Robinson [**17**, Theorem VII] to matrix-summability methods where the entries are linear (not necessarily bounded) operators $A_{n,k} : X \to Y$, where X and Y are Banach spaces. In the statement of the next Theorem, the *group norm* of a row of linear operators $T = (T_k)_{k\geq 0}$ is defined by

$$\|T\| := \sup\Big\{ \Big\| \sum_{k=0}^{n} T_k x_k \Big\|_Y \ : n \in \mathbb{N} \text{ and } \|x_k\|_X \leq 1 \Big\}.$$

Also, the acronym "SOT" means the strong operator topology.

THEOREM 4.2 (Robinson's Theorem). *Let $(A, c_A(X), \lim_A)$ be a matrix-summability method where the matrix is given by $(A_{m,n})_{m,n\geq 0}$, with $A_{n,k} : X \to Y$ linear operators. The method $\mathcal{A}$ is regular if and only if there exists an integer $N \geq 0$ such that*

(1) $\sup_{m\geq 0} \|(A_{m,n})_{k\geq N}\| < \infty$*;*
(2) $\lim_{n\to\infty} A_{m,n} = 0$ *(SOT);*

(3) $\sum_{n=0}^{\infty} A_{m,n} = A_m$ *exists (SOT) for any* $n \geq 0$*; and*
(4) $\lim_{m\to\infty} A_m = I$ *(SOT).*

The group norm enjoys many useful properties (see [**12**, Proposition 2.3]). One of them is the following: if $T_k = a_k I$ where $I : X \to Y$ is the identity operator and $a_k \in \mathbb{C}$, then $\|(T_k)_{k\geq 0}\| = \sum_{k\geq 0} |a_k|$. The last property tells us that the group norm of a sequence of complex numbers is simply its ℓ^1-norm. Because of this last property of the group norm, we see how condition (1) in Theorem 4.2 (Robinson's Theorem) is the right generalization of condition (1) in Theorem 1.1 (Silverman–Toeplitz Theorem).

Robinson proved this result using a "gliding hump" argument adapted from Hardy's book *Divergent Series* [**5**]. There is a functional analytic proof of Robinson's result in Maddox's book (see [**12**, Theorem 4.2]).

Recently, Leonneti proved a generalized version of Robinson's result. Instead of using the standard lim operator of sequences to construct the $\lim_A$ in the triplet $(A, c_A(X), \lim_A)$, he replaced it by the notion of ideal-convergence. The word "ideal" refers to a family $\mathcal{I}$ of subsets of $\mathbb{N}$ with desirable properties. For more details on this recent result, the reader may consult [**10**]. Leonniti's proof uses a generalization of the Banach–Steinhaus Theorem using the notion of limit superior with respect to an ideal of nonnegative integers.

There is a version of the Silverman–Toeplitz Theorem for sequence-to-function summability methods $(A, c_A(X), \lim_A)$ given by a sequence of functions (a_n). For more details on that, see [**5**, Theorem 5] and [**15**, Theorem 2.1] when the functions $(a_n)_{n\geq 0}$ are scalar-valued and see [**17**, Theorem X] when the functions $(a_n)_{n\geq 0}$ are operator-valued.

4.2. Vector-valued kernel-summability methods. Because of the different ways the kernel function $a : F \times E \to \mathbb{C}$ may behave compared to the more predictable behavior of a kernel $a : \mathbb{N} \times E \to \mathbb{C}$ (matrix or sequence-to-function summability methods), Hardy [**5**, p. 50] mentioned that the expected conditions in the Silverman–Toeplitz Theorem are less "symmetric" than those in the other versions of the Silverman–Toeplitz Theorem. He therefore confined himself to state sufficient conditions. He does, however, work out potential necessary and sufficient conditions later in his book [**5**, pp. 61–63]. We present here an adaptation of these conditions for kernel-summability methods as defined in Definition 3.11. Recall that there is a sequence of compact sets $(K_n)_{n\geq 0}$ such that $E = \cup_{n\geq 0} K_n$, $K_n \subset K_{n+1}$, and $\mu(K_n) < \infty$.

THEOREM 4.3. *Let* $\mathcal{A} = (A, c_A(X), \lim_A)$ *be a kernel-summability method with measurable function* $a : F \times E \to \mathbb{C}$. *The method* $\mathcal{A}$ *is regular if and only if*

(1) *the map* $t \mapsto a(r,t)$ *is in* $L^1_{\mathbb{C}}(E)$ *for any* $r \in F$*;*
(2) *the map* $r \mapsto \int_E |a(r,t)|\, d\mu(t) \in L^\infty(F, \mathbb{C})$*;*
(3) *for any compact set* $K \subset E$, $\displaystyle\int_K |a(r,t)|\, d\mu(t) \to 0$ *as* $r \to \infty$*; and*
(4) $\displaystyle\lim_{r\to\infty} \int_E a(r,t)\, d\mu(t) = 1.$

PROOF. Suppose that the conditions (1)–(4) are satisfied. We have to show that $\mathcal{A}$ is regular. Therefore, we have to show the following:

(a) $c(E, X) \subset D_A$;

(b) $A(c(E,X)) \subset c(F,X)$; and
(c) $\lim_{r\to\infty} A_v(r) = \lim_{t\to\infty} v(t)$ for any $v \in c(E,X)$.

We start by showing (a). Let $v \in c(E,X)$. In particular, the function v is strongly measurable and bounded. Therefore, the function $t \mapsto a(r,t)v(t)$ is strongly measurable and the Bochner integral

$$A_r(v) := A(v)(r) = \int_E a(r,t)v(t)\, d\mu(t)$$

exists by condition (1) and by Property 2.5 (II). In particular, $v \in D_A$.

We now show (b), that is $A(c(E,X)) \subset c(F,X)$. By condition (2), the map $r \mapsto A_r(v)$ is strongly measurable. To show that $\lim_{r\to\infty} A_r(v)$ exists, we will show directly that (c) holds. With the previously introduced notation, we have to show that

$$\lim_{r\to\infty} A_r(v) = \lim_{t\to\infty} v(t),$$

for any $v \in c(E,X)$. Let $v \in c(E,X)$ and let $x := \lim_{t\to\infty} v(t)$. We have

$$\begin{aligned}\|A_r(v) - x\|_X &= \Big\| \int_E a(r,t)v(t)\, d\mu(t) - x \Big\|_X \\ &= \Big\| \int_E a(r,t)(v(t)-x)\, d\mu(t) + \Big(\int_E a(r,t)\, d\mu(t) - 1 \Big) x \Big\|_X .\end{aligned}$$

Let $\varepsilon > 0$. There is a compact set $K_0 \subset E$ such that $\|v(t) - x\|_X < \varepsilon$ for $t \notin K_0$. Then, we have

$$\begin{aligned}\Big\| \int_E a(r,t)(v(t)-x)\, d\mu(t) \Big\|_X &\le \Big\| \int_{K_0} a(r,t)(v(t)-x)\, d\mu(t) \Big\|_X \\ &\quad + \Big\| \int_{E\setminus K_0} a(r,t)(v(t)-x)\, d\mu(t) \Big\|_X \\ &\le \sup_{t\in K_0} \|v(t)-x\|_X \int_{K_0} |a(r,t)|\, d\mu(t) + M\varepsilon.\end{aligned}$$

Therefore, taking a compact set $K \subset F$ sufficiently big, we see that for any $r \notin K$, we have

$$\|A_r(v) - x\|_X < \sup_{t\in K_0} \|v(t)-x\|_X \varepsilon + M\varepsilon + \varepsilon\|x\|_X,$$

which means $\lim_A v(t) = \lim v(t)$.

We will now show that the four conditions are necessary. We first start by showing that condition (1) is necessary. Given a unit vector $x \in X$, the function $v(t) = x$, for every $t \in E$, belongs to $c(E,X)$. Therefore, since $c(E,X) \subset D_A$, the (weak) integral

$$\int_E a(r,t)x\, d\mu(t)$$

exist, for every $r \in F$. However, for a fixed $r \in F$, using the Hahn–Banach Theorem, we can find a continuous linear functional $\phi \in X^*$ such that $\phi(x) = \|x\|_X = 1$. Therefore, from the definition of the weak integral, we see that $\phi(a(r,t)x) = a(r,t)$ must be in $L^1_{\mathbb{C}}(E)$. Furthermore, we have $\lim_{t\to\infty} v(t) = x$ and therefore, from the regularity of $\mathcal{A}$,

$$\lim_{r\to\infty} \int_E a(r,t)v(t)\, d\mu(t) = x.$$

Using the fact that $\phi(x) = \|x\|_X = 1$, we conclude that

$$\lim_{r\to\infty} \int_E a(r,t)\, d\mu(t) = 1.$$

Thus, condition (4) is also necessary.

Notice that since we now know that $t \mapsto a(r,t) \in L^1_{\mathbb{C}}(E)$ for every $r \in F$, then the Bochner integral of $t \mapsto a(r,t)v(t)$ exists, for every $r \in F$ and every $v \in c(E,X)$.

To show that condition (2) is necessary, we will introduce some notation. Let $\|\cdot\|_\infty$ be the essential sup-norm on $L^\infty(E,X)$. Therefore, after identifying the functions that agree (μ)-a.e. on E, the space $L^\infty(E,X)$ becomes a Banach space and $c(E,X)$ becomes a Banach space under the same norm. We define the linear operator $A_r : c(E,X) \to X$ as follows:

$$A_r(v) := \int_E a(r,t)v(t)\, d\mu(t).$$

For each $r \in F$, the linear operator $A_r : c(E,X) \to X$ is bounded. Indeed, given $v \in c(E,X)$ and using Property 2.5 (III),

$$\|A_r(v)\|_X \leq \int_E |a(r,t)| \|v(t)\|_X\, d\mu(t) \leq \Big(\int_E |a(r,t)|\, d\mu(t) \Big) \|v\|_\infty.$$

For each $v \in c(E,X)$, the function $r \mapsto A_r(v)$ belongs to $c(F,X)$ because the method $\mathcal{A}$ is regular. Therefore, for each $v \in c(E,X)$, we have $\sup_{r\in F} \|A_r(v)\|_X < \infty$. By the Banach–Steinhaus Theorem, we conclude that

$$\sup_{r\in F} \|A_r\|_{c(E,X)\to X} < \infty.$$

It remains to compute the norm of A_r. Define the bounded linear operator $A_r^n : c(E,X) \to X$ as follows:

$$A_r^n(v) := \int_{K_n} a(r,t)v(t)\, d\mu(t).$$

Let $x \in X$ be a unit vector and let $E' := \{t \in E : a(r,t) \neq 0\}$. Define the function $v_n : E \to X$ by

$$v_n := \chi_{E'\cap K_n} a(r,\cdot)/\overline{a(r,\cdot)}. \tag{4.1}$$

Then we have $v_n \in c_\infty(E,X)$, $\|v_n\|_\infty = 1$, and moreover v_n is Bochner–integrable. Applying A_r^n and using Property 2.5 (V), we obtain

$$\|A_r^n(v_n)\|_X = \int_{K_n} |a(r,t)|\, d\mu(t) \|x\|_X = \int_{K_n} |a(r,t)|\, d\mu(t).$$

Therefore, the norm of A_r^n is the right-hand side of the last equality. We then obtain

$$\|A_r^n(v) - A_r(v)\|_X \leq \int_{E\setminus K_n} |a(r,t)|\, d\mu(t) \quad (\|v\|_\infty \leq 1)$$

and using the Lebesgue Dominated Convergence Theorem in $L^1_{\mathbb{C}}(E)$

$$\lim_{n\to\infty} \|A_r^n - A_r\|_{c(E,X)\to X} = 0.$$

Thus, we conclude that

$$\|A_r\|_{c(E,X)\to X} = \lim_{n\to\infty} \int_{K_n} |a(r,t)|\, d\mu(t) = \int_E |a(r,t)|\, d\mu(t).$$

Finally, to prove condition (3), define the function $v_K : E \to X$ as in (4.1), where $K \subset E$ is compact and $x \in X$ with $\|x\|_X = 1$. Since $v_K \in c(E, X)$ with $\lim_{t\to\infty} v_K(t) = 0$ and $\mathcal{A}$ is regular, we conclude that

$$\lim_{r\to\infty} \int_K |a(r,t)| x \, d\mu(t) = 0.$$

Applying Property 2.5 (V), we conclude that condition (3) must be satisfied and this ends the proof. □

5. Inclusion between summability methods

Regularity for a summability method $\mathcal{A} = (A, c_A(X), \lim_A)$ can also be expressed in the following way: The domain of the summability method of strong convergence is included in the domain of the mapping A and the restrictions of the operators lim and $\lim_A$ to the space $c(E, X)$ are equal. We generalize this idea of inclusion to two arbitrary summability methods.

DEFINITION 5.1. Let $\mathcal{A} := (A, c_A(X), \lim_A)$ and $\mathcal{B} := (B, c_B(X), \lim_B)$ be two summability methods and let $M \subset \mathcal{F}(E, X)$.

(1) The method $\mathcal{A}$ is *M-included* in the method $\mathcal{B}$ if the following conditions are satisfied:
  (a) $M \cap D_A \neq \emptyset$;
  (b) $M \cap D_A \subset D_B$; and
  (c) $M \cap c_A(X) \subseteq c_B(X)$ and $\lim_A v(t) = \lim_B v(t)$ for any $v \in c_A(X)$.
  We denote this by $A \subseteq_M B$.
(2) The methods $\mathcal{A}$ and $\mathcal{B}$ are *M-equivalent* if $A \subseteq_M B$ and $B \subseteq_M A$. In this case, we write $A \sim_M B$.

When $c_A(X) \subset M$, then $\mathcal{A}$ is M-included in $\mathcal{B}$ becomes simply $\mathcal{A}$ is *included* in the method $\mathcal{B}$ and we denote this by $A \subseteq B$. Also, when $c_A(X) \cup c_B(X) \subset M$, then $\mathcal{A}$ and $\mathcal{B}$ are M-equivalent becomes $\mathcal{A}$ and $\mathcal{B}$ are *equivalent* and we write $A \sim B$.

For $X = \mathbb{C}$, the Cesàro summability method $(C, c_C(\mathbb{C}), \lim_C)$ is included in the Abel summability method $(A^0, c_{A^0}(\mathbb{C}), \lim_{A^0})$. A reference for this result is [**1**, Theorem 3.6.11(a)]. The method of proof can be easily adapted to sequences in a Banach space. Therefore, we can claim that the Cesàro summability method $(C, c_C(X), \lim_C)$ is included in the Abel summability method $(A^0, c_{A^0}(X), \lim_{A^0})$.

For $X = \mathbb{C}$, the Cesàro summability method is equivalent to the Riesz summability method. We won't elaborate here on the Riesz summability method, but for more details, we refer to [**7**, **16**].

To deduce the Silverman–Toeplitz Theorem for the matrix-summability methods, we used the fact that the original Silverman–Toeplitz Theorem was true for matrix-summability methods applied to complex-valued sequences. Therefore, it seems that when a summability method is included in another for complex-valued sequences, then the first method is included in the second one for vector-valued sequences. This depends, however, strongly on the proof of the inclusion of one summability method into another one for scalar-valued sequences. In [**15**], the authors found a way to generalize a result about inclusion of summability methods valid for complex-valued sequences to vector-valued sequences. Before recalling this result, we introduce some vocabulary.

We denote by $M_x := \{\lambda x : \lambda \in \mathbb{C}\}$, where $x \in X$ is a unit vector. The set of complex numbers is isomorphic to M_x and the set $\mathcal{F}(E, M_x)$ is called the *space of scalar-valued functions on* E. This avoids inconsistencies in the next definition.

DEFINITION 5.2. Let $\mathcal{A} := (A, c_A(X), \lim_A)$ and $\mathcal{B} := (B, c_B(X), \lim_B)$ be two summability methods. We say that $\mathcal{A}$ is *scalar-included* in $\mathcal{B}$, denoted by $A \subseteq_{\mathbb{C}} B$ if there exists a unit vector $x \in X$ such that $\mathcal{A}$ is $\mathcal{F}(E, M_x)$-included in $\mathcal{B}$.

Definition 5.2 is simply Definition 5.1 with $M = \mathcal{F}(E, M_x)$. In the rest of this paper, since the kernel-summability methods are determined by complex-valued kernels, we will identify M_x with $\mathbb{C}$. We can now state the result obtained in [**15**] for sequence-to-function summability methods.

THEOREM 5.3 ([**15**, Theorem 5.1]). *Let $\mathcal{A}$ and $\mathcal{B}$ be two regular sequence-to-function summability methods. Let X and Y be Banach spaces, and let $S : X \to Y$ and $S_n : X \to Y$ $(n \geq 0)$ be bounded linear operators. Suppose that:*

(1) *$S_n(x) \to S(x)$ for all $x \in W$, where W is a dense subset of X;*
(2) *$(S_n(X))_{n \geq 0}$ is $\mathcal{A}$-summable to $S(x)$ for all $x \in X$; and*
(3) *$\mathcal{A}$ is scalar-included in $\mathcal{B}$.*

Then $(S_n(x))_{n \geq 0}$ is $\mathcal{B}$-summable to $S(x)$ for all $x \in X$.

We will recall an important step in the proof of this result to clarify why we are using the weak integral over the Bochner integral in the definition of a kernel-summability method. Denote by $b_n(r)$ the function defining the summability method $\mathcal{B}$ in the statement of the theorem. The proof is split into two parts: (a) to prove that the series $\sum_{n \geq 0} b_n(r) S_n(x)$ exists and (b) to prove that, as $r \to \infty$, $\sum_{n \geq 0} b_n(r) S_n(x)$ converges to $S(x)$. The reason to use the weak integral over the Bochner integral comes from the proof of part (a). To prove that the series is convergent, it is shown that the sequence of partial sums $(\sum_{n=0}^{N} b_n(r) S_n(x))_{N \geq 0}$ is Cauchy in Y. The limit y of this Cauchy sequence satisfies the properties of the weak integral of the function $n \mapsto b_n(r) S_n(x)$ on E, that is, for all $\phi \in X^*$,

$$\phi(y) = \sum_{n \geq 0} b_n(r) \phi(S_n(x)).$$

Unfortunately, the proof cannot be modified to show that $\sum_{n \geq 0} b_n(r) S_n(x)$ is in $\ell^1(\mathbb{N}, Y) := L^1(\mathbb{N}, Y)$, which would be the Bochner integral of the function $n \mapsto b_n(r) S_n(x)$. This is why the weak integral is used in the definition of the domain of a kernel-summability method.

We now generalize the last result to kernel-summability methods. Recall that E is σ-compact. We will also assume that F is metrizable and its Alexandrov compactification [**13**] denoted by F_∞ is metrizable. An example of an F with these properties is an interval $[0, R)$ for some $R \in (0, \infty]$.

THEOREM 5.4. *Let X and Y be two Banach spaces. Let $\mathcal{A} = (A, c_A(Y), \lim_A)$ and $\mathcal{B} = (B, c_B(Y), \lim_B)$ be two kernel-summability methods. Let $S_t : X \to Y$, $S : X \to Y$ be bounded linear operators for each $t \in E$. Assume that*

(1) *$t \mapsto S_t(x)$ is continuous for any $x \in X$;*
(2) *$\lim_{t \to \infty} S_t(w) = S(w)$ for any w in a dense subset W of X;*
(3) *$t \mapsto S_t(x)$ is $\mathcal{A}$-summable to $S(x)$ for any $x \in X$;*

(4) *$\mathcal{B}$ is regular; and*
(5) *$\mathcal{A}$ is scalar-included in $\mathcal{B}$.*

Then, for any $x \in X$, the function $t \mapsto S_t(x)$ is $\mathcal{B}$-summable to $S(x)$.

PROOF. Let $(K_n)_{n\geq 0}$ be the sequence of compact sets such that $K_n \subset K_{n+1}$ and $E = \cup_n K_n$. Let $a : F\times E \to \mathbb{C}$ and $b : F\times E \to \mathbb{C}$ be the two $F\times E$-measurable summability kernels associated with the methods $\mathcal{A}$ and $\mathcal{B}$, respectively.

To show that $t \mapsto S_t(x)$ is $\mathcal{B}$-summable for any $x \in X$, we have to show two things:

(a) the function $t \mapsto S_t(x) \in D_B$, where D_B is the domain of $\mathcal{B}$; and
(b) the function $t \mapsto S_t(x) \in c_B(Y)$ and

$$\lim_B S_t(x) = \lim_A S_t(x).$$

We first show part (a). We therefore have to show that the weak integral of $t \mapsto b(r,t)S_t(x)$ exists for every $r \in F$ and every $x \in X$. Let $r \in F$ and $x \in X$ be fixed. Since the map $t \mapsto S_t(x)$ is $\mathcal{A}$-convergent by condition (3), it implies that the weak integral

$$A_r(x) := \int_E a(r,t)S_t(x)\, d\mu(t)$$

exists. In particular, the map $t \mapsto a(r,t)\phi(S_t(x))$ is integrable for every $\phi \in Y^*$ and

$$\phi(A_r(x)) = \int_E a(r,t)\phi(S_t(x))\, d\mu(t).$$

Since $\mathcal{A}$ is scalar-included in $\mathcal{B}$, this means, in particular, that

$$\int_E b(r,t)\phi(S_t(x))\, d\mu(t)$$

exists. From Proposition 2.4, the function $t \mapsto S_t(x)$ is strongly measurable. Also, since $\mathcal{B}$ is regular, from Theorem 4.3, the map $t \mapsto b(r,t)$ belongs to $L^1_{\mathbb{C}}(E)$. Therefore, the map $t \mapsto b(r,t)S_t(x)$ is strongly measurable. Also, by continuity, the map $t \mapsto S_t(x)$ is bounded on each K_n and therefore the restriction of the map $t \mapsto a(r,t)S_t(x)$ to the compact set K_n is Bochner–integrable by Property 2.5 (II), that is, the Bochner integral

$$\int_{K_n} b(r,t)S_t(x)\, d\mu(t)$$

exists. By the Lebesgue Dominated Convergence Theorem (for $L^1_{\mathbb{C}}(E)$)

$$\begin{aligned}\int_E b(r,t)\phi(S_t(x))\, d\mu(t) &= \lim_{n\to\infty} \int_{K_n} b(r,t)\phi(S_t(x))\, d\mu(t) \\ &= \lim_{n\to\infty} \phi\Big(\int_{K_n} b(r,t)S_t(x)\, d\mu(t)\Big).\end{aligned}$$

and, in particular,

$$\sup_{n\geq 0} \Big|\phi\Big(\int_{K_n} b(t,r)S_t(x)\, d\mu(t)\Big)\Big| < \infty.$$

This is true for any $\phi \in Y^*$. Therefore, by the Banach–Steinhaus Theorem and the Hahn–Banach Theorem,

$$\sup_{n\geq 0}\Big\| \int_{K_n} b(r,t)S_t(x)\,d\mu(t)\Big\|_Y < \infty. \tag{5.1}$$

For any integer $n \geq 0$, define the linear operator $B_{r,n} : X \to Y$ by

$$B_{r,n}(x) := \int_{K_n} b(r,t)S_t(x)\,d\mu(t).$$

Then $B_{r,n}$ is well-defined and continuous. Indeed, by the assumption (1) and the fact that K_n is compact, for any $x \in X$,

$$\sup_{t\in K_n} \|S_t(x)\|_Y < \infty.$$

Therefore, the Banach–Steinhaus Theorem gives

$$M_n := \sup_{t\in K_n} \|S_t\|_{X\to Y} < \infty.$$

So, if $x \in X$, then

$$\|B_{r,n}(x)\|_Y \leq \int_{K_n} |b(r,t)|\|S_t(x)\|_Y\,d\mu(t) \leq M_n\Big(\int_E |b(r,t)|\,d\mu(t)\Big)\|x\|_X.$$

Now, using (5.1) and a third time the Banach–Steinhaus Theorem, we obtain

$$M := \sup_{n\geq 0} \|B_{n,r}\|_{X\to Y} < \infty. \tag{5.2}$$

Let $\varepsilon > 0$. Use the density of W in assumption (2) to choose a $w \in W$ such that $\|x - w\|_X < \varepsilon$ and such that $S_t(w) \to S(w)$ as $t \to \infty$. In particular, the map $t \mapsto S_t(w)$ is bounded on E. This last fact combined with the fact that the map $t \mapsto b(r,t)$ is integrable implies that the function $t \mapsto b(r,t)S_t(w)$ is Bochner–integrable on E. Therefore, the sequence $(B_{r,n}(w))_{n\geq 0}$ is a Cauchy sequence in Y. This means that there is an N such that whenever $m, n \geq N$,

$$\|B_{r,m}(w) - B_{r,n}(w)\|_Y < \varepsilon.$$

Therefore, we have

$$\begin{aligned}\|B_{r,m}(x) - B_{r,n}(x)\|_Y &\leq 2M\|x-w\|_X + \|B_{r,m}(w) - B_{r,n}(w)\|_Y\\ &\leq (2M+1)\varepsilon.\end{aligned}$$

From this, we conclude that $(B_{r,n}(x))_{n\geq 0}$ is a Cauchy sequence and since Y is complete, $\lim_{n\to\infty} B_{r,n}(x)$ exists, say it is $I^B(x)$. The vector $I^B(x)$ is the weak integral of the map $t \mapsto b(r,t)S_t(x)$ because, for any $\phi \in Y^*$, we have

$$\phi(I^B(x)) = \lim_{n\to\infty} \int_{K_n} \phi(b(r,t)S_t(x))\,d\mu(t) = \int_E \phi(b(r,t)S_t(x))\,d\mu(t).$$

This concludes the proof of part (a).

Let's turn our attention to part (b). Fix $x \in X$. We will show that

$$\lim_{r\to\infty} \int_E b(r,t)S_t(x)\,d\mu(t) = \lim_{r\to\infty} \int_E a(r,t)S_t(x)\,d\mu(t) = S(x).$$

For $r \in F$, define the linear operator $B_r : X \to Y$ by

$$B_r(x) = \int_E b(r,t)S_t(x)\,d\mu(t),$$

where the integral on the right-hand side is the weak integral $I^B(x)$ from part (a). The linear operator B_r is also continuous because $B_r(x) = \lim_{n\to\infty} B_{r,n}(x)$, where $B_{r,n}$ are the continuous linear operators defined in part (a). Using the characterization of limits in the metric space F_∞ in terms of sequences, let $r_n \to \infty$ be an arbitrary sequence converging to ∞ as $n \to \infty$. As a consequence of assumption (3), we have that $\phi(A_r(x)) \to \phi(S(x))$ as $r \to \infty$ for any $\phi \in Y^*$. Therefore, $\phi(B_{r_n}(x)) \to \phi(S(x))$ as $n \to \infty$ and, in particular,

$$\sup_{n\geq 0} |\phi(B_{r_n}(x))| < \infty.$$

Using the Banach–Steinhaus Theorem twice, we conclude that

$$M' := \sup_{n\geq 0} \|B_{r_n}\|_{X\to Y} < \infty.$$

Let $\varepsilon > 0$. Choose $w \in W$ so that $\|x - w\|_X < \varepsilon$ and $S_t(w) \to S(w)$. Since $\mathcal{B}$ is regular, $B_r(w) \to S(w)$, as $r \to \infty$. Therefore, choose N such that $\|B_{r_n}(w) - S(w)\|_Y < \varepsilon$ for any $n \geq N$. We then have, for $n \geq N$,

$$\begin{aligned}
\|B_{r_n}(x) - S(x)\|_Y &\leq \|B_{r_n}(x-w)\|_Y + \|S(x-w)\|_Y + \|B_{r_n}(w) - S(w)\|_Y \\
&\leq \|B_{r_n}\|_{X\to Y}\|x-w\|_X + \|S\|_{X\to Y}\|x-w\|_X + \varepsilon \\
&\leq (M' + \|S\|_{X\to Y} + 1)\varepsilon.
\end{aligned}$$

It follows that $\lim_{n\to\infty} B_{r_n}(x) = S(x)$. Since the sequence $(r_n)_{n\geq 0}$ was arbitrary, we conclude that $\lim_{r\to\infty} B_r(x) = S(x)$ as required. This ends the proof. □

If we put $E := \mathbb{N}$ with the discrete topology, then we recover [**15**, Theorem 5.1]. This comes from the fact that the first assumption is automatically satisfied when $E = \mathbb{N}$ and the fourth assumption is included in the prelude of the Theorem. Notice that in the previous result, we didn't assume $\mathcal{A}$ to be regular, in contrast to what was assumed in [**15**, Theorem 5.1]. Note also that we assumed implicitly that the methods $\mathcal{A}$ and $\mathcal{B}$ have the same parameter space F. In fact, the kernel functions defining $\mathcal{A}$ and $\mathcal{B}$ can have different parameters space in their first argument.

Furthermore, the last result requires the existence of a dense subspace W and uses a function with values in $B(X,Y)$, the Banach space of bounded linear operators from X into Y. We might ask if those assumptions can be dropped in the statement. As a careful inspection of the proof of Theorem 5.4 shows, one will finds out that the hypothesis (2) is crucial. Indeed, in the proof, we need this assumption to show that the weak integral defining $B_r(x)$ exists for each $r \in F$. Otherwise, the best conclusion that we can draw from this hypothesis is that the sequence $(B_{r,n}(x))_{n\geq 0}$ is weakly Cauchy in Y. In this situation, unfortunately, the weak limit might not exist. Examples of Banach spaces Y where weakly Cauchy sequences do not have limit are $L^\infty([0,1])$ and $A(\mathbb{D})$ (the disk Algebra). For reference, see [**3**, Example 4.5]. However, if more restrictions are added on the Banach space X, we can obtain the following result.

THEOREM 5.5. *Let X be a Banach space. Let $\mathcal{A}$, $\mathcal{B}$ be two kernel-summability methods. Assume further that*

(1) *$\mathcal{A}$ is scalar-included in $\mathcal{B}$;*
(2) *$\mathcal{B}$ is regular; and*
(3) *X is reflexive.*

*Then the method $\mathcal{A}$ is **weakly-included** in the method $\mathcal{B}$.*

In Theorem 5.5, a summability method $\mathcal{A}$ is weakly included in another summability method $\mathcal{B}$ if the following conditions are satisfied:

(a) $D_A \subset D_B$; and
(b) if $t \mapsto v(t)$ is $\mathcal{A}$-summable weakly to $x \in X$ as $t \to \infty$, then $t \mapsto v(t)$ is $\mathcal{B}$-summable weakly to x.

PROOF OF THEOREM 5.5. Let (K_n), F_∞, $a, b : F \times E \to \mathbb{C}$ be as in the preceding proof. To show that $\mathcal{A}$ is weakly-included in the method $\mathcal{B}$, we have to show that

(a) $D_A \subset D_B$; and
(b) if $\lim_A \phi(v(t)) = \phi(x)$ for every $\phi \in X^*$, then $\lim_B \phi(v(t)) = \phi(x)$.

We first prove part (a). Let $v \in D_A$. Then, this means that

$$A_r(v) = \int_E a(r,t)v(t)\, d\mu(t)$$

exists for every $r \in F$. From the properties of the integral, if $\phi \in X^*$, then

$$\phi(A_r(v)) = \int_E a(r,t)\phi(v(t))\, d\mu(t).$$

Since $\mathcal{A}$ is scalar-included in $\mathcal{B}$, we find, in particular, that the map $t \mapsto b(r,t)\phi(v(t))$ is integrable for every $r \in F$ and every $\phi \in X^*$.

Fix $r \in F$ for the moment. Define the linear operator $T_r : X^* \to L^1_{\mathbb{C}}(E)$ by

$$T_r(\phi) := b(r,t)\phi(v(t)) \quad (\phi \in X^*).$$

By the Closed-Graph Theorem, the operator T_r is bounded. Its adjoint is then also defined and $T_r^* : (L^1_{\mathbb{C}}(E))^* \to X^{**}$. Since E is a σ-finite measure space, its dual is isomorphic to $L^\infty_{\mathbb{C}}(E)$ and the isomorphism is given by

$$\psi(u) := \int_E u(t)w(t)\, d\mu(t) \quad (u \in L^1_{\mathbb{C}}(E),\, w \in L^\infty_{\mathbb{C}}(E))$$

with ψ in the dual of $L^1_{\mathbb{C}}(E)$ (see [**18**, Theorem 6.16]). The space X is also reflexive by assumption (3) and therefore $T_r^* : L^\infty_{\mathbb{C}}(E) \to X$, under the appropriate identification. The vector $T_r^*(\chi_E)$ is then the weak integral of $t \mapsto b(r,t)v(t)$ as one can check using the isomorphism of $(L^1_{\mathbb{C}}(E))^*$ with $L^\infty_{\mathbb{C}}(E)$ and the isomorphism of X with X^{**}.

Part (b) is immediate from the assumption (1). □

The above result may be sufficient in certain applications. For example, it can be applied to the convergence/divergence of Taylor series of holomorphic functions in Hilbert spaces as in [**14, 15**], instead of Theorem 5.4.

6. Consequences on summability in holomorphic Banach spaces

Recall that a Banach space $X \subset \mathrm{Hol}(\mathbb{D})$ is a *Banach holomorphic function space on the unit disk* if the inclusion map $\mathfrak{i} : X \to \mathrm{Hol}(\mathbb{D})$ is continuous. We will use the abbreviation BHFD to refer to these spaces. Examples of BHFD are the Hardy spaces H^p $(0 < p \leq \infty)$, the disk algebra, the Wiener algebra, the weighted Dirichlet spaces, Bloch spaces, the Bergman spaces A^p $(0 < p < \infty)$, the de Branges–Rovnyak spaces $\mathcal{H}(b)$ associated to an element b of the closed unit ball of H^∞, and more.

Suppose that the set of polynomials is dense in X, so in particular any polynomial is in X. We define maps $S_n : X \to X$ by

$$S_n(f)(z) := \sum_{k=0}^{n} a_k z^k \quad (f \in X,\, z \in \mathbb{D}).$$

Each S_n maps a function $f(z) = \sum_{m=0}^{\infty} a_m z^m$ in X to the n-th partial sums of the power series of f. Given a summability method $\mathcal{A} = (A, c_A(X), \lim_A)$, the Taylor series of a function $f \in X$ is $\mathcal{A}$-summable to some $g \in X$ if the sequence $(S_n(f))_{n\geq 0}$ is $\mathcal{A}$-summable to g.

COROLLARY 6.1. *Let X be a BHFD and assume the set of polynomials is dense in X. Let $\mathcal{A} = (A, c_A(X), \lim_A)$ and $\mathcal{B} = (B, c_B(X), \lim_B)$ be two sequence-to-function summability methods such that $\mathcal{A}$ is scalar-included in $\mathcal{B}$ and $\mathcal{B}$ is regular. If the Taylor series is $\mathcal{A}$-summable to f for every $f \in X$, then it is also $\mathcal{B}$-summable to f for every $f \in X$.*

PROOF. Let $E = \mathbb{N}$ and let F be as in Section 2. To obtain the result, we only need to verify the assumptions of Theorem 5.4.

Let W be the set of polynomials. We know that W is dense in X. Define $S(f) = f$, the identity operator on X. We first show that $S_n : X \to X$ is continuous for any $n \geq 0$. Fix $n \geq 0$ and let $f \in X$ with $f(z) = \sum_{m\geq 0} a_m z^m$. Since $a_m = \frac{f^{(m)}(0)}{m!}$, we have

$$\|S_n(f)\|_X \leq \sum_{k=0}^{n} \left|\frac{f^{(n)}(0)}{n!}\right| \|z^k\|_X.$$

The maps $D_k : \mathrm{Hol}(\mathbb{D}) \to \mathbb{C}$ defined by $D_k(f) := f^{(k)}(0)$ are continuous and linear. Since the inclusion $\mathfrak{i} : X \to \mathrm{Hol}(\mathbb{D})$ is continuous, the linear functionals $D_k|_X : X \to \mathbb{C}$ are continuous on X. Therefore there are constants C_k $(k \geq 0)$ such that

$$|f^{(k)}(0)| \leq C_k \|f\|_X.$$

From this last estimate, we obtain

$$\|S_n(f)\|_X \leq \Big(\sum_{k=0}^{n} \frac{C_k}{k!} \|z^k\|_X\Big) \|f\|_X$$

and thus $S_n : X \to X$ is continuous.

Secondly, for each polynomial p with $N := \deg p$, we have $S_n(p) = p$, whenever $n \geq N$. Therefore $\lim_{n\to\infty} S_n(p) = p$ for any polynomial p. Finally, the method $\mathcal{A}$ is scalar-included in the method $\mathcal{B}$. Assumptions (1), (2), (4) and (5) in Theorem 5.4 are then satisfied.

Suppose that the Taylor series of any $f \in X$ is $\mathcal{A}$-summable to f. Then the assumption (4) of Theorem 5.4 is satisfied and we can conclude that the Taylor series of any $f \in X$ is $\mathcal{B}$-summable to f. $\square$

We also have a similar result for the Abel means (or the radial dilates) modulo a certain additional assumption. For $0 < r < 1$, let $A_r : \mathrm{Hol}(\mathbb{D}) \to \mathrm{Hol}(\mathbb{D})$ be defined as follows:

$$A_r(f) := (1 - r) \sum_{m\geq 0} S_m(f) r^n \quad (f \in X).$$

This is a continuous linear map on X if we assume that $\limsup_{k\to\infty} \|z^k\|_X^{1/k} \leq 1$. We say f is $\mathcal{B}$-Abel summable to some $g \in X$ if $A_r(f)$ is $\mathcal{B}$-summable to g.

COROLLARY 6.2. *Let X be a BHFD and assume the set of polynomials is dense in X with* $\limsup_{k\to\infty} \|z^k\|_X^{1/k} \leq 1$. *Let $\mathcal{B} = (B, c_B(X), \lim_B)$ and $\mathcal{D} = (D, c_D(X), \lim_D)$ be two kernel-summability methods such that $\mathcal{B}$ is scalar-included in $\mathcal{D}$ and $\mathcal{D}$ is regular. If any $f \in X$ is $\mathcal{B}$-Abel summable to f, then for any $f \in X$, it is $\mathcal{D}$-Abel summable to f.*

PROOF. Since $\limsup_{k\to\infty} \|z^k\|_X \leq 1$, each map $A_r : X \to X$ is a bounded linear operator. The proof then follows the same scheme as the preceding proof. □

It is important to note that Corollary 6.2 does not apply to certain de Branges–Rovnyak spaces for which the set of polynomials is dense. This is mainly due to the fact that, for certain elements b of the closed unit ball of H^∞, the linear map A_r may not be bounded on $\mathcal{H}(b)$ (see [**4**]).

Notice that $A_r(f)$ can be replaced by any other summability method. For example, we can use the logarithmic summability methods and define

$$L_r(f)(z) := -\frac{1}{\log(1-r)} \int_0^r \frac{A_r(f)(z)}{1-t}\,dt.$$

In Corollary 6.2, the $\mathcal{B}$-Abel summability is then replaced by the $\mathcal{B}$-logarithmic summability of the power series of f.

Acknowledgments

The author would like to thank the anonymous referee for his careful reading of the manuscript. His comments greatly improved the presentation of the paper. The author also thanks Christopher Felder and William Verreault for their careful reading of an earlier version of this manuscript.

The results in Section 6 are a generalization of a collection of results which appeared in the author's PhD thesis (in French) that was not published anywhere. Therefore, the author would like to thank Thomas Ransford for pointing out a mistake in the proof of Corollary 6.1 while the author was in the process of writing his PhD thesis.

References

[1] J. Boos, *Classical and modern methods in summability*, Oxford Mathematical Monographs, Oxford University Press, Oxford, 2000. Assisted by Peter Cass; Oxford Science Publications. MR1817226

[2] D. Borwein, *On a scale of Abel-type summability methods*, Proc. Cambridge Philos. Soc. **53** (1957), 318–322. MR86158

[3] J. B. Conway, *A course in functional analysis*, 2nd ed., Graduate Texts in Mathematics, vol. 96, Springer-Verlag, New York, 1990. MR1070713

[4] O. El-Fallah, E. Fricain, K. Kellay, J. Mashreghi, and T. Ransford, *Constructive approximation in de Branges-Rovnyak spaces*, Constr. Approx. **44** (2016), no. 2, 269–281, DOI 10.1007/s00365-015-9312-4. MR3544000

[5] G. H. Hardy, *Divergent Series*, Oxford, at the Clarendon Press, 1949. MR30620

[6] G. F. Kangro, *Theory of summability of sequences and series*, J. Math. Sci. **5** (1976), 1–45.

[7] B. Kuttner, *On discontinuous Riesz means of type n*, J. London Math. Soc. **37** (1962), 354–364, DOI 10.1112/jlms/s1-37.1.354. MR142953

[8] F. León-Saavedra, M. del P. Romero de la Rosa, and A. Sala, *Schur lemma and uniform convergence of series through convergence methods*, Mathematics **8** (2020), no. 10, p. 11.

[9] P. Leonetti, *A characterization of Cesàro convergence*, Amer. Math. Monthly **128** (2021), no. 6, 559–562, DOI 10.1080/00029890.2021.1898876. MR4265483

[10] P. Leonetti, *Regular matrices of unbounded linear operators*, 2022, arXiv:2201.13059.

[11] A. D. Lewis, *Integrable and absolutely continuous vector-valued functions*, Rocky Mountain J. Math. **52** (2022), no. 3, 925–947, DOI 10.1216/rmj.2022.52.925. MR4441105

[12] I. J. Maddox, *Infinite matrices of operators*, Lecture Notes in Mathematics, vol. 786, Springer, Berlin, 1980. MR568707

[13] M. Mandelkern, *Metrization of the one-point compactification*, Proc. Amer. Math. Soc. **107** (1989), no. 4, 1111–1115, DOI 10.2307/2047675. MR991703

[14] J. Mashreghi, P.-O. Parisé, and T. Ransford, *Cesàro summability of Taylor series in weighted Dirichlet spaces*, Complex Anal. Oper. Theory **15** (2021), no. 1, Paper No. 7, 8, DOI 10.1007/s11785-020-01058-3. MR4179962

[15] J. Mashreghi, P.-O. Parisé, and T. Ransford, *Power-series summability methods in de Branges–Rovnyak spaces*, Integral Equations Operator Theory **94** (2022), no. 2, Paper No. 20, 17. MR4418300

[16] M. Riesz, *Sur L'Equivalence de Certaines Methodes de Sommation*, Proc. London Math. Soc. (2) **22** (1924), 412–419, DOI 10.1112/plms/s2-22.1.412. MR1575717

[17] A. Robinson, *On functional transformations and summability*, Proc. London Math. Soc. (2) **52** (1950), 132–160, DOI 10.1112/plms/s2-52.2.132. MR37371

[18] W. Rudin, *Real and complex analysis*, 3rd ed., McGraw-Hill Book Co., New York, 1987. MR924157

Department of Mathematics, University of Hawai'i at Mānoa, Honolulu, Hawai'i 96822.

Email address: parisepo@hawaii.edu

Contemporary Mathematics
Volume **799**, 2024
https://doi.org/10.1090/conm/799/16024

On Rhaly operators on l^2

George Popescu and Gabriel T. Prăjitură

Abstract. We study the class of Rhaly operators and discuss their property on l^2. In particular we studied when such operators are bounded, and offered estimates for the norm, when they are compact and they are hyponormal. We also discussed some of their spectral properties.

1. Introduction

Rhaly operators are one of the possible generalizations and abstractizations of the Cesàro operator.

As far as we know the name and the concept was introduced by Gerald Leibowitz in [**7**]. Rhaly himself, who initially studied only some particular forms, always called them terraced matrices. He still does.

Most of the properties of Rhaly operators come down to conditions about series or sequences of complex numbers and thus give the impression that they should be easy. Unfortunately they seem to be related to open problems about series and so really hard to solve. We restricted ourselves to the case of l^2, where all fundamental questions are still open. We tried to keep our approach as simple as possible.

2. Definition

Let (a_n) be a sequence of complex numbers. The Rhaly operator defined by this sequence is

$$R_{(a_n)} = \begin{pmatrix} a_1 & 0 & 0 & 0 & \dots \\ a_2 & a_2 & 0 & 0 & \dots \\ a_3 & a_3 & a_3 & 0 & \dots \\ a_4 & a_4 & a_4 & a_4 & \dots \\ \dots & & & & \end{pmatrix}.$$

It can be viewed as an operator on any sequence space but we will only discuss it on l^2 in this paper, not just because most of the questions are still open but also because we believe we know what the answers are.

In particular, when $a_n = \frac{1}{n}$ one gets the Cesàro operator that we will denote by C. For the properties of C we recommend [**3**] and [**15**].

2020 *Mathematics Subject Classification.* Primary 47A05, 47A10, 47B07, 47B20.

Key words and phrases. Rhaly operators, Cesàro operator, compact, hyponormal, spectrum.

This work was partially done while the second author was a Fulbright scholar visiting the Polytechnic University of Bucharest.

DEFINITION 2.1. We will call a sub–Cesàro operator a Rhaly operator for which a_n equals either $\frac{1}{n}$ or 0. We will denote such an operator by C_S where

$$S = \{n \geq 1 : a_n = 0\}.$$

3. Related matrices

It is good to keep in mind some of the related operators. Thus

$$R^*_{(a_n)} = \begin{pmatrix} \overline{a}_1 & \overline{a}_2 & \overline{a}_3 & \overline{a}_4 & \dots \\ 0 & \overline{a}_2 & \overline{a}_3 & \overline{a}_4 & \dots \\ 0 & 0 & \overline{a}_3 & \overline{a}_4 & \dots \\ 0 & 0 & 0 & \overline{a}_4 & \dots \\ \dots \end{pmatrix}$$

$$R_{(a_n)}R^*_{(a_n)} = \begin{pmatrix} |a_1|^2 & a_1\overline{a}_2 & a_1\overline{a}_3 & a_1\overline{a}_4 & \dots \\ \overline{a}_1 a_2 & 2|a_2|^2 & 2a_2\overline{a}_3 & 2a_2\overline{a}_4 & \dots \\ \overline{a}_1 a_3 & 2\overline{a}_2 a_3 & 3|a_3|^2 & 3a_3\overline{a}_4 & \dots \\ \overline{a}_1 a_4 & 2\overline{a}_2 a_4 & 3\overline{a}_3 a_4 & 4|a_4|^2 & \dots \\ \dots \end{pmatrix}$$

and

$$R^*_{(a_n)}R_{(a_n)} = \begin{pmatrix} \sum_{j=1}^{\infty}|a_j|^2 & \sum_{j=2}^{\infty}|a_j|^2 & \sum_{j=3}^{\infty}|a_j|^2 & \sum_{j=4}^{\infty}|a_j|^2 & \dots \\ \sum_{j=2}^{\infty}|a_j|^2 & \sum_{j=2}^{\infty}|a_j|^2 & \sum_{j=3}^{\infty}|a_j|^2 & \sum_{j=4}^{\infty}|a_j|^2 & \dots \\ \sum_{j=3}^{\infty}|a_j|^2 & \sum_{j=3}^{\infty}|a_j|^2 & \sum_{j=3}^{\infty}|a_j|^2 & \sum_{j=4}^{\infty}|a_j|^2 & \dots \\ \sum_{j=4}^{\infty}|a_j|^2 & \sum_{j=4}^{\infty}|a_j|^2 & \sum_{j=4}^{\infty}|a_j|^2 & \sum_{j=4}^{\infty}|a_j|^2 & \dots \\ \dots \end{pmatrix}.$$

Notice that the entries of $R^*_{(a_n)}R_{(a_n)}$ are infinite series whose convergence is implied by the boundedness of $R_{(a_n)}$. This and the boundedness of $R^*_{(a_n)}R_{(a_n)}$ will provide some necessary conditions for the boundedness of $R_{(a_n)}$. Moreover, notice that $R^*_{(a_n)}R_{(a_n)}$ is a reversed L shaped matrix.

Also of interest is to keep in mind how a Rhaly operator acts on a vector

$$\begin{pmatrix} a_1 & 0 & 0 & 0 & \dots \\ a_2 & a_2 & 0 & 0 & \dots \\ a_3 & a_3 & a_3 & 0 & \dots \\ a_4 & a_4 & a_4 & a_4 & \dots \\ \dots \end{pmatrix} \begin{pmatrix} x_1 \\ x_2 \\ x_3 \\ x_4 \\ \dots \end{pmatrix} = \begin{pmatrix} a_1 x_1 \\ a_2(x_1 + x_2) \\ a_3(x_1 + x_2 + x_3) \\ a_4(x_1 + x_2 + x_3 + x_4) \\ \dots \end{pmatrix}.$$

4. Factorizations

In general, if (b_n) is a sequence of complex numbers, we will denote by $D_{(b_n)}$ the diagonal operator of diagonal (b_n). Notice that

$$D_{(b_n)}R_{(a_n)} = R_{(b_n a_n)}.$$

Some particular forms of this equation are of relevance for some of the properties of the Rhaly operators.

With the convention that $\frac{|a_n|}{a_n} = 1$ and $\frac{a_n}{|a_n|} = 1$ when $a_n = 0$,

$$D_{(|a_n|/a_n)}R_{(a_n)} = R_{(|a_n|)} \quad \text{and} \quad D_{(a_n/|a_n|)}R_{(|a_n|)} = R_{(a_n)}.$$

This implies that in general it suffices to consider only the case of positive a_n.

Other particular relations of this type are

$$D_{(na_n)}C = R_{(a_n)} \quad \text{and} \quad D_{(1/na_n)}R_{(a_n)} = C.$$

where the second assumes that $a_n \neq 0$ for all $n \geq 1$.

It is also interesting to look at the product of two Rhaly operators.

$$R_{(b_n)}R_{(a_n)} = \begin{pmatrix} b_1a_1 & 0 & 0 & 0 & \dots \\ b_2(a_1+a_2) & b_2a_2 & 0 & 0 & \dots \\ b_3(a_1+a_2+a_3) & b_3(a_2+a_3) & b_3a_3 & 0 & \dots \\ b_4(a_1+a_2+a_3+a_4) & b_4(a_2+a_3+a_4) & b_4(a_3+a_4) & b_4a_4 & \dots \\ \dots \end{pmatrix}.$$

5. Boundedness

First we will show that it suffices to consider positive sequences.

PROPOSITION 5.1. *$R_{(a_n)}$ is bounded if and only if $R_{(|a_n|)}$ is bounded and when that is the case, $\|R_{(a_n)}\| = \|R_{(|a_n|)}\|$.*

PROOF. It is clear that if at least for one n we have $a_n \neq 0$, $\|D_{(a_n/|a_n|)}\| = \|D_{(|a_n|/a_n)}\| = 1$.

The factorizations $R_{(|a_n|)} = D_{(|a_n|/a_n)}R_{(a_n)}$ and $R_{(a_n)} = D_{(a_n/|a_n|)}R_{(|a_n|)}$ imply that if one operator is bounded, the other one is also bounded. Moreover, we have that

$$\|R_{(|a_n|)}\| \leq \|D_{(|a_n|/a_n)}\| \cdot \|R_{(a_n)}\| = \|R_{(a_n)}\|$$

and

$$\|R_{(a_n)}\| \leq \|D_{(a_n/|a_n|)}\| \cdot \|R_{(|a_n|)}\| = \|R_{(|a_n|)}\|.$$

□

Thus, from now on in this section, we will only consider positive sequences.

PROPOSITION 5.2. *Let (a_n) and (b_n) be two positive sequences such that there is some positive constant M with the property that $b_n \leq Ma_n$ for every $n \geq 1$. If $R_{(a_n)}$ is bounded, then $R_{(b_n)}$ is bounded and $\|R_{(b_n)}\| \leq M\|R_{(a_n)}\|$.*

PROOF. It is easy to see from the way a Rhaly operator acts on a vector that for every $x \in l^2$ we have

$$\|R_{(b_n)}x\| \leq M\|R_{(a_n)}x\|.$$

□

The next result summarizes some sufficient conditions for boundedness of Rhaly operators or for the lack of it.

THEOREM 5.3. *Let (a_n) be a positive sequence.*

(1) *If the sequence (na_n) is bounded, then the Rhaly operator is bounded and*

$$\|R_{(a_n)}\| \leq 2\sup_n na_n.$$

(2) *If the series $\sum_n na_n^2$ is convergent, then the Rhaly operator is bounded and*

$$\|R_{(a_n)}\| \leq \sqrt{\sum_n na_n^2}.$$

(3) *If $\lim_n na_n = \infty$, then the Rhaly operator is unbounded.*

PROOF.

(1) Since

$$R_{(a_n)} = D_{(na_n)}C,$$

the Rhaly operator is bounded being the product of two bounded operators.

Moreover,

$$\|R_{(a_n)}\| \leq \|D_{(na_n)}\| \cdot \|C\| = 2 \sup_n na_n.$$

Here we used the fact that $\|C\| = 2$. See [**3**, Theorem 2 part (1)].

(2) If $x = (x_n) \in l^2$, then

$$\|R_{(a_n)}\|^2 = \sum_n a_n^2 |x_1 + \dots x_n|^2 \leq \sum_n a_n^2 \cdot n(|x_1|^2 + \dots |x_n|^2) \leq \sum_n na_n^2 \|x\|^2$$
$$= \left(\sum_n na_n^2 \right) \|x\|^2$$

which implies the conclusion.

(3) Suppose that $R_{(a_n)}$ is bounded.

Since $\lim_n na_n = \infty$, the set $S = \{m; a_m = 0\}$ is finite.

Let

$$b_n = \begin{cases} \frac{1}{na_n} & \text{if} \quad a_n \neq 0 \\ 0 & \text{if} \quad a_n = 0. \end{cases}$$

Since $\lim_n b_n = 0$, the operator $D_{(b_n)}$ is compact, and because

$$D_{(b_n)} R_{(a_n)} = C_S,$$

so is C_S.

However, $C - C_S$ is a finite rank operator which implies that C is compact, which is a contradiction (for example because, by [**3**, Theorem 2 part (6)], the spectrum has nonempty interior). Therefore the Rhaly operator is unbounded.

□

REMARK 5.4. Neither (1) nor (2) of Theorem 5.3 are necessary for the Rhaly operator to be bounded.

It is easy to see that C fails the condition in (2), which, as we will see later, it is considerably restrictive.

Let $\alpha \in (1/2, 1)$ and $A \subset \mathbb{N}$, infinite, such that

$$\sum_{n \in A} \frac{1}{n^{2\alpha - 1}}$$

is convergent.

Let

$$a_n = \begin{cases} \frac{1}{n^\alpha} & \text{if} \quad n \in A \\ 0 & \text{if} \quad n \notin A. \end{cases}$$

Then $\limsup_n na_n = \lim_n n^{1-\alpha} = \infty$. However,

$$\sum_n na_n^2 = \sum_{n \in A} n \frac{1}{n^{2\alpha}} = \sum_{n \in A} \frac{1}{n^{2\alpha - 1}}$$

which is convergent. Therefore $R_{(a_n)}$ is bounded although the sequence (na_n) is not bounded. This shows that condition (1) in the theorem is not necessary for boundedness.

REMARK 5.5. It makes sense to look at the sequence (na_n) because in other spaces (c_0 and l^∞, for example) the Rhaly operator is bounded if and only if this sequence is bounded (see [7] for details). We saw that here it is not necessary, but the behavior of the sequence is still related to the boundedness of the operator.

We have a clear case in which the operator is bounded, (1), and another one in which the operator is unbounded, (3). In between there is the case of a sequence for which

$$\limsup_n na_n = \infty \quad \text{and} \quad \liminf_n na_n < \infty.$$

In this case we believe that if the operator is bounded it is possible to decompose the sequence into a disjoint union of two subsequences, $(l_n a_{l_n})$ and $(k_n a_{k_n})$ such that the first is bounded and the second has limit infinity. At a quick look this may seem obvious, but it is not always true that a sequence can be decomposed into two subsequences, one bounded and one with limit infinity. For example if we consider the sequence of positive rationals, any subsequence with limit infinity contains only a finite number of terms from every interval of the type $[n, n+1]$ and thus its complement is not bounded.

When this decomposition is possible, let $S = \{l_1, l_2, l_3, \dots\}$ and D a diagonal operator as in the proof of (3). The proof works the same way and if the Rhaly operator is bounded it implies that C_S is compact. Thus the relevance of the next question.

QUESTION 5.6. For which sets S of natural numbers is the operator C_S compact?

We will give a partial answer to this question in Section 6.

The next theorem summarizes some necessary conditions for the boundedness of the Rhaly operators.

THEOREM 5.7. *If the Rhaly operator $R_{(a_n)}$ is bounded, then*

(1)
$$\lim_{n\to\infty} a_n = 0.$$

(2)
$$\lim_{n\to\infty} a_n \left(1 + \frac{1}{2} + \cdots + \frac{1}{n}\right) = 0.$$

(3)
$$\lim_{n\to\infty} a_n \left(x_1 + x_2 + \cdots + x_n\right) = 0$$

for every vector $x = (x_1, x_2, x_3, \dots) \in l^2$.

(4) *The sequence (na_n^2) is bounded.*

(5) *The sequence (na_n) is bounded if the sequence (a_n) is decreasing.*

(6) *The sequence*
$$\frac{1}{n}(a_1 + 2a_2 + \cdots + na_n)$$
is bounded.

(7) *The series*
$$\sum_{n=1}^{\infty} a_n^2$$

is convergent and

$$\|R_{(a_n)}\| \geq \sqrt{\sum_{n=1}^{\infty} a_n^2}.$$

PROOF.

(1) This follows from the convergence of the series that are entries of $R^*_{(a_n)}R_{(a_n)}$.

(2) Let $x = (1, 1/2, 1/3, 1/4 \dots)$. If $R_{(a_n)}$ is bounded, then $R_{(a_n)}x$ is in l^2 and thus the series

$$\sum_n a_n^2 \left(1 + \frac{1}{2} + \dots + \frac{1}{n}\right)^2$$

is convergent.

(3) Same as (2). The property is relevant only if the vector is not in l^1, otherwise the result follows from (1).

(4) Let x be the vector with the first n entries equal to 1 and the rest equal to 0. Then $\|x\| = \sqrt{n}$.

We have $\|R_{(a_n)}x\| \leq \|R_{(a_n)}\| \cdot \|x\|$ which means

$$\sum_{k=1}^{n} k^2 a_k^2 + n^2 \sum_{k=n+1}^{\infty} a_k^2 \leq \|R_{(a_n)}\|^2 \cdot n.$$

In particular,

$$n^2 a_n^2 \leq \|R_{(a_n)}\|^2 \cdot n$$

which implies the result.

(5) Let x be as in (4). Since

$$a_n^2 \sum_{k=1}^{n} k^2 \leq \sum_{k=1}^{n} k^2 a_k^2,$$

we get

$$a_n^2 \frac{n(n+1)(2n+1)}{6} \leq \|R_{(a_n)}\|^2 \cdot n$$

which implies the result.

(6) Let x be as in (4). We will look at $CR_{(a_n)}x$, which is bounded.

$$\|CR_{(a_n)}x\|^2 = \sum_{k=1}^{n} \frac{1}{k^2}(a_1 + 2a_2 + \dots + ka_k)^2 + \sum_{k=n+1}^{\infty} \frac{1}{k^2}(a_1 + \dots + na_n + \dots + na_k)^2.$$

Therefore

$$\sum_{k=n+1}^{2n} \frac{1}{k^2}(a_1 + \dots + na_n + \dots + na_k)^2 \leq \|CR_{(a_n)}x\|^2 \leq \|CR_{(a_n)}\|^2 \cdot n.$$

For each $n+1 \leq k \leq 2n$,

$$\frac{1}{4n^2}(a_1 + 2a_2 + \dots + na_n)^2 \leq \frac{1}{k^2}(a_1 + \dots + na_n + \dots + na_k)^2$$

and thus

$$\frac{1}{4n}(a_1 + 2a_2 + \dots + na_n)^2 \leq \|CR_{(a_n)}\|^2 \cdot n$$

which implies the result.

(7) This is because

$$\|R^*_{(a_n)}R_{(a_n)}\| \geq \sum_{n=1}^{\infty} a_n^2.$$

□

Corollary 5.8. *If* (a_n) *is a decreasing sequence, then* $R_{(a_n)}$ *is bounded if and only if the sequence* (na_n) *is bounded.*

Remark 5.9. We believe that the boundedness of a Rhaly operator in the case when the sequence (na_n) is not bounded depends on the density in $\mathbb{N}$ of the sequence k_n as well as on the rate of growth to ∞ of the sequence $k_n a_{k_n}$. The next definition comes by analogy with the statistical convergence of the positive sequences.

Definition 5.10. We will call a positive sequence (x_n) statistically bounded if the sequence

$$\frac{x_1 + x_2 + \cdots + x_n}{n}$$

is bounded.

We will continue with what we believe characterizes the boundedness of *Rhaly* operators.

Conjecture 5.11. $R_{(a_n)}$ *is bounded if and only if the sequence* (na_n) *is statistically bounded.*

Notice that the only if part was proved in (6) of Theorem 5.7.

We will end this section with one more reduction of the problem of boundedness.

Remark 5.12. Let (na_n) be an unbounded sequence such that the corresponding Rhaly operator is bounded and the disjoint decomposition $(l_n a_{l_n})$ and $(k_n a_{k_n})$ as in Remark 5.5. We define

$$b_n = \begin{cases} a_{l_m} & \text{if there is } m \text{ such that } n = l_m \\ 0 & \text{if there is } m \text{ such that } n = k_m \end{cases}$$

and

$$c_n = \begin{cases} a_{k_m} & \text{if there is } m \text{ such that } n = k_m \\ 0 & \text{if there is } m \text{ such that } n = l_m. \end{cases}$$

Notice that $R_{(b_n)}$ is bounded and that

$$R_{(a_n)} = R_{(b_n)} + R_{(c_n)}.$$

Therefore $R_{(a_n)}$ is bounded if and only if $R_{(c_n)}$ is bounded.

6. A better estimation of the norm

Following [**7**] we will show how the method in [**3**] can be adjusted to Rhaly operators to provide, in some cases, a better estimation of the norm than in Theorem 5.3.

As before, we can restrict the computation to the case of positive sequences.

We assume that (na_n) is bounded.

Since

$$D_{(na_n)} - R_{(a_n)} = \begin{pmatrix} 0 & 0 & 0 & 0 & \dots \\ -a_2 & a_2 & 0 & 0 & \dots \\ -a_3 & -a_3 & 2a_3 & 0 & \dots \\ -a_4 & -a_4 & -a_4 & 3a_4 & \dots \\ \dots \end{pmatrix},$$

it is easy to check that

$$(D_{(na_n)} - R_{(a_n)})(D_{(na_n)} - R_{(a_n)})^* = D_{((n^2-n)a_n^2)}.$$

Then

$$\begin{aligned} \|D_{(na_n)} - R_{(a_n)}\|^2 &= \|(D_{(na_n)} - R_{(a_n)})(D_{(na_n)} - R_{(a_n)})^*\| = \|D_{((n^2-n)a_n^2)}\| \\ &= \sup_n (n^2 - n)a_n^2. \end{aligned}$$

Therefore

$$\|D_{(na_n)} - R_{(a_n)}\| = \sup_n \sqrt{n^2 - n}a_n$$

and thus

$$\|R_{(a_n)}\| \leq \|D_{(na_n)}\| + \|D_{(na_n)} - R_{(a_n)}\| = \sup_n na_n + \sup_n \sqrt{n^2 - n}a_n.$$

In the same time,

$$\begin{aligned} &(D_{(na_n)} - R_{(a_n)})(D_{(na_n)} - R_{(a_n)})^* = D_{((n^2-n)a_n^2)} \\ &\iff D^2_{(na_n)} + R_{(a_n)}R^*_{(a_n)} = D_{(na_n)}R^*_{(a_n)} + R_{(a_n)}D_{(na_n)} + D_{((n^2-n)a_n^2)}. \end{aligned}$$

Hence

$$\begin{aligned} \|D^2_{(na_n)} + R_{(a_n)}R^*_{(a_n)}\| &= \|D_{(na_n)}R^*_{(a_n)} + R_{(a_n)}D_{(na_n)} + D_{((n^2-n)a_n^2)}\| \\ &\leq \|D_{(na_n)}\| \cdot \|R^*_{(a_n)}\| + \|R_{(a_n)}\| \cdot \|D_{(na_n)}\| + \|D_{((n^2-n)a_n^2)}\| \\ &= 2\|R_{(a_n)}\| \cdot \|D_{(na_n)}\| + \|D_{((n^2-n)a_n^2)}\|. \end{aligned}$$

Since

$$\|D^2_{(na_n)}\| - \|R_{(a_n)}R^*_{(a_n)}\| \leq \|D^2_{(na_n)} + R_{(a_n)}R^*_{(a_n)}\|,$$

we conclude that

$$\begin{aligned} &\|D^2_{(na_n)}\| - \|R_{(a_n)}R^*_{(a_n)}\| \leq 2\|R_{(a_n)}\| \cdot \|D_{(na_n)}\| + \|D_{((n^2-n)a_n^2)}\| \\ &\iff \|D_{(na_n)}\|^2 - \|R_{(a_n)}\|^2 \leq 2\|R_{(a_n)}\| \cdot \|D_{(na_n)}\| + \|D_{((n^2-n)a_n^2)}\| \\ &\iff \|R_{(a_n)}\|^2 + 2\|R_{(a_n)}\| \cdot \|D_{(na_n)}\| + \|D_{((n^2-n)a_n^2)}\| - \|D_{(na_n)}\|^2 \geq 0. \end{aligned}$$

This is a quadratic inequality with positive discriminant having one positive root and one negative root. Therefore $\|R_{(a_n)}\|$ is greater than the positive root, which means that

$$\begin{aligned} \|R_{(a_n)}\| &\geq \sqrt{2\|D_{na_n}\|^2 - \|D_{((n^2-n)a_n^2)}\|} - \|D_{(na_n)}\| \\ &= \sqrt{2\sup_n n^2a_n^2 - \sup_n (n^2-n)a_n^2} - \sup_n na_n. \end{aligned}$$

Thus we have established the following.

THEOREM 6.1. *If the sequence (na_n) is bounded, then*

$$\sqrt{2\sup_n n^2a_n^2 - \sup_n (n^2-n)a_n^2} - \sup_n na_n \leq \|R_{(a_n)}\| \leq \sup_n na_n + \sup_n \sqrt{n^2-n}a_n.$$

REMARK 6.2. Of course, if $\sup_n na_n = \sup_n \sqrt{n^2 - n}a_n$, as it is the case for the Cesàro operator, Theorem 6.1 does not say anything new.

7. Compactness

The argument used in the proposition about boundedness can also be used to justify the following.

PROPOSITION 7.1. *$R_{(a_n)}$ is compact if and only if $R_{(|a_n|)}$ is compact.*

Thus we can restrict again the discussion to positive sequences.

We will continue with the only result where the characterization is complete. It is part of [**7**, Proposition 4.2].

PROPOSITION 7.2. *The Rhaly operator $R_{(a_n)}$ is Hilbert–Schmidt if and only if $\sum_n na_n^2$ is convergent.*

PROOF. It is known that an operator is Hilbert–Schmidt if and only if in its matrix representation the sum of the squares of the absolute values of its entries is finite. See [**4**, p. 267].

In our case this condition comes exactly to $\sum_n na_n^2$ to be convergent. □

REMARK 7.3. The condition (2) in Theorem 5.3 is way to powerful for boundedness because it is equivalent to Hilbert–Schmidt.

REMARK 7.4. The second example in Remark 5.4 is a Hilbert–Schmidt operator.

COROLLARY 7.5. *C_S is a Hilbert–Schmidt operator if and only if $\sum_{n \notin S} \frac{1}{n}$ is convergent.*

REMARK 7.6. There is a one-to-one correspondence between the Hilbert–Schmidt sub–Cesàro operators and the convergent subseries of the harmonic series.

We do not know a sub–Cesàro operator which is compact but not Hilbert–Schmidt. We have the following.

CONJECTURE 7.7. *A sub–Cesàro operator is either Hilbert–Schmidt or noncompact.*

There is another $0-1$ law regarding compactness and Cesàro related operators. Rhaly studied in [**12**] the operators $R_{(1/n^p)}$ for positive p and determined that for $p < 1$ they are unbounded, for $p = 1$ bounded noncompact (the Cesàro operator) while for $p > 1$ they are Hilbert–Schmidt. These facts can be easily checked using (3) of Theorem 5.3 and Proposition 7.2.

The next result summarizes the known facts about compactness.

THEOREM 7.8. *Let (a_n) be a positive sequence.*

(1) *If*
$$\lim_n na_n = 0,$$
then the Rhaly operator is compact.

(2) *If*
$$\liminf_n na_n > 0,$$
then the Rhaly operator is not compact.

(3) *If the Rhaly operator is compact, then*

$$\lim_n \sqrt{n} a_n = 0.$$

PROOF.

(1) Since $R_{(a_n)} = D_{(na_n)}C$ and the diagonal operator is compact the Rhaly operator is compact. Here we use the fact that a diagonal operator is compact if and only if its diagonal converges to 0 (see [**6**, Problem 171]).

(2) Suppose that $R_{(a_n)}$ is compact. We use the same factorization as in (3) of Theorem 5.3, except that this time the Rhaly operator is compact and the condition implies that the diagonal operator is bounded. It leads to the same contradiction.

(3) If the Rhaly operator is compact, then its adjoint is compact. Let e_n be the standard orthonormal basis of l^2. Since e_n converges weakly to 0, $R^*_{(a_n)}e_n$ converges to 0 in norm. The result follows from

$$\sqrt{n} a_n = \|R^*_{(a_n)}e_n\|.$$

□

REMARK 7.9. The condition in (1) is not necessary. The second example in Remark 5.4 shows that the operator may be even Hilbert–Schmidt without the sequence (na_n) to be at least bounded.

REMARK 7.10. There are compact non–Hilbert–Schmidt Rhaly operators. For example, if $a_1 = 1$ and for $n \geq 2$

$$a_n = \frac{1}{n\sqrt{\ln n}}.$$

This example comes from [**7**].

REMARK 7.11. Even though $lim_n na_n$ does not have to be 0 for the Rhaly operator to be compact, a subsequence of (na_n) must have limit 0. We believe that its complement must have density 0 for the operator to be compact. In other words, the sequence (na_n) converges to 0 statistically.

Thus we have the following.

CONJECTURE 7.12. $R_{(a_n)}$ *is compact if and only if*

$$\lim_n \frac{a_1 + 2a_2 + \cdots + na_n}{n} = 0.$$

Recall the classical Olivier Theorem: if $\sum_n x_n$ is convergent and x_n is a sequence decreasing to 0, then

$$\lim_n nx_n = 0.$$

A more general version of it was proved in [**16**] using the concept of statistical convergence.

PROPOSITION 7.13. *If for some* $k \geq 1$ *the series* $\sum_n n^{k-1}a_n^k$ *is convergent, then*

$$\lim_n \frac{a_1 + 2a_2 + \cdots + na_n}{n} = 0.$$

PROOF. Since $\sum_n n^{k-1}a_n^k$ is convergent by the statistical Olivier Theorem (see [**16**]), the sequence $n^k a_n^k$ converges to zero except for a rare subsequence (that is, a subsequence of the density 0 in the counting measure).

Therefore the sequence na_n converges to 0 except for a rare subsequence which implies the conclusion (see [**8**]). □

COROLLARY 7.14. *If the Rhaly operator is Hilbert–Schmidt, then*

$$\lim_n \frac{a_1 + 2a_2 + \cdots + na_n}{n} = 0.$$

REMARK 7.15. In the case of operators C_S this follows directly from the classical Olivier Theorem. If (k_n) is a subsequence of the sequence of natural numbers such that

$$\sum_n \frac{1}{k_n}$$

converges, since we have a decreasing sequence, we get that $\lim_n n\frac{1}{k_n} = 0$ which says exactly that the subsequence is rare.

Therefore a negative answer to Conjecture 7.7 implies a negative answer to Conjecture 7.12.

REMARK 7.16. Similar to the discussion at the end of Section 5, we can reduce the problem to the case in which the subsequence of (na_n) convergent to 0 is replaced by zeros.

8. Point spectrum

We will look first at $R_{(a_n)}$.

PROPOSITION 8.1. *If $R_{(a_n)}$ is a Rhaly operator, then*

$$\sigma_p(R_{(a_n)}) \subset \{a_1, a_2, a_3, \ldots\}.$$

PROOF. Let $\lambda \in \sigma_p(R_{(a_n)})$ and $x = (x_1, x_2, \ldots) \neq 0$ such that $R_{(a_n)}x = \lambda x$. This implies that

$$a_1x_1 = \lambda x_1$$

$$a_2x_1 + a_2x_2 = \lambda x_2$$

$$a_3x_1 + a_3x_2 + a_3x_3 = \lambda x_3$$

and so on.

Since $x \neq 0$, there is j such that $x_1 = x_2 = \ldots x_{j-1} = 0$ and $x_j \neq 0$.

Thus the jth equation becomes

$$a_jx_j = \lambda x_j$$

which implies that $\lambda = a_j$. □

REMARK 8.2. It is very hard to determine a condition to imply when some a_k is in fact an eigenvalue of the Rhaly operator, unless the sequence (a_n) is one-to-one. In this case we can solve the system in the previous proof and get a vector. Then a_k is an eigenvalue if and only if this vector is in l^2.

REMARK 8.3. The Cesàro operator does not have eigenvalues (this is [**3**, Theorem 2 part (3)]) but some Rhaly operators may have. Let S be the set of odd numbers. Then 0 is an eigenvalue of C_S and the corresponding eigenspace has infinite dimension. To see this notice that for every $n \geq 1$, the vector having the first $2n$ entries equal to 0 and after that $1, -1, 1/2, -1/2, 1/3, -1/3, \ldots$ is an eigenvector of C_S.

Now we will turn our attention to the adjoint.

PROPOSITION 8.4. $\lambda \in \sigma_p(R^*_{(a_n)}) \iff$ *there is a nonzero vector* $x = (x_1, x_2, x_3, \ldots) \in l^2$ *such that for every* n

$$\overline{a}_n x_n = \lambda(x_n - x_{n+1}).$$

PROOF. $R^*_{(a_n)}x = \lambda x$ implies that

$$\sum_{k=1}^{\infty} \overline{a}_k x_k = \lambda x_1$$

$$\sum_{k=2}^{\infty} \overline{a}_k x_k = \lambda x_2$$

$$\sum_{k=3}^{\infty} \overline{a}_k x_k = \lambda x_3$$

and so on.

Subtracting each sum from the previous one gives the relations in the statement.

For the reciprocal, adding all relations in the statement for $k \geq j$ gives

$$\sum_{k=j}^{\infty} \overline{a}_k x_k = \lambda x_j.$$

□

PROPOSITION 8.5. $\{\overline{a}_1, \overline{a}_2, \overline{a}_3, \ldots\} \subset \sigma_p(R^*_{(a_n)})$.

PROOF. We will just look at the initial system from the previous proof.

Let a_m be term of the sequence. We will show that $\overline{a}_m$ is an eigenvalue of $R^*_{(a_n)}$ by induction.

If $x = (1, 0, 0, 0, \ldots)$, it is obvious that $R^*_{(a_n)}x = a_1 x$.

Suppose that $\overline{a}_k$ are eigenvalues of $R^*_{(a_n)}$ for all $1 \leq k \leq m-1$. If there is such k with the property that $a_m = a_k$, then $\overline{a}_m$ is an eigenvalue.

If for all $1 \leq k \leq m-1$, $a_k \neq a_m$, we will consider x a vector for which $x_m = 1$ and $x_k = 0$ for all $k > m$. To show that $\overline{a}_m$ is an eigenvalue, notice that the original system in the previous proof is, in this case, equivalent to

$$\sum_{k=1}^{m} \overline{a}_k x_k = \overline{a}_m x_1$$

$$\sum_{k=2}^{m} \overline{a}_k x_k = \overline{a}_m x_2$$

$$\vdots$$

$$\overline{a}_{m-1} x_{m-1} + \overline{a}_m = \overline{a}_m x_{m-1}$$

or

$$(\overline{a}_1 - \overline{a}_m)x_1 + \sum_{k=2}^{m-1} \overline{a}_k x_k = -\overline{a}_m$$

$$(\overline{a}_2 - \overline{a}_m)x_2 + \sum_{k=3}^{m-1} \overline{a}_k x_k = -\overline{a}_m$$

$$\vdots$$

$$(\overline{a}_{m-1} - \overline{a}_m)x_{m-1} = -\overline{a}_m.$$

This is an upper diagonal system with nonzero entries on the main diagonal and thus it has solutions. □

COROLLARY 8.6. $\{a_1, a_2, a_3, \dots\} \subset \sigma(R_{(a_n)})$.

COROLLARY 8.7. $\|R_{(a_n)}\| \geq \sup_n |a_n|$.

QUESTION 8.8. If $a_n = \frac{i^n}{n}$, what is the spectrum of $R_{(a_n)}$?

COROLLARY 8.9. *If $R_{(a_n)}$ is a compact operator, then $\sigma(R_{(a_n)}) = \{0\} \cup \{a_1, a_2, a_3, \dots\}$.*

COROLLARY 8.10. *If $R_{(a_n)}$ is a compact operator and $a_m \neq 0$, then a_m is an eigenvalue of $R_{(a_n)}$.*

Next we will look into what other eigenvalues, if any, can the adjoint have.

PROPOSITION 8.11. $0 \in \sigma_p(R^*_{(a_n)}) \iff$ *there is j such that $a_j = 0$.*

PROOF. Let $x = (x_1, x_2, x_3, \dots) \in l^2$ be a nonzero vector such that $R^*_{(a_n)}x = 0$. Then, from the previous proposition,

$$\overline{a}_n x_n = 0$$

for every n. Since at least one x_n is nonzero, at least one a_n must be zero.

Conversely, if there is n such that $a_n = 0$ then let $x_n = 1$ and $x_k = 0$ for all $k \neq n$. It is easy to see that the vector x with these entries satisfies the condition $R^*_{(a_n)}x = 0$. □

PROPOSITION 8.12.

$$\lambda \notin \{0\} \cup \{\overline{a}_1, \overline{a}_2, \overline{a}_3, \dots\}$$

is an eigenvalue of $R^*_{(a_n)}$ if and only if the series

$$\sum_{n=1}^{\infty} \left| \Pi_{k=1}^{n} \left(1 - \frac{\overline{a}_k}{\lambda}\right) \right|^2$$

is convergent.

Moreover, when this is the case, the corresponding eigenspace has dimension 1.

PROOF. The condition in Proposition 8.4 can be rewritten as

$$x_{n+1} = \left(1 - \frac{\overline{a}_n}{\lambda}\right) x_n$$

which implies that

$$x_{n+1} = \Pi_{k=1}^{n} \left(1 - \frac{\overline{a}_k}{\lambda}\right)$$

and so

$$\|x\|^2 = |x_1|^2 \left(1 + \sum_{n=1}^{\infty} \left|\Pi_{k=1}^{n}\left(1 - \frac{\overline{a}_k}{\lambda}\right)\right|^2\right)$$

which implies the conclusion. □

REMARK 8.13. It is possible for $R^*_{(a_n)}$ to have eigenvalues other than a_m. For example, if $a_n = \frac{1}{n+1}$ and $x = (1, 1/2, 1/3, \dots)$, then $R^*_{(a_n)}x = x$.

REMARK 8.14. Let

$$\lambda \notin \{0\} \cup \{\overline{a}_1, \overline{a}_2, \overline{a}_3, \dots\}.$$

Then

$$\sum_{n=1}^{\infty} \left|\Pi_{k=1}^{n}\left(1 - \frac{\overline{a}_k}{\lambda}\right)\right|^2$$

is convergent if and only if for some $m \geq 1$

$$\sum_{n=m}^{\infty} \left|\Pi_{k=1}^{n}\left(1 - \frac{\overline{a}_k}{\lambda}\right)\right|^2$$

is convergent.

PROPOSITION 8.15. *Let (a_n) be any sequence of complex numbers.*

(1) *If λ is an eigenvalue for $R^*_{(a_n)}$, then $|\lambda|$ is an eigenvalue for $R^*_{|a_n|}$.*
(2) *If λ is an eigenvalue for $R^*_{(a_n)}$ and $0 < |\mu| < |\lambda|$, then $|\mu|$ is an eigenvalue for $R^*_{|a_n|}$.*

PROOF.

(1) If $|\lambda| = |a_m|$ for some m, then this follows from Proposition 8.5.
If $\lambda = 0$, then this follows from Proposition 8.11.
If $|\lambda \notin \{0\} \cup \{|a_1|, |a_2|, |a_3|, \dots\}$, then

$$\lambda \notin \{0\} \cup \{\overline{a}_1, \overline{a}_2, \overline{a}_3, \dots\}$$

and thus, by Proposition 8.12, the series

$$\sum_{n=1}^{\infty} \left|\Pi_{k=1}^{n}\left(1 - \frac{\overline{a}_k}{\lambda}\right)\right|^2$$

is convergent.

Since

$$\left|1 - \frac{|a_n|}{|\lambda|}\right| \leq \left|1 - \frac{\overline{a}_n}{\lambda}\right|,$$

it follows that

$$\sum_{n=1}^{\infty} \left|\Pi_{k=1}^{n}\left(1 - \frac{|a_k|}{|\lambda|}\right)\right|^2$$

is convergent and so, by Proposition 8.12, $|\lambda|$ is an eigenvalue of $R_{(|a_n|)}$.

(2) If $|\mu| = |a_m|$ for some m, then this follows from Proposition 8.5.
Otherwise, since $\lim_n a_n = 0$, there is m such that $\frac{|a_n|}{|\mu|} < 1$ for all $n \geq m$. Then

$$\left|1 - \frac{|a_n|}{|\mu|}\right| = 1 - \frac{|a_n|}{|\mu|} < 1 - \frac{|a_n|}{|\lambda|} = \left|1 - \frac{|a_n|}{|\lambda|}\right|,$$

and the result follows from part (1) and Proposition 8.12.

□

9. The spectrum

The approach in this section follows [**11**] but the sequences we are considering are way more general.

Besides what we showed in the previous section, there is not much known about the spectrum of a general Rhaly operator. However, we can be more specific if we assume that

$$\lim_n na_n = L$$

exists and is finite.

If $L = 0$, then the operator is compact, and its spectrum was determined in Section 8.

THEOREM 9.1.

(1) *If $L \neq 0$ and for every n, $a_n \neq 0$, then*

$$\begin{aligned}\{a_1, a_2, a_3, \dots\} \cup \{\lambda : |\lambda - \overline{L}| < |L|\} &\subset \sigma_p(R^*_{(a_n)}) \\ &\subset \{a_1, a_2, a_3, \dots\} \cup \{\lambda : |\lambda - \overline{L}| \leq |L|\} \setminus \{0\}.\end{aligned}$$

(2) *If $L \neq 0$ and for some n, $a_n = 0$, then*

$$\begin{aligned}\{a_1, a_2, a_3, \dots\} \cup \{\lambda : |\lambda - \overline{L}| < |L|\} &\subset \sigma_p(R^*_{(a_n)}) \\ &\subset \{a_1, a_2, a_3, \dots\} \cup \{\lambda : |\lambda - \overline{L}| \leq |L|\}.\end{aligned}$$

PROOF. We already know that in this case each a_n is an eigenvalue and that 0 is not in the first case and it is in the second.

We will try to determine the convergence of the series in Proposition 8.11.

We will use the Raabe test (see [**2**, p. 48]):

$$\begin{aligned}\lim_n n\left(1 - \frac{\left|\Pi_{k=1}^{n+1}\left(1 - \frac{\overline{a}_k}{\lambda}\right)\right|^2}{\left|\Pi_{k=1}^{n}\left(1 - \frac{\overline{a}_k}{\lambda}\right)\right|^2}\right) &= \lim_n n\left(1 - \left|1 - \frac{\overline{a}_{n+1}}{\lambda}\right|^2\right) \\ &= \lim_n n\left(1 - 1 + \frac{\overline{a}_{n+1}}{\lambda} + \frac{a_{n+1}}{\overline{\lambda}} - \frac{|a_{n+1}|^2}{|\lambda|^2}\right) \\ &= \lim_n \left(\frac{n\overline{a}_{n+1}}{\lambda} + \frac{na_{n+1}}{\overline{\lambda}} - \frac{n|a_{n+1}|^2}{|\lambda|^2}\right) = \frac{\overline{L}}{\lambda} + \frac{L}{\overline{\lambda}} \\ &= 1 + \left|\frac{\overline{L}}{\lambda}\right|^2 - \left|1 - \frac{\overline{L}}{\lambda}\right|^2 = 1 + \frac{1}{|\lambda|^2}(|\overline{L}|^2 - |\lambda - \overline{L}|^2).\end{aligned}$$

Therefore, whenever $|\lambda - \overline{L}| < |\overline{L}|$, the limit is > 1 and the series is convergent; and whenever $|\lambda - \overline{L}| > |\overline{L}|$, the limit is < 1 and the series is divergent.

For the case when the limit is 1, the Raabe test is inconclusive. □

REMARK 9.2. In the case of the points on the circle, that is, when

$$\frac{\overline{L}}{\lambda} + \frac{L}{\overline{\lambda}} = 1,$$

they may or may not be eigenvalues for the adjoint of the Rhaly operator.

To see this, we will use a strengthening of Raabe test (see [**9**, p. 396])

We are computing

$$\lim_n \ln n \left(1 - n\left(1 - \frac{\left|\Pi_{k=1}^{n+1}\left(1-\frac{\overline{a}_k}{\lambda}\right)\right|^2}{\left|\Pi_{k=1}^{n}\left(1-\frac{\overline{a}_k}{\lambda}\right)\right|^2}\right)\right)$$

$$= \lim_n \ln n \left(\frac{n\overline{a}_{n+1}}{\lambda} + \frac{na_{n+1}}{\overline{\lambda}} - \frac{n|a_{n+1}|^2}{|\lambda|^2} - 1\right)$$

$$= \lim_n \ln n \left(\frac{n\overline{a}_{n+1}}{\lambda} + \frac{na_{n+1}}{\overline{\lambda}} - 1\right) - \lim_n \frac{\ln n}{n}\frac{n^2|a_n|^2}{|\lambda|^2}$$

$$= \lim_n \ln n \left(\frac{n\overline{a}_{n+1}}{\lambda} + \frac{na_{n+1}}{\overline{\lambda}} - \frac{\overline{L}}{\lambda} - \frac{L}{\overline{\lambda}}\right)$$

$$= \lim_n \ln n \left(\frac{n\overline{a}_{n+1} - \overline{L}}{\lambda} + \frac{na_{n+1} - L}{\overline{\lambda}}\right).$$

Let $p > 0$ and the sequence $a_1 = a_2 = 1$ and for $n \geq 3$,

$$a_n = \frac{1}{n-1}\left(\frac{p}{\ln n} + L\right).$$

Then $lim_n na_n = L$ and

$$\lim_n \ln n \left(\frac{n\overline{a}_{n+1} - \overline{L}}{\lambda} + \frac{na_{n+1} - L}{\overline{\lambda}}\right) = p\left(\frac{1}{\lambda} + \frac{1}{\overline{\lambda}}\right).$$

If this limit is greater than 1, then the series is divergent and so the points on the circumference are not eigenvalues; while if this limit is < 1, the series is convergent and the points on the circumference are eigenvalues, except possibly for 0.

REMARK 9.3. According to [**11**], the previous example is due to Gary Jones, but no reference was given and we could not find one.

COROLLARY 9.4. *If $\lim_n na_n$ exists, is finite, and equals $L \neq 0$, then the closed disk of radius $|L|$ centered at L is included in the spectrum of the corresponding Rhaly operator. When this is the case, the spectral radius of the Rhaly operator is $\geq 2|L|$ and thus $\|R_{(a_n)}\| \geq 2|L|$.*

COROLLARY 9.5. *If $\lim_n na_n$ exists, is finite, and equals $L \neq 0$ and if $n|a_n| \leq |L|$, then $\|R_{(a_n)}\| = 2|L|$.*

We will return now the possible eigenvalues of a Rhaly operator.

THEOREM 9.6. *Let $R = R_{(a_n)}$.*

(1) *If $|a_k - L| > |L|$, then $a_k \in \sigma_p(R)$.*
(2) *If a_k is an eigenvalue of R, then $|a_n - L| \leq |L|$.*

PROOF. By Proposition 8.1, if λ is an eigenvalue of R, then there is k such that $\lambda = a_k$. Moreover, any eigenvector must satisfy $x_1 = x_2 = \ldots x_{k-1} = 0$ and, for all $n \geq k$,

$$a_n(x_k + x_{k+1} + \cdots + x_n) = a_k x_n.$$

Notice that $\lim_n na_n = L \neq 0$ implies that $a_n = 0$ for at most a finite number of indices. Thus there is m such that $a_n \neq 0$ for all $n \geq m$. Then, for $n \geq \max\{k, m\}$ the condition above is equivalent to

$$x_k + x_{k+1} + \cdots + x_n = \frac{a_k}{a_n}x_n.$$

Subtracting each such such relation from the next one we get

$$x_{n+1} = \frac{a_k}{a_{n+1}}x_{n+1} - \frac{a_k}{a_n}x_n \iff \frac{x_n}{x_{n+1}} = \frac{a_n(a_k - a_{n+1})}{a_k a_{n+1}}.$$

We need to see if the vector thus determined is in l^2.

Let $p_n = 1/(n|a_n|^2)$. We will use Raabe test (see [**9**, p. 395]).

$$\begin{aligned}
\lim_n &\left(p_n \frac{|x_n|^2}{|x_{n+1}|^2} - p_{n+1}\right) \\
&= \lim_n \left(\frac{1}{n|a_n|^2}\frac{|a_n|^2|a_k - a_{n+1}|^2}{|a_k|^2|a_{n+1}|^2} - \frac{1}{(n+1)|a_{n+1}|^2}\right) \\
&= \lim_n \frac{1}{n^2|a_{n+1}|^2|a_k|^2}\left(n|a_k - a_{n+1}|^2 - \frac{n^2}{n+1}|a_k|^2)\right) \\
&= \frac{1}{|L|^2 \cdot |a_k|^2} \lim_n \left(\frac{n}{n+1}|a_k|^2 - n a_k \overline{a}_{n+1} - n\overline{a}_k a_{n+1} + n|a_{n+1}|^2\right) \\
&= \frac{1}{|L|^2 \cdot |a_k|^2}(|a_k|^2 - a_k\overline{L} - \overline{a}_k L) = \frac{1}{|L|^2 \cdot |a_k|^2}(|a_k - L|^2 - |L|^2).
\end{aligned}$$

Raabe test implies now the conclusion. Notice that the second part requires the divergence of the series $\sum_n n|a_n|^2$, which is true in this case because $\lim_n na_n = L \neq 0$ implies that the operator is not compact and hence not Hilbert–Schmidt; thus, the series is divergent. □

THEOREM 9.7. *If $lim_n na_n = L \neq 0$ and $R = R_{(a_n)}$, then*

$$\sigma(R) = \{\lambda : |\lambda - L| \leq |L|\} \cup \{a_1, a_2, a_3, \dots\}.$$

PROOF. We already know that the right hand side is included in the left-hand side.

Let $\varepsilon > 0$ and K such that for $k \geq K$ we have $(k+1)|a_k| < |L| + \varepsilon$.

Let D_K be the diagonal matrix with the first K diagonal entries equal to 0 and the rest equal to na_n. We call $R_K = D_K C$ which is $R_{(a_n)}$ with the first K rows replaced by zeros.

Let L_K be the first K diagonal entries are equal to 0 and the remaining diagonal entries are equal to L.

Let $\lambda \in \sigma(R) \setminus \sigma_p(R)$. Then $L - \lambda \in \sigma(L - R) \setminus \sigma_p(L - R)$.

Notice that $(L - R) - (L_K - R_K)$ is a compact operator and thus, by Weyl's Theorem (see [**6**, Problem 82]), $L - \lambda \in \sigma(L_K - R_K)$. Also, recall from [**3**] that $\|I - C\| = 1$.

Therefore

$$\begin{aligned}
|L - \lambda| &\leq \|L_K - R_K\| \leq \|L_K - D_K\| + \|D_K - R_K\| \\
&= \|L_K - D_K\| + \|D_K(I - C)\| < \varepsilon + \|D_K\| \cdot \|I - C\| < |L| + 2\varepsilon.
\end{aligned}$$

Since this is true for every $\varepsilon > 0$, it follows that $|L - \lambda| \leq |L|$ which completes the proof. □

COROLLARY 9.8. *Under the same hypothesis, the spectral radius of R is*

$$r(R) = \sup\{2|L|, |a_1|, |a_2|, |a_3|, \dots\}.$$

10. Hyponormality

If a Rhaly operator is normal, then, by the equality of the $(1,1)$ entries of $R_{(a_n)}R^*_{(a_n)}$ and $R^*_{(a_n)}R_{(a_n)}$, we get that

$$|a_1|^2 = \sum_{j=1}^{\infty} |a_j|^2.$$

Thus it is easy to see that a Rhaly operator is normal if and only if $a_n = 0$ for all $n \geq 2$.

It is by far more interesting to see when such an operator is hyponormal or subnormal. Unfortunately the known conditions are very restrictive.

We will assume that the Rhaly operators are bounded. Recall that an operator T is hyponormal if $T^*T - TT^* \geq 0$.

THEOREM 10.1. *If $R_{(a_n)}$ is hyponormal, then*

(1) *For every n, $\sum_{j=n+1}^{\infty} |a_j|^2 \geq (n-1)|a_n|^2$.*
(2) *The sequence $(|a_n|)$ is decreasing.*
(3) *Either $a_n = 0$ for every $n \geq 2$ or $a_n \neq 0$ for every $n \geq 2$ and*

$$\lim_n \frac{|a_{n+1}|}{|a_n|} = 1.$$

(4) *If the sequence $(a_n) \subset \mathbb{R}$, then all numbers have the same sign.*

PROOF. Let (e_n) be the standard orthonormal basis of l^2. Suppose that $R_{(a_n)}$ is hyponormal.

(1)

$$\begin{aligned}\left\langle \left(R^*_{(a_n)}R_{(a_n)} - R_{(a_n)}R^*_{(a_n)}\right) e_n, e_n \right\rangle \geq 0 &\iff \sum_{j=n}^{\infty} |a_j|^2 - n|a_n|^2 \geq 0 \\ &\iff \sum_{j=n+1}^{\infty} |a_j|^2 \geq (n-1)|a_n|^2\end{aligned}$$

(2)

$$\begin{aligned}&\left\langle \left(R^*_{(a_n)}R_{(a_n)} - R_{(a_n)}R^*_{(a_n)}\right) (e_n - e_{n+1}), e_n - e_{n+1} \right\rangle \geq 0 \\ &\iff n(a_n\overline{a}_{n+1} + \overline{a}_n a_{n+1}) - (n-1)|a_n|^2 - (n+1)|a_{n+1}|^2 \geq 0 \\ &\iff n(a_n\overline{a}_{n+1} + \overline{a}_n a_{n+1}) \geq (n-1)|a_n|^2 + (n+1)|a_{n+1}|^2\end{aligned}$$

In particular this implies that

$$n(a_n\overline{a}_{n+1} + \overline{a}_n a_{n+1}) \geq 0$$

and thus

$$n(a_n\overline{a}_{n+1} + \overline{a}_n a_{n+1}) = n|a_n\overline{a}_{n+1} + \overline{a}_n a_{n+1}| \leq n(|a_n\overline{a}_{n+1}| + |\overline{a}_n a_{n+1}|) = 2n|a_n| \cdot |a_{n+1}|.$$

Combining this with the previous inequality, we get that

$$2n|a_n|\cdot|a_{n+1}| \geq (n-1)|a_n|^2 + (n+1)|a_{n+1}|^2 \iff |a_n|^2 - |a_{n+1}|^2 \geq n(|a_n| - |a_{n+1}|)^2$$

which implies the conclusion.

(3) We will get back to

$$2n|a_n| \cdot |a_{n+1}| \geq (n-1)|a_n|^2 + (n+1)|a_{n+1}|^2.$$

It is easy to see from this inequality that $a_n = 0 \iff a_{n+1} = 0$. This shows the first statement.

For the rest of this part we will assume that $a_n \neq 0$ for every n.

In the same time

$$\begin{aligned} &2n|a_n| \cdot |a_{n+1}| \geq (n-1)|a_n|^2 + (n+1)|a_{n+1}|^2 \\ &\qquad \iff (|a_{n+1}| - |a_n|)((n+1)|a_{n+1}| - (n-1)|a_n|) \geq 0 \end{aligned}$$

We already know that the first factor is positive. If $|a_{n+1}| - |a_n| > 0$, then

$$(n+1)|a_{n+1}| - (n-1)|a_n| \geq 0 \iff |a_{n+1}| \geq \frac{n-1}{n+1}|a_n|.$$

If $|a_{n+1}| - |a_n| = 0$, then

$$|a_{n+1}| \geq \frac{n-1}{n+1}|a_n|$$

is still true.

Therefore

$$1 \geq \frac{|a_{n+1}|}{|a_n|} \geq \frac{n-1}{n+1}$$

from where the conclusion follows.

(4) From

$$n(a_n\overline{a}_{n+1} + \overline{a}_n a_{n+1}) \geq (n-1)|a_n|^2 + (n+1)|a_{n+1}|^2$$

we get that, in this case,

$$2na_n a_{n+1} \geq (n-1)a_n^2 + (n+1)a_{n+1}^2$$

and thus, for every n, a_n and a_{n+1} have the same sign.

□

Corollary 10.2. *Suppose that the sequence $(a_n) \subset \mathbb{R}$ and that the Rhaly operator is hyponormal.*

(1) *If $a_1 > 0$, then the sequence is positive and decreasing to 0.*
(2) *If $a_1 < 0$, then the sequence is negative and increasing to 0.*

Remark 10.3. The conditions (2)–(4) in Theorem 10.1 are not sufficient to imply hyponormality. One can see that when $a_n = \frac{1}{n^2}$.

We will show some sufficient conditions next.

PROPOSITION 10.4. *Let (a_n) be a sequence of complex numbers and $R = R_{(a_n)}$.*

(1) *If there is $(t_n) \subset [0,1]$ such that*
$$R^* D_{(t_n)} R - RR^* \geq 0,$$
then R is hyponormal.

(2) *If there is $(t_n) \subset [1,m]$ for some $m > 1$ such that*
$$R^* R - R D_{(t_n)} R^* \geq 0,$$
then R is hyponormal.

PROOF.

(1) The condition tells us that $I - D_{(t_n)} \geq 0$.

Since
$$R^* R - RR^* = R^* R - R^* D_{(t_n)} R + R^* D_{(t_n)} R - RR^*,$$
the result follows by noticing that for every $x \in l^2$,
$$\langle (R^* R - R^* D_{(t_n)} R)x, x\rangle = \langle R^*(I - D_{(t_n)})Rx, x\rangle = \langle (I - D_{(t_n)})Rx, Rx\rangle \geq 0.$$
□

Operators T for which there is a positive operator P such that
$$TT^* = T^* P T$$
are called posinormal and were introduced by Rhaly in [**13**] and studied by him and many others. They are another possible generalization and abstractization of the Cesàro operator.

REMARK 10.5. Since T is hyponormal $\iff$ αT is hyponormal, it suffices to consider only Rhaly operators for which $|a_1| \leq 1$.

PROPOSITION 10.6. *Let (a_n) be a sequence of nonzero complex numbers such that $|a_1| \leq 1$ and for every n*
$$|a_n|(1 - |a_n|) \leq |a_{n+1}| \leq |a_n|$$
and $R = R_{(a_n)}$.

Then R is hyponormal if the operator
$$\begin{pmatrix} |a_1| - |a_1|^2 & |a_2| - a_1\overline{a}_2 & |a_3| - a_1\overline{a}_3 & |a_4| - a_1\overline{a}_4 & \dots \\ |a_2| - \overline{a}_1 a_2 & |a_2| - 2|a_2|^2 & |a_3| - 2a_2\overline{a}_3 & |a_4| - 2a_2\overline{a}_4 & \dots \\ |a_3| - \overline{a}_1 a_3 & |a_3| - 2\overline{a}_2 a_3 & |a_3| - 3|a_3|^2 & |a_4| - 3a_3\overline{a}_4 & \dots \\ |a_4| - \overline{a}_1 a_4 & |a_4| - 2\overline{a}_2 a_4 & |a_4| - 3\overline{a}_3 a_4 & |a_4| - 4|a_4|^2 & \dots \\ \dots & & & & \end{pmatrix}$$
is positive.

PROOF. Let
$$t_n = \frac{|a_n| - |a_{n+1}|}{|a_n|^2}.$$
The hypothesis means that $0 \leq t_n \leq 1$. By (1) of Proposition 10.4, R is hyponormal if $R^* D_{(t_n)} R - RR^*$ is positive. In this case $R^* D_{(t_n)} R - RR^*$ is exactly the operator in the statement. □

REMARK 10.7. Notice that in the matrix in the previous proposition when all the numbers are real and positive, if we multiply the first n entries on the nth row by a_{n+1}/a_n, we get the first n entries of the $(n+1)$th row.

PROPOSITION 10.8. *Let a_n be a positive sequence such that $a_1 \le 1$ and for every n*

$$a_n(1-a_n) \le a_{n+1} \le \frac{a_n}{1+a_n}.$$

Then $R_{(a_n)}$ is hyponormal.

PROOF. We need to show that the operator in the previous proposition is positive.

For this it suffices to show that for each n, the $n \times n$ upper-left corner minor has positive determinant.

In this determinant, if we multiply row $n-1$ by a_n/a_{n-1} and subtract it from row n, then multiply row $n-2$ by a_{n-1}/a_{n-2} and subtract it from row $n-1$ and so on, we get, by the previous remark, an upper triangular determinant with first diagonal entry $a_1(1-a_1)$ and for $k>1$ with kth diagonal entry

$$\frac{a_k(1+a_{k-1})}{a_{k-1}}\left(\frac{a_{k-1}}{1+a_{k-1}} - a_k\right).$$

Therefore the determinant is positive. □

REMARK 10.9. Since the function $f(x) = \frac{x}{1+x}$ is increasing for $x>0$, if $a_n \le \frac{1}{n}$, then $f(a_n) \le f(1/n) = 1/(n+1)$. Since $a_{n+1} \le f(a_n)$, we get $a_{n+1} \le 1/(n+1)$. The condition $a_1 \le 1$ implies then that $a_n \le 1/n$ for every n. This means that the Rhaly operator $R_{(a_n)}$ is bounded since its entries are dominated by the entries of the Cesàro operator.

COROLLARY 10.10. *If (a_n) is a positive sequence such that for every n*

$$a_n\left(1-\frac{a_n}{a_1+1}\right) \le a_{n+1} \le \frac{a_n}{1+\frac{a_n}{a_1+1}},$$

then the corresponding Rhaly operator is hyponormal.

PROOF. We multiply the operator by $1/(1+a_1)$ and apply Proposition 10.8. □

If $R = R_{(a_n)}$ is hyponormal and compact, then, by [**1**, Corollary 2], it is normal. Therefore except for the trivial case discussed at the beginning of this section, a hyponormal operator is not compact. This implies that if $\lim_n na_n = L \ne 0$ and finite, the spectrum consists of a disc of radius $|L|$ and at most a discrete set. Thus, by Putnam inequality, (see [**10**])

$$\|R^*R - RR^*\| \le \frac{1}{\pi}\mu(\sigma(R)) = |L|^2.$$

Here μ denoted the 2-dimensional Lebesgue measure.

The upper-left corner entry of $R^*R - RR^*$ is $\sum_{n=2}^{\infty} |a_n|^2$ which is obviously $\le \|R^*R - RR^*\|$. Therefore we have proved the following theorem.

THEOREM 10.11. *If $\lim_n na_n = L \ne 0$ and the corresponding Rhaly operator is hyponormal, then*

$$\sum_{n=2}^{\infty} |a_n|^2 \le |L|^2.$$

Remark 10.12. For $n \geq 2$, by (1) of Theorem 10.1,

$$n|a_n|^2 \leq \sum_{j=n}^{infty} |a_j|^2 \leq \sum_{j=2}^{\infty} |a_j|^2 \leq |L|^2$$

which implies that

$$|a_n| \leq \frac{|L|}{\sqrt{n}}$$

in this case.

However, this inequality is of importance only for smaller values of n because eventually

$$|a_n| \leq \frac{|L|+1}{n},$$

which is a better inequality.

Acknowledgments

The second author would like to thank the Fulbright Scholar Program, Fulbright Commission Romania, Mathematical Institute of the Romanian Academy, and the Department of Applied Sciences of the Polytechnic University of Bucharest for their support.

Note added in proof

Conjectures 5.11, 7.7, and 7.12 have been disproved in [**5**].

References

[1] T. Andô, *On hyponormal operators*, Proc. Amer. Math. Soc. **14** (1963), 290–291, DOI 10.2307/2034629. MR145353

[2] D. D. Bonar and M. Khoury Jr., *Real infinite series*, Classroom Resource Materials Series, Mathematical Association of America, Washington, DC, 2006. MR2193802

[3] A. Brown, P. R. Halmos, and A. L. Shields, *Cesàro operators*, Acta Sci. Math. (Szeged) **26** (1965), 125–137. MR187085

[4] J. B. Conway, *A course in functional analysis*, 2nd ed., Graduate Texts in Mathematics, vol. 96, Springer-Verlag, New York, 1990. MR1070713

[5] P. Galanopoulos, D. Girela, and G. Prăjitură, *Rhaly operators acting on l^p spaces*, Submitted January 2024.

[6] P. R. Halmos, *A Hilbert space problem book*, 2nd ed., Encyclopedia of Mathematics and its Applications, vol. 17, Springer-Verlag, New York-Berlin, 1982. Graduate Texts in Mathematics, 19. MR0675952

[7] G. Leibowitz, *Rhaly matrices*, J. Math. Anal. Appl. **128** (1987), no. 1, 272–286, DOI 10.1016/0022-247X(87)90230-7. MR915983

[8] C. P. Niculescu and G. T. Prăjitură, *Some open problems concerning the convergence of positive series*, Ann. Acad. Rom. Sci. Ser. Math. Appl. **6** (2014), no. 1, 92–107. MR3241141

[9] J. M. H. Olmsted, *Advanced calculus*, Appleton–Century–Crofts, New York, 1961.

[10] C. R. Putnam, *An inequality for the area of hyponormal spectra*, Math. Z. **28** (1971), 473–477.

[11] H. C. Rhaly Jr., *Terraced matrices*, Bull. London Math. Soc. **21** (1989), no. 4, 399–406, DOI 10.1112/blms/21.4.399. MR998639

[12] H. C. Rhaly Jr., *p-Cesàro matrices*, Houston J. Math. **15** (1989), no. 1, 137–146. MR1002086

[13] H. C. Rhaly Jr., *Posinormal operators*, J. Math. Soc. Japan **46** (1994), no. 4, 587–605, DOI 10.2969/jmsj/04640587. MR1291108

[14] H. C. Rhaly Jr., *Hyponormal terraced matrices*, Far East J. Math. Sci. **5** (1997), no. 3, 425–428. MR1619791

[15] W. T. Ross, *The Cesàro operator*, Recent progress in function theory and operator theory, Contemp. Math., vol. 799, Amer. Math. Soc., Providence, RI, 2024.

[16] T. Šalát and V. Toma, *A classical Olivier's theorem and statistical convergence* (English, with English and French summaries), Ann. Math. Blaise Pascal **10** (2003), no. 2, 305–313. MR2031274

Department of Applied Mathematics, University of Craiova, Craiova, Romania
Email address: grgpop@gmail.com

Department of Mathematics, State University of New York Brockport, Brockport, New York 14420
Email address: gprajitu@brockport.edu

Contemporary Mathematics
Volume **799**, 2024
https://doi.org/10.1090/conm/799/16025

The Cesàro operator

William T. Ross

Abstract. This paper surveys the various aspects of the Cesàro operator with a special emphasis on the Hilbert space setting of ℓ^2. We include a discussion of summability methods of Fourier series, the Cesàro matrix and integral operator, and the Cesàro operator on the Hilbert space ℓ^2. In this setting of ℓ^2 we cover the spectral properties of the Cesàro operator as well as a treatment of its hyponormality (via posinormal operators) and subnormality (via a multiplication operator of Kriete and Trutt). Using the results of Kriete and Trutt, we describe the commutant of the Cesàro operator as well as its bounded square roots. The Cesàro operator has a rich lattice of invariant subspaces whose description remains unknown and we suspect might never be fully understood. We survey some results which display the complexity of these invariant subspaces and provide the reader with several paths forward for further discussion. Though the Cesàro operator was originally explored in the ℓ^2 and Hardy space settings, it has been explored in various other settings such as the general Hardy spaces as well as the Bergman spaces. Finally, we explore some of the so-called generalized Cesàro operator which are classes of integral operators on the Hardy spaces that connect to many areas of classical function theory.

1. Introduction

There has been renewed interest in the classical Cesàro operator and its generalizations as of late [**1, 17, 28, 32, 38, 44**] so perhaps it is a good time to put together an extended survey of what is currently known about this operator. Most of us in analysis know the name Cesàro from his summability method for infinite series and the important role this plays in summing the Fourier series of an integrable function. We will outline some of the aspects of Cesàro's life and the historical development of his summability method in Section 2.

Graduating from Cesàro summability, which involves creating the sequence of Cesàro averages

$$\Big(\frac{a_0 + a_1 + a_2 + \cdots + a_N}{N+1}\Big)_{N \geqslant 0}$$

from the sequence $(a_n)_{n \geqslant 0}$ of complex numbers, to considering the linear transformation

$$(a_n)_{n \geqslant 0} \mapsto \Big(\frac{a_0 + a_1 + a_2 + \cdots + a_N}{N+1}\Big)_{N \geqslant 0}$$

2020 *Mathematics Subject Classification.* 47B35, 47B02, 47A05.
Key words and phrases. Cesàro summability, Cesàro matrix, Hardy spaces, Cesàro operator.

on the sequence space ℓ^2, was taken up by Brown, Halmos, and Shields [**8**]. By representing this linear transformation with respect to the standard orthonormal basis $(\mathbf{e}_n)_{n\geqslant 0}$ for ℓ^2, as the Cesàro matrix

$$C := \begin{bmatrix} 1 & 0 & 0 & 0 & 0 & \cdots \\ \frac{1}{2} & \frac{1}{2} & 0 & 0 & 0 & \cdots \\ \frac{1}{3} & \frac{1}{3} & \frac{1}{3} & 0 & 0 & \cdots \\ \frac{1}{4} & \frac{1}{4} & \frac{1}{4} & \frac{1}{4} & 0 & \cdots \\ \frac{1}{5} & \frac{1}{5} & \frac{1}{5} & \frac{1}{5} & \frac{1}{5} & \cdots \\ \vdots & \vdots & \vdots & \vdots & \vdots & \ddots \end{bmatrix},$$

they were able to describe its (operator) norm $\|C\|$ and its spectrum $\sigma(C)$. We survey this in Sections 5 and 6. Along the way, we discuss the Cesàro matrix as a linear transformation on the set of *all* complex sequences in Section 3.

Brown, Halmos, and Shields also showed that the Cesàro operator on ℓ^2 is hyponormal, i.e., $C^*C - CC^* \geqslant 0$ (with, of course, normal being equality). By representing the Cesàro operator as a multiplication operator on the closure of the polynomials in an $L^2(\mu)$ space, Kriete and Trutt [**30**], in one of the gems in the subject, showed that the Cesàro operator satisfies the stronger condition of subnormality (a normal operator restricted to one of its invariant subspaces). This will be outlined in Sections 7 and 8.

As a consequence of the Kriete–Trutt result, and some known results concerning the commutant of a multiplication operator on a reproducing kernel Hilbert space of analytic functions, one can describe the bounded operators on ℓ^2 that commute with the Cesàro operator. By a result of Shields and Wallen [**43**], these operators coincide with the closure in the weak operator topology of $p(C)$, where $p \in \mathbb{C}[z]$. This is discussed in Section 9—with a corresponding discussion in Section 3 of the linear transformations on the space of all complex sequences that commute with the Cesàro matrix. We use this discussion in Section 10 to show that the Cesàro operator has exactly two bounded square roots and many other unbounded ones.

The Cesàro operator has several interesting and useful connections to strongly continuous semigroups of composition operators on the Hardy space H^2. We explore some of these connections in Section 11 and then again in Section 12 when we discuss the complexity of the invariant subspaces of the Cesàro operator.

We end this survey with a selective survey in Sections 14 and 15 of the Cesàro operator on other spaces of analytic functions along with generalizations of the Cesàro operator that are inspired by its integral representation on the Hardy space H^2.

2. Cesàro summability

Many people who work on the Cesàro operator are not familiar with its namesake Ernesto Cesàro who was born in Naples, Italy on March 12, 1859 and died tragically, while trying to save his son from drowning, in Torre Annunziata, Italy, on September 12, 1906. He began his university studies in Ècole des Mines of Liège (in Belgium) under Eugène Catalan and continued his studies in Rome. He held university positions in Palermo, Naples, and Bologna.

Though readers of this paper probably know Cesàro for his concept of summability of infinite series, one might not know that Cesàro was also a prolific author with varied interests in several hundred papers in analysis, geometry, number theory, and mathematical physics.

Traditionally, one sums an infinite series $a_0 + a_1 + a_2 + \cdots$ of complex numbers by considering

$$\lim_{N \to \infty} S_N, \quad S_N := \sum_{j=0}^{N} a_j, \tag{2.1}$$

the limit of the partial sums S_N. When the limit exists, we say the series converges and its value is denoted by $\sum_{n=0}^{\infty} a_n$. When the limit $\lim_{N \to \infty} S_N$ fails to exist, we declare the series $a_0 + a_1 + a_2 + \cdots$ to be divergent. In some sense, this definition of convergence is rather restrictive (and perhaps capricious) in that one runs into trouble with the very natural problem of summing a Fourier series. It is well known that for an $f \in L^1 := L^1(\mathbb{T}, m)$ there are technical problems in summing the Fourier series

$$\sum_{n=-\infty}^{\infty} \widehat{f}(n)\xi^n.$$

In the above, $\mathbb{T} = \{\xi \in \mathbb{C} : |\xi| = 1\}$ is the unit circle in the complex plane $\mathbb{C}$, m is standard normalized Lebesgue measure (i.e., $d\theta/2\pi$) on $\mathbb{T}$, and

$$\widehat{f}(n) := \int_{\mathbb{T}} f(\xi)\overline{\xi}^n dm(\xi), \quad n \in \mathbb{Z},$$

is the nth Fourier coefficient of f. An example of du Bois-Reymond from 1873 [**14**] produces a continuous function $\mathbb{T}$ whose Fourier series diverges at $\xi = 1$. An even more pathological example of Kolmogorov from 1923 [**29**] produces an L^1 function whose Fourier series diverges at *every* point of $\mathbb{T}$. In spite of these pathologies, one still needs to somehow relate a function with its Fourier series in a meaningful way. One way out of this conundrum (there are several) involves a summation method of Cesàro [**10**].

DEFINITION 2.1. For a sequence $(a_n)_{n \geqslant 0}$ of complex numbers, we say the infinite series $\sum_{n=0}^{\infty} a_n$ is *Cesàro summable* to L if

$$L = \lim_{N \to \infty} \frac{1}{N+1} \sum_{n=0}^{N} S_n.$$

Notice how L is the limit of the averages of the partial sums. The reader might wonder why we are indexing our sequences starting with a_0 instead of a_1. As we shall see momentarily, when we recast the Cesàro operator from the setting of vector spaces of sequences to that of vector spaces of analytic functions, we will identify the sequence $(a_n)_{n \geqslant 0}$ with the power series $a_0 + a_1 z + a_2 z^2 + \cdots$.

The reader can verify that the famous *Grandi* series

$$1 - 1 + 1 - 1 + 1 - \cdots$$

diverges in the "usual sense" of (2.1) but is Cesàro summable to $\frac{1}{2}$. A nice real analysis exercise will confirm the following.

PROPOSITION 2.2. *If $(a_n)_{n \geqslant 0}$ is a sequence of complex numbers for which $S_N \to L$, then the series $\sum_{n=0}^{\infty} a_n$ is Cesàro summable to L.*

Well known theorems of Fejér and Lebesgue [**23**] show that although the Fourier series of an L^1 function need not converge in the usual sense, it does converge in the sense of Cesàro.

THEOREM 2.3. *The Fourier series of an $f \in L^1$ is Cesàro summable to f almost everywhere on $\mathbb{T}$.*

The proof of this theorem is often the starting point of just about any harmonic analysis course. Indeed, if

$$(S_n f)(\xi) := \sum_{k=-n}^{n} \widehat{f}(k)\xi^k$$

is the nth partial sum of the Fourier series for f, then

$$(S_n f)(\xi) = \int_{\mathbb{T}} f(w)\Big(\sum_{k=-n}^{n} (\xi\overline{w})^k\Big)dm(w).$$

Thus,

$$(\sigma_N f)(\xi) := \frac{1}{N+1}\sum_{n=0}^{N}(S_n f)(\xi),$$

the Nth Cesàro mean of the Fourier series of f, becomes

$$\int_{\mathbb{T}} f(w)\Big(\frac{1}{N+1}\sum_{n=0}^{N}\sum_{k=-n}^{n} (\xi\overline{w})^k\Big)dm(w).$$

The quantity

$$\frac{1}{N+1}\sum_{n=0}^{N}\sum_{k=-n}^{n} (\xi\overline{w})^k$$

in the previous integral is known as the Fejér kernel and has the requisite properties needed to prove Theorem 2.3.

Before moving on, let us mention that properties of the Fejér kernel show that if f is continuous on $\mathbb{T}$, then $\sigma_N(f) \to f$ uniformly (which gives a proof of the Weierstrass theorem on the density of the trigonometric polynomials in the Banach space of continuous functions on $\mathbb{T}$). Furthermore, if $f \in L^p$, $1 \leqslant p < \infty$, then $\sigma_N(f) \to f$ in L^p norm. If $f \in L^\infty$, then $\sigma_N(f) \to f$ weak-$*$.

There are higher order summability methods for a series $a_0 + a_1 + a_2 + \cdots$. Define

$$H_N^0 := \sum_{j=0}^{N} a_j,$$

$$H_N^1 := \frac{H_0^0 + H_1^0 + H_2^0 + \cdots + H_N^0}{N+1},$$

$$H_N^2 := \frac{H_0^1 + H_1^1 + H_2^1 + \cdots + H_N^1}{N+1},$$

and for $r \in \mathbb{N}$,

$$H_N^r := \frac{H_0^{r-1} + H_1^{r-1} + H_2^{r-1} + \cdots + H_N^{r-1}}{N+1}.$$

We say a series $a_0 + a_1 + a_2 + \cdots$ is *Hölder summable of order* r, written (H, r) summable, to L if

$$\lim_{N\to\infty} H_N^r = L.$$

See [**24**] for the original reference.

Let

$$S_N^0 := a_0 + a_1 + a_2 + \cdots + a_N,$$

$$S_N^1 := S_0^0 + S_1^0 + S_2^0 + \cdots + S_N^0,$$

$$S_N^2 := S_0^1 + S_1^1 + S_2^1 + \cdots + S_N^1,$$

and for $r \in \mathbb{N}$,

$$S_N^r := S_0^{r-1} + S_1^{r-1} + S_2^{r-1} + \cdots + S_N^{r-1}.$$

We say a series $a_0 + a_1 + a_2 + \cdots$ is *Cesàro summable of order* r, written (C, r) summable, to L if

$$\lim_{N\to\infty} \frac{r! S_N^r}{(N+1)^r} = L.$$

Note that $(H, 1)$ and $(C, 1)$ summability are the same. It is also known that if a series is (H, s) summable to L for some s, then it is (H, r) summable to L for all $r \geqslant s$ (similarly for (C, r) summability). In fact, a series is (H, r) summable to L if and only if it is (C, r) is summable to L. We will revisit these summability methods as matrices momentarily.

For example, the $(C, p+1)$ sum of the series

$$1^p - 2^p + 3^p - 4^p + \cdots$$

is equal to

$$\frac{2^{p+1} - 1}{p+1} B_{p+1},$$

where B_{p+1} is the corresponding Bernoulli number. Note that when $p = 0$, we have $B_1 = \frac{1}{2}$ and $(C, 1)$ sum of the series $1 - 1 + 1 - 1 + 1 + \cdots$ is $\frac{1}{2}$ which we discussed earlier. When $p = 1$, we have $B_2 = \frac{1}{6}$ and the $(C, 2)$ sum of the series $1 - 2 + 3 - 4 + 5 - \cdots$ is $\frac{1}{4}$. In fact, Cesàro found these formulas curious and said [**9**]: "I say that, although these formulas are false, they may be used as a base of a theory, which shall not be more absurd than the theory of imaginaries."

For further reading on all things concerning both summability methods of series of constants as well as Fourier series, we direct the reader to two well-known texts. The first is Hardy's classic text [**20**] on various aspects of divergent series and summability methods. The second is Zygmund's thorough treatment of Fourier series [**47**]. For a well-documented history behind what Cesàro was thinking when he developed his summability method, the paper [**16**] is an excellent resource.

3. The Cesàro matrix

For a sequence $(a_n)_{n\geqslant 0}$ of complex numbers, the Cesàro averages

$$\sigma_N := \frac{1}{N+1}(S_0 + S_1 + S_2 + \cdots + S_N)$$

can be obtained via a linear transformation on the vector space $\mathcal{V}$ of all complex sequences as follows. By means of the *Cesàro matrix*

$$C := \begin{bmatrix} 1 & 0 & 0 & 0 & 0 & \cdots \\ \frac{1}{2} & \frac{1}{2} & 0 & 0 & 0 & \cdots \\ \frac{1}{3} & \frac{1}{3} & \frac{1}{3} & 0 & 0 & \cdots \\ \frac{1}{4} & \frac{1}{4} & \frac{1}{4} & \frac{1}{4} & 0 & \cdots \\ \frac{1}{5} & \frac{1}{5} & \frac{1}{5} & \frac{1}{5} & \frac{1}{5} & \cdots \\ \vdots & \vdots & \vdots & \vdots & \vdots & \ddots \end{bmatrix},$$

observe that

$$\begin{bmatrix} 1 & 0 & 0 & 0 & 0 & \cdots \\ \frac{1}{2} & \frac{1}{2} & 0 & 0 & 0 & \cdots \\ \frac{1}{3} & \frac{1}{3} & \frac{1}{3} & 0 & 0 & \cdots \\ \frac{1}{4} & \frac{1}{4} & \frac{1}{4} & \frac{1}{4} & 0 & \cdots \\ \frac{1}{5} & \frac{1}{5} & \frac{1}{5} & \frac{1}{5} & \frac{1}{5} & \cdots \\ \vdots & \vdots & \vdots & \vdots & \vdots & \ddots \end{bmatrix} \begin{bmatrix} S_0 \\ S_1 \\ S_2 \\ S_3 \\ S_4 \\ \vdots \end{bmatrix} = \begin{bmatrix} S_0 \\ \frac{1}{2}(S_0 + S_1) \\ \frac{1}{3}(S_0 + S_1 + S_2) \\ \frac{1}{4}(S_0 + S_1 + S_2 + S_3) \\ \frac{1}{5}(S_0 + S_1 + S_2 + S_3 + S_4) \\ \vdots \end{bmatrix} = \begin{bmatrix} \sigma_0 \\ \sigma_1 \\ \sigma_2 \\ \sigma_3 \\ \sigma_4 \\ \vdots \end{bmatrix}. \tag{3.1}$$

Not only does the matrix C produce the Cesàro partial sums, it is the starting point in studying the linear transformation $\mathbf{a} \mapsto C\mathbf{a}$ on $\mathcal{V}$. Let us first explore the eigenvalues and eigenvectors of C on $\mathcal{V}$. For each $m \in \mathbb{N}_0 := \mathbb{N} \cup \{0\}$, define the sequence (as a column vector)

$$\mathbf{b}_m = \begin{bmatrix} 0 & 0 & 0 & \cdots & 0 & \binom{m+0}{m} & \binom{m+1}{m} & \binom{m+2}{m} & \binom{m+3}{m} & \cdots \end{bmatrix}^T. \tag{3.2}$$

PROPOSITION 3.1. *For each $m \in \mathbb{N}_0$,*

$$C\mathbf{b}_m = \frac{1}{m+1}\mathbf{b}_m.$$

Moreover, each eigenspace is one dimensional.

We will outline a proof of this in the next section when we reimagine the Cesàro operator as an integral operator on the space of all analytic functions on $\mathbb{D}$. Next observe that the Cesàro matrix takes $\mathcal{V}$, the vector space of all complex sequences, *onto* $\mathcal{V}$.

PROPOSITION 3.2. *The Cesàro matrix defines an invertible linear transformation of $\mathcal{V}$ onto $\mathcal{V}$ with*

$$C^{-1} = \begin{bmatrix} 1 & 0 & 0 & 0 & 0 & \cdots \\ -1 & 2 & 0 & 0 & 0 & \cdots \\ 0 & -2 & 3 & 0 & 0 & \cdots \\ 0 & 0 & -3 & 4 & 0 & \cdots \\ 0 & 0 & 0 & -4 & 5 & \cdots \\ \vdots & \vdots & \vdots & \vdots & \vdots & \ddots \end{bmatrix}.$$

One proves this by referring to (3.1) and solving the infinite triangular system

$$\begin{aligned}\sigma_0 &= S_0\\ \sigma_1 &= \tfrac{1}{2}(S_0 + S_1)\\ \sigma_2 &= \tfrac{1}{3}(S_0 + S_1 + S_2)\\ \sigma_3 &= \tfrac{1}{4}(S_0 + S_1 + S_2 + S_3)\end{aligned}$$

and so on, for $S_0, S_1, S_2, \ldots$

Interestingly, the Cesàro matrix connects with some other famous matrices. These factorizations where pointed out by Bennett [**4, 5**] as a way or estimating, and even computing, the norms of various well-known matrices and their generalizations. For example, if H is the classical *Hilbert matrix*

$$H = \begin{bmatrix} 1 & \frac{1}{2} & \frac{1}{3} & \frac{1}{4} & \cdots\\ \frac{1}{2} & \frac{1}{3} & \frac{1}{4} & \frac{1}{5} & \cdots\\ \frac{1}{3} & \frac{1}{4} & \frac{1}{5} & \frac{1}{6} & \cdots\\ \vdots & \vdots & \vdots & \vdots & \ddots \end{bmatrix},$$

i.e., $H = [H_{jk}]_{j,k \geqslant 0}$, where

$$H_{jk} = \frac{1}{j+k+1},$$

and

$$B = \begin{bmatrix} \frac{1}{2} & \frac{1}{3} & \frac{1}{4} & \frac{1}{5} & \frac{1}{6} & \frac{1}{7} & \cdots\\ \frac{1}{6} & \frac{1}{6} & \frac{3}{20} & \frac{2}{15} & \frac{5}{42} & \frac{3}{28} & \cdots\\ \frac{1}{12} & \frac{1}{10} & \frac{1}{10} & \frac{2}{21} & \frac{5}{56} & \frac{1}{12} & \cdots\\ \frac{1}{20} & \frac{1}{15} & \frac{1}{14} & \frac{1}{14} & \frac{5}{72} & \frac{1}{15} & \cdots\\ \frac{1}{30} & \frac{1}{21} & \frac{3}{56} & \frac{1}{18} & \frac{1}{18} & \frac{3}{55} & \cdots\\ \frac{1}{42} & \frac{1}{28} & \frac{1}{24} & \frac{2}{45} & \frac{1}{22} & \frac{1}{22} & \cdots\\ \vdots & \vdots & \vdots & \vdots & \vdots & \vdots & \ddots \end{bmatrix},$$

i.e., $B = [B_{jk}]_{j,k \geqslant 0}$, where

$$B_{jk} = \frac{k+1}{(j+k+1)(j+k+2)},$$

one can show [**4**] that

$$H = BC \tag{3.3}$$

which links the Hilbert matrix with the Cesàro matrix.

If one considers the transpose

$$\begin{bmatrix} 1 & \frac{1}{2} & \frac{1}{3} & \frac{1}{4} & \frac{1}{5} & \cdots\\ 0 & \frac{1}{2} & \frac{1}{3} & \frac{1}{4} & \frac{1}{5} & \cdots\\ 0 & 0 & \frac{1}{3} & \frac{1}{4} & \frac{1}{5} & \cdots\\ 0 & 0 & 0 & \frac{1}{4} & \frac{1}{5} & \cdots\\ 0 & 0 & 0 & 0 & \frac{1}{5} & \cdots\\ \vdots & \vdots & \vdots & \vdots & \vdots & \ddots \end{bmatrix}$$

of C and defines

$$L := \begin{bmatrix} 1 & \frac{1}{2} & \frac{1}{3} & \frac{1}{4} & \cdots \\ \frac{1}{2} & \frac{1}{2} & \frac{1}{3} & \frac{1}{4} & \cdots \\ \frac{1}{3} & \frac{1}{3} & \frac{1}{3} & \frac{1}{4} & \cdots \\ \frac{1}{4} & \frac{1}{4} & \frac{1}{4} & \frac{1}{4} & \cdots \\ \vdots & \vdots & \vdots & \vdots & \ddots \end{bmatrix},$$

in other words, $L = [L_{ij}]_{i,j \geqslant 0}$, with

$$L_{ij} := \frac{1}{\max\{i, j\} + 1},$$

a computation will show that $L = CC^*$. The previous matrix is a special case of an L-*matrix* and was explored recently in [**6**, **7**].

As with any matrix, one can ask about the infinite matrices T for which $TC = CT$, i.e., the commutant of C as a linear transformation on $\mathcal{V}$. These involve the *Hausdorff matrices* which are constructed as follows. Define the matrix

$$W := \begin{bmatrix} 1 & 0 & 0 & 0 & 0 & 0 & \cdots \\ 1 & -1 & 0 & 0 & 0 & 0 & \cdots \\ 1 & -2 & 1 & 0 & 0 & 0 & \cdots \\ 1 & -3 & 3 & -1 & 0 & 0 & \cdots \\ 1 & -4 & 6 & -4 & 1 & 0 & \cdots \\ 1 & -5 & 10 & -10 & 5 & -1 & \cdots \\ \vdots & \vdots & \vdots & \vdots & \vdots & \vdots & \ddots \end{bmatrix}.$$

Notice how each row consists of the alternating binomial coefficients. A matrix calculation (which is allowed since W is lower triangular) shows that

$$W^2 = I.$$

Furthermore, and this will be used in Section 9, one can verify the matrix identity [**22**]

$$C = WDW, \tag{3.4}$$

where D is the diagonal matrix

$$D := \begin{bmatrix} 1 & 0 & 0 & 0 & \cdots \\ 0 & \frac{1}{2} & 0 & 0 & \cdots \\ 0 & 0 & \frac{1}{3} & 0 & \cdots \\ 0 & 0 & 0 & \frac{1}{4} & \cdots \\ \vdots & \vdots & \vdots & \vdots & \ddots \end{bmatrix}.$$

If $TC = CT$, then the matrix $T' := WTW$ commutes with D and, since D is a diagonal matrix with distinct entries, one can argue that

$$T' = \begin{bmatrix} z_0 & 0 & 0 & 0 & \cdots \\ 0 & z_1 & 0 & 0 & \cdots \\ 0 & 0 & z_2 & 0 & \cdots \\ 0 & 0 & 0 & z_3 & \cdots \\ \vdots & \vdots & \vdots & \vdots & \ddots \end{bmatrix}$$

for some sequence $(z_n)_{n\geqslant 0}$ of complex numbers. Thus,

$$T = W \begin{bmatrix} z_0 & 0 & 0 & 0 & \cdots \\ 0 & z_1 & 0 & 0 & \cdots \\ 0 & 0 & z_2 & 0 & \cdots \\ 0 & 0 & 0 & z_3 & \cdots \\ \vdots & \vdots & \vdots & \vdots & \ddots \end{bmatrix} W.$$

This yields the following.

THEOREM 3.3 (Hurwitz–Silverman, [**26**]). *A matrix T commutes with the Cesàro matrix C if and only if*

$$T = W \begin{bmatrix} z_0 & 0 & 0 & 0 & \cdots \\ 0 & z_1 & 0 & 0 & \cdots \\ 0 & 0 & z_2 & 0 & \cdots \\ 0 & 0 & 0 & z_3 & \cdots \\ \vdots & \vdots & \vdots & \vdots & \ddots \end{bmatrix} W$$

for some complex sequence $(z_n)_{n\geqslant 0}$.

The matrix T as defined in Theorem 3.3 is known as a *Hausdorff matrix.* An interesting class of Hausdorff matrices are the *Euler matrices* E_λ, where $\lambda \in \mathbb{C}$ and

$$E_\lambda := W \begin{bmatrix} 1 & 0 & 0 & 0 & \cdots \\ 0 & \lambda & 0 & 0 & \cdots \\ 0 & 0 & \lambda^2 & 0 & \cdots \\ 0 & 0 & 0 & \lambda^3 & \cdots \\ \vdots & \vdots & \vdots & \vdots & \ddots \end{bmatrix} W. \tag{3.5}$$

We will see these again in Section 9. The above Hausdorff operators have various generalizations that were explored in [**42**].

Before leaving this section, let us connect these Hausdorff matrices with higher order Hölder and Cesàro summability discussed earlier. For an infinite series $a_0 + a_1 + a_2 + \cdots$, recall the quantities

$$H_N^0 := \sum_{j=0}^{N} a_j,$$

$$H_N^r := \frac{H_0^{r-1} + H_1^{r-1} + H_2^{r-1} + \cdots + H_N^{r-1}}{N+1}, \quad r \in \mathbb{N}.$$

Also recall that a series $a_0 + a_1 + a_2 + \cdots$ is Hölder summable or order r, written (H, r) summable, to L if

$$\lim_{N\to\infty} H_N^r = L.$$

The matrix that corresponds to (H,r) summability, as the Cesàro matrix corresponded to Cesàro summability in (3.1), is

$$W\begin{bmatrix} f(1) & 0 & 0 & 0 & \cdots \\ 0 & f(\frac{1}{2}) & 0 & 0 & \cdots \\ 0 & 0 & f(\frac{1}{3}) & 0 & \cdots \\ 0 & 0 & 0 & f(\frac{1}{4}) & \cdots \\ \vdots & \vdots & \vdots & \vdots & \ddots \end{bmatrix} W,$$

where $f(z) = z^r$.

Similarly, recall the quantities

$$S_N^0 := a_0 + a_1 + a_2 + \cdots + a_N,$$

$$S_N^r := S_0^{r-1} + S_1^{r-1} + S_2^{r-1} + \cdots + S_N^{r-1}, \quad r \in \mathbb{N}.$$

A series $a_0 + a_1 + a_2 + \cdots$ is Cesàro summable or order r, written (C,r) summable, to L if

$$\lim_{N \to \infty} \frac{r! S_N^r}{(N+1)^r} = L.$$

The matrix that corresponds to (C,r) summability is

$$W\begin{bmatrix} g(1) & 0 & 0 & 0 & \cdots \\ 0 & g(\frac{1}{2}) & 0 & 0 & \cdots \\ 0 & 0 & g(\frac{1}{3}) & 0 & \cdots \\ 0 & 0 & 0 & g(\frac{1}{4}) & \cdots \\ \vdots & \vdots & \vdots & \vdots & \ddots \end{bmatrix} W,$$

where

$$g(z) = \frac{r! z^r}{(1+z)(1+2z)\cdots(1+(r-1)z)}.$$

4. The Cesàro operator on the space of analytic functions

Let $\mathcal{O}(\mathbb{D})$ denote the vector space of all analytic functions on the open unit disk $\mathbb{D}$. It is well known that $\mathcal{O}(\mathbb{D})$ is a Fréchet space when endowed with the topology arising from uniform convergence on compact subsets of $\mathbb{D}$. For $f \in \mathcal{O}(\mathbb{D})$ define

$$(4.1) \qquad (Cf)(z) = \frac{1}{z}\int_0^z \frac{f(\xi)}{1-\xi} d\xi, \quad z \in \mathbb{D},$$

where the integral is along any path from 0 to z in $\mathbb{D}$. Basic theory of integrals shows that $Cf \in \mathcal{O}(\mathbb{D})$. In fact, we have more.

PROPOSITION 4.1. *The map $f \mapsto Cf$ is a linear isomorphism of $\mathcal{O}(\mathbb{D})$.*

PROOF. The linearity of the map $f \mapsto Cf$ comes from the linearity of the integral. This map is injective since

$$\frac{1}{z}\int_0^z \frac{f(\xi)}{1-\xi} d\xi \equiv 0 \implies \int_0^z \frac{f(\xi)}{1-\xi} d\xi \equiv 0 \implies \frac{f(z)}{1-z} \equiv 0 \implies f \equiv 0.$$

This map is surjective since if $g \in \mathcal{O}(\mathbb{D})$, then $f(z) = (1-z)(zg'(z) + g(z))$ belongs to $\mathcal{O}(\mathbb{D})$ and one can quickly check that $Cf = g$. □

COROLLARY 4.2. *The inverse of $C : \mathcal{O}(\mathbb{D}) \to \mathcal{O}(\mathbb{D})$ is given by*

$$(C^{-1}f)(z) = (1-z)\frac{d}{dz}(zf(z)).$$

Let us connect this seemingly "new" linear transformation on $\mathcal{O}(\mathbb{D})$, defined by an integral, with the Cesàro matrix as a linear transformation on $\mathcal{V}$. We do this by looking at the action of C on the functions $(z^n)_{n\geqslant 0}$. One can check via integration that

$$C1 = -\frac{1}{z}\log(1-z) = 1 + \frac{z}{2} + \frac{z^2}{3} + \frac{z^3}{4} + \cdots,$$

$$Cz = -\frac{z+\log(1-z)}{z} = \frac{z}{2} + \frac{z^2}{3} + \frac{z^3}{4} + \frac{z^4}{5} + \cdots,$$

$$Cz^2 = -\Big(1 + \frac{z}{2} + \frac{\log(1-z)}{z}\Big) = \frac{z^2}{3} + \frac{z^3}{4} + \frac{z^4}{5} + \frac{z^5}{6} + \cdots,$$

and so on. This says that the matrix representation of $C : \mathcal{O}(\mathbb{D}) \to \mathcal{O}(\mathbb{D})$ with respect to the functions $(z^n)_{n\geqslant 0}$ is the Cesàro matrix

$$\begin{bmatrix} 1 & 0 & 0 & 0 & 0 & \cdots \\ \frac{1}{2} & \frac{1}{2} & 0 & 0 & 0 & \cdots \\ \frac{1}{3} & \frac{1}{3} & \frac{1}{3} & 0 & 0 & \cdots \\ \frac{1}{4} & \frac{1}{4} & \frac{1}{4} & \frac{1}{4} & 0 & \cdots \\ \frac{1}{5} & \frac{1}{5} & \frac{1}{5} & \frac{1}{5} & \frac{1}{5} & \cdots \\ \vdots & \vdots & \vdots & \vdots & \vdots & \ddots \end{bmatrix}.$$

We have already seen the eigenvalues and eigenvectors of the Cesàro matrix from Proposition 3.1. Let us see them as functions. To do this, we fix a $\lambda \in \mathbb{C}$ and solve the functional equation $Cf = \lambda f$. This yields

$$\frac{1}{z}\int_0^z \frac{f(\xi)}{1-\xi}\,d\xi = \lambda f(z).$$

Multiplying through by z and then differentiating gives us

$$\frac{f(z)}{1-z} = \lambda z f'(z) + \lambda f(z).$$

This differential equation has the solution

$$f(z) = z^{\frac{1-\lambda}{\lambda}}(1-z)^{-\frac{1}{\lambda}}.$$

Since this function needs to be analytic on $\mathbb{D}$, it must be the case that the exponent on z must be a nonnegative integer and so

$$\frac{1-\lambda}{\lambda} \in \mathbb{N}_0.$$

Solving for λ produces the eigenvalues

$$\frac{1}{n+1}, \quad n \in \mathbb{N}_0, \tag{4.2}$$

and the corresponding eigenfunctions

$$f_n(z) = \frac{z^n}{(1-z)^{n+1}}. \tag{4.3}$$

Notice how this corresponds, via the identification of the function $z^m \in \mathcal{O}(\mathbb{D})$ with the basis vector $\mathbf{e}_m \in \mathcal{V}$ to the eigenvalues and eigenvectors seen in (3.2). Indeed,

$$f_n(z) = \sum_{k=n}^{\infty} \binom{k}{n} z^n$$

and the Taylor coefficients of f_n are the entries of the vector $\mathbf{b}_n$ from (3.2).

We end this section with a re-examination of (3.3) which relates the Hilbert and Cesàro matrices. A change of variable in (4.1) will show that the integral form of C on $\mathcal{O}(\mathbb{D})$ can be written as

$$(Cf)(z) = \int_0^1 f(tz) \frac{1}{1 - tz} dt. \tag{4.4}$$

Since the matrix entries of the Hilbert matrix H are

$$H_{jk} = \frac{1}{j + k + 1}, \quad j, k \geqslant 0,$$

we can use this to define a linear transformation H on $\mathcal{O}(\mathbb{D})$ by

$$(Hf)(z) = \sum_{n=0}^{\infty} \Big(\sum_{k=0}^{\infty} \frac{a_k}{n + k + 1} \Big) z^n, \quad f(z) = \sum_{k=0}^{\infty} a_k z^k.$$

This can be written as integral form as

$$(Hf)(z) = \int_0^1 f(t) \frac{1}{1 - tz} dt.$$

Notice in the integral representation of C from (4.4) that $f(tz)$ appears in the integrand while $f(t)$ (without the z) appears in the integrand for H. For the functions

$$e_n(z) = z^n, \quad n \geqslant 0,$$

observe that

$$(Ce_n)(z) = z^n (He_n)(z). \tag{4.5}$$

This says that one can transform the Hilbert matrix to the Cesàro matrix as follows: move the nth column of H down n places and fill in the n empty slots with zeros. In an analogous way, one can transform C to H.

5. The Cesàro operator is bounded on ℓ^2

The Cesàro matrix defines a linear transformation $\mathbf{a} \mapsto C\mathbf{a}$ on the vector space $\mathcal{V}$ of all complex sequences $\mathbf{a} = (a_n)_{n \geqslant 0}$. In fact, it defines an isomorphism of $\mathcal{V}$ onto itself. Now consider the well known Hilbert space

$$\ell^2 := \Big\{ \mathbf{a} = (a_n)_{n \geqslant 0} : \|\mathbf{a}\| := \Big(\sum_{n=0}^{\infty} |a_n|^2 \Big)^{\frac{1}{2}} < \infty \Big\}$$

with inner product

$$\langle \mathbf{a}, \mathbf{b} \rangle = \sum_{n=0}^{\infty} a_n \overline{b_n}$$

and corresponding norm $\|\mathbf{a}\| = \sqrt{\langle \mathbf{a}, \mathbf{a} \rangle}$. Is the linear transformation $\mathbf{a} \mapsto C\mathbf{a}$ defined/bounded on ℓ^2? What is

$$\|C\| = \sup_{\|\mathbf{a}\|=1} \|C\mathbf{a}\|,$$

the norm of this linear transformation on ℓ^2?

THEOREM 5.1 (Brown–Halmos–Shields, [**8**]). *The linear transformation* $\mathbf{a} \mapsto C\mathbf{a}$*, denoted simply by* C*, defines a bounded operator on* ℓ^2 *with* $\|C\| = 2$.

PROOF. Let us outline the original proof from [**8**]. The boundedness of C comes from Hardy's inequality [**19**] (see also [**21**]):

$$\sum_{n=1}^{\infty} \left| \frac{b_0 + b_1 + b_2 + \cdots + b_n}{N+1} \right|^2 \leqslant 4 \sum_{n=1}^{\infty} |b_n|^2. \tag{5.1}$$

One can also get the boundedness of C from Schur's theorem. This shows that C has a meaningful (bounded) Hilbert space adjoint

$$\begin{bmatrix} 1 & 0 & 0 & 0 & 0 & \cdots \\ \frac{1}{2} & \frac{1}{2} & 0 & 0 & 0 & \cdots \\ \frac{1}{3} & \frac{1}{3} & \frac{1}{3} & 0 & 0 & \cdots \\ \frac{1}{4} & \frac{1}{4} & \frac{1}{4} & \frac{1}{4} & 0 & \cdots \\ \frac{1}{5} & \frac{1}{5} & \frac{1}{5} & \frac{1}{5} & \frac{1}{5} & \cdots \\ \vdots & \vdots & \vdots & \vdots & \vdots & \ddots \end{bmatrix}^* = \begin{bmatrix} 1 & \frac{1}{2} & \frac{1}{3} & \frac{1}{4} & \frac{1}{5} & \cdots \\ 0 & \frac{1}{2} & \frac{1}{3} & \frac{1}{4} & \frac{1}{5} & \cdots \\ 0 & 0 & \frac{1}{3} & \frac{1}{4} & \frac{1}{5} & \cdots \\ 0 & 0 & 0 & \frac{1}{4} & \frac{1}{5} & \cdots \\ 0 & 0 & 0 & 0 & \frac{1}{5} & \cdots \\ \vdots & \vdots & \vdots & \vdots & \vdots & \ddots \end{bmatrix}. \tag{5.2}$$

The computation of $\|C\|$ goes as follows. A matrix calculation shows that

$$(I-C)(I-C)^* = \begin{bmatrix} 0 & 0 & 0 & 0 & \cdots \\ 0 & \frac{1}{2} & 0 & 0 & \cdots \\ 0 & 0 & \frac{2}{3} & 0 & \cdots \\ 0 & 0 & 0 & \frac{3}{4} & \cdots \\ \vdots & \vdots & \vdots & \vdots & \ddots \end{bmatrix}.$$

Now use the C^*-algebra identity for bounded operators (i.e., $\|A^*A\| = \|A\|^2$) on Hilbert space to obtain

$$\|I - C\|^2 = \|(I-C)(I-C)^*\| = \sup_{n \geqslant 0} \frac{n}{n+1} = 1. \tag{5.3}$$

Note the use of the fact that the norm of a diagonal operator is the supremum of the absolute values of its entries. The triangle inequality now shows that

$$\|C\| = \|I - (I - C)\| \leqslant \|I\| + \|I - C\| = 1 + 1 = 2.$$

To prove that $\|C\| = 2$, it is enough to prove that $\|C^*\| = 2$. Certainly $\|C^*\| = \|C\| \leqslant 2$. One can argue, as was done in [**8**], that for each $a > \frac{1}{2}$ we have

$$\|C^*(\tfrac{1}{(n+1)^a})_{n \geqslant 0}\| \geqslant \frac{1}{a} \|(\tfrac{1}{(n+1)^a})_{n \geqslant 0}\|.$$

Letting $a \to \frac{1}{2}$ proves the result. □

Though the focus of this paper will be on the Cesàro operator in the Hilbert space setting of ℓ^2 (and an associated space of analytic functions called the Hardy space and described in the next section), we would be remiss if we did not make a few remarks about an extension of the previous discussion to the more general setting of

$$\ell^p := \Big\{\mathbf{a} = (a_n)_{n\geqslant 0} : \|\mathbf{a}\|_p := \Big(\sum_{n=0}^{\infty} |a_n|^p\Big)^{\frac{1}{p}} < \infty\Big\}, \quad 1 < p < \infty.$$

The boundedness of C on ℓ^p for $1 < p < \infty$ comes from another version of Hardy's inequality:

$$\sum_{n=0}^{\infty} \Big|\frac{b_0 + b_1 + b_2 + \cdots + b_n}{n+1}\Big|^p \leqslant q^p \sum_{n=0}^{\infty} |b_n|^p.$$

In this inequality, $q = p/(p-1)$ is the conjugate index to p. Furthermore, the constant q^p is optimal [**21**] and thus,

$$\|C\|_{\ell^p \to \ell^p} = q.$$

There are results concerning the pesky cases $p = 1$ and $p = \infty$ [**40**].

As a bit of a diversion, let us unpack the fact that

$$\|I - C\|_{\ell^2 \to \ell^2} = 1$$

used in the proof of Theorem 5.1. What is $\|I - C\|_{\ell^p \to \ell^p}$? This question was posed in [**5**] and answered recently in [**44**] with the following:

$$\|I - C\|_{\ell^p \to \ell^p} = \begin{cases} \dfrac{1}{p-1} & \text{if } 1 < p \leqslant 2, \\[2mm] m_p^{-\frac{1}{p}} & \text{if } 2 < p < \infty, \end{cases}$$

where $m_p = \min\{pt^{p-1} + (1-t)^p - t^p : 0 \leqslant t \leqslant \frac{1}{2}\}$.

6. The spectral properties of the Cesàro operator

The spectrum of C on ℓ^2, denoted by $\sigma(C)$, is the set of $\lambda \in \mathbb{C}$ such that $C - \lambda I$ is not invertible in the bounded operators on ℓ^2. Recall that $\sigma(C)$ is a nonempty compact subset of $\mathbb{C}$ and that $\sigma(C^*) = \{\overline{\lambda} : \lambda \in \sigma(C)\}$. The point spectrum, denoted by $\sigma_p(C)$, is the set of eigenvalues of C.

THEOREM 6.1 (Brown–Halmos–Shields, [**8**]). *The following hold for the Cesàro operator on ℓ^2.*

(1) $\sigma_p(C) = \varnothing$.
(2) $\sigma_p(C^*) = \{z : |z - 1| < 1\}$.
(3) $\sigma(C) = \{z : |z - 1| \leqslant 1\}$.

There is a direct proof of this theorem, using only functional analysis and sequence spaces, in the Brown–Halmos–Shields paper. An alternative proof involves function theory on the disk. To do this, we introduce the *Hardy space*

$$H^2 := \Big\{f(z) = \sum_{n=0}^{\infty} a_n z^n : (a_n)_{n\geqslant 0} \in \ell^2\Big\}.$$

Routine estimates show that every such power series $f \in H^2$ has a radius of convergence at least one and thus defines an analytic function on $\mathbb{D}$. The Hardy space is a Hilbert space of analytic functions on $\mathbb{D}$ with the norm and inner product it naturally inherits by identifying a power series $f(z) = \sum_{n=0}^{\infty} a_n z^n$ in H^2 with its Taylor coefficients $(a_n)_{n \geqslant 0}$ in ℓ^2. In other words,

$$\|f\| = \Big(\sum_{n=0}^{\infty} |a_n|^2\Big)^{\frac{1}{2}} \quad \text{and} \quad \langle f, g\rangle = \sum_{n=0}^{\infty} a_n \overline{b_n},$$

where $(b_n)_{n \geqslant 0}$ is the sequence of Taylor coefficients of $g \in H^2$. From (4.1) there is the integral form of the Cesàro operator via

$$(Cf)(z) = \frac{1}{z} \int_0^z \frac{f(\xi)}{1-\xi} d\xi.$$

This shows that if $f(z) = \sum_{n=0}^{\infty} a_n z^n$ belongs to H^2, then

$$(Cf)(z) = \sum_{n=0}^{\infty} \Big(\frac{1}{n+1} \sum_{j=0}^{n} a_j\Big) z^n.$$

By Hardy's inequality from (5.1), the Taylor coefficients of Cf belong to ℓ^2 and thus $Cf \in H^2$. We also see that the linear transformation $f \mapsto Cf$, denoted simply by C, defines a bounded operator on H^2 with $\|C\| = 2$. Essentially, we are recasting what we know about the Cesàro operator on ℓ^2, written in terms of sequences, as in operator on a Hilbert space of analytic functions on $\mathbb{D}$, written in terms of Taylor coefficients.

What we gain from changing viewpoints is an enlightening way of proving Theorem 6.1. For example, we can show that $\sigma_p(C) = \varnothing$. Indeed, if $f \in H^2$ and $Cf = \lambda f$ for some $\lambda \in \mathbb{C}$, then, as discovered in (4.2) and (4.3),

$$\lambda = \frac{1}{n+1} \quad \text{and} \quad f(z) = k \frac{z^n}{(1-z)^{n+1}}.$$

For any $n \in \mathbb{N}_0$, one can check that

$$\frac{1}{(1-z)^{n+1}} = 1 + (n+1)z + \tfrac{1}{2}(n+2)(n+1)z^2 + \tfrac{1}{6}(n+3)(n+2)(n+1)z^3 + \cdots$$

and so the sequence of Taylor coefficients for f does not belong to ℓ^2. Thus, the only way for f to belong to H^2 is for $k = 0$. In other words, the Cesàro operator on ℓ^2 has no eigenvalues.

Continuing with our identification of the Cesàro operator as a matrix operator on ℓ^2 with an integral operator on H^2, one can discover from a matrix calculation using the adjoint formula from (5.1) that

$$(C^* f)(z) = \sum_{n=0}^{\infty} \Big(\sum_{j=n}^{\infty} \frac{a_j}{j+1}\Big) z^n = \frac{1}{1-z} \int_z^1 f(\xi) d\xi. \tag{6.1}$$

Thus, if we want to compute the eigenvalues of C^*, we need to find the $\lambda \in \mathbb{C}$ such that $C^* f = \lambda f$ for some $f \in H^2 \setminus \{0\}$. This leads us to the integral equation

$$\frac{1}{1-z} \int_z^1 f(\xi) d\xi = \lambda f(z)$$

which, in a similar way as before, has solutions

$$f(z) = k(1-z)^{\frac{1-\lambda}{\lambda}}. \tag{6.2}$$

These functions will belong to H^2, i.e., have Taylor coefficients which belong to ℓ^2, precisely when

$$\Re\Big(\frac{1-\lambda}{\lambda}\Big) > -\tfrac{1}{2}.$$

A little algebra will show that

$$\Re\Big(\frac{1-\lambda}{\lambda}\Big) > -\tfrac{1}{2} \iff |1-\lambda| < 1.$$

Thus $\sigma_p(C^*) = \{z : |z-1| < 1\}$ as advertised in Theorem 6.1.

From (5.3) it follows that $\sigma(I-C) \subseteq \{z : |z| \leqslant 1\}$ and thus $\sigma(C) \subseteq \{z : |z-1| \leqslant 1\}$. Hence, since $\{z : |z-1| < 1\} = \sigma_p(C^*) \subseteq \sigma(C^*) = \{\overline{w} : w \in \sigma(C)\}$, we obtain the containment

$$\{z : |z-1| < 1\} \subseteq \sigma(C) \subseteq \{z : |z-1| \leqslant 1\}.$$

Equality follows by taking closures. This proves Theorem 6.1.

The analog of Theorem 6.1 for the Cesàro operator on ℓ^p is the following.

THEOREM 6.2. *The spectrum of the Cesàro operator on ℓ^p for $1 < p < \infty$ is the closed disk $\{z : |z - \frac{q}{2}| \leqslant \frac{q}{2}\}$.*

7. The Cesàro operator is hyponormal

A bounded operator A on a Hilbert space $\mathcal{H}$ is *hyponormal* if $A^*A - AA^* \geqslant 0$, meaning that $\langle (A^*A - AA^*)\mathbf{x}, \mathbf{x}\rangle \geqslant 0$ for all $\mathbf{x} \in \mathcal{H}$. The Brown, Halmos, Shields paper [**8**] contains a proof that the Cesaro operator C on ℓ^2 is hyponormal. Their proof relies on a classical result of Sylvester concerning positivity of square matrices and is somewhat technical. Here is a simpler proof by Rhaly which connects to the interesting topic of *posinormal operators* [**41**].

PROPOSITION 7.1. *The Cesàro operator on ℓ^2 is hyponormal.*

PROOF. Consider the diagonal operator

$$D = \begin{bmatrix} \frac{1}{2} & 0 & 0 & 0 & \cdots \\ 0 & \frac{2}{3} & 0 & 0 & \cdots \\ 0 & 0 & \frac{3}{4} & 0 & \cdots \\ 0 & 0 & 0 & \frac{4}{5} & \cdots \\ \vdots & \vdots & \vdots & \vdots & \ddots \end{bmatrix}.$$

Notice that $I - D \geqslant 0$ (diagonal with positive entries).

A matrix computation will show that $C^*DC = CC^*$ and so for all $\mathbf{a} \in \ell^2$,

$$\begin{aligned} \langle (C^*C - CC^*)\mathbf{a}, \mathbf{a}\rangle &= \langle (C^*C - C^*DC)\mathbf{a}, \mathbf{a}\rangle \\ &= \langle (C - DC)\mathbf{a}, C\mathbf{a}\rangle \\ &= \langle (I - D)C\mathbf{a}, C\mathbf{a}\rangle \\ &\geqslant 0. \end{aligned}$$

This says $C^*C - CC^* \geqslant 0$, i.e., C is hyponormal. □

A bounded operator A on a Hilbert space $\mathcal{H}$ is *posinormal* if $AA^* = A^*PA$ for some positive operator P (called the interrupter). We have already seen that C is posinormal. It turns out that C^* is posinormal as well. Indeed, if S denotes the *unilateral shift* on ℓ^2, i.e.,

$$S := \begin{bmatrix} 0 & 0 & 0 & 0 & \cdots \\ 1 & 0 & 0 & 0 & \cdots \\ 0 & 1 & 0 & 0 & \cdots \\ 0 & 0 & 1 & 0 & \cdots \\ \vdots & \vdots & \vdots & \vdots & \ddots \end{bmatrix},$$

one can check that the positive operator

$$P = (C^* - S)(C - S^*)$$

satisfies $C^*C = CPC^*$ which makes C^* posinormal.

General properties of posinormal operators explored in [**41**] applied to the Cesàro operator and its adjoint, prove the following.

PROPOSITION 7.2. *For the Cesàro operator on* ℓ^2, *we have* $\operatorname{Ran} C = \operatorname{Ran} C^*$ *and hence* $\operatorname{Ran} C$ *is dense in* ℓ^2.

One can also see the density of the range of C on ℓ^2 more directly from the simple facts, using the matrix representation of C, that

$$\mathbf{e}_0 = C\mathbf{e}_0 - C\mathbf{e}_1,$$

$$\tfrac{1}{2}\mathbf{e}_1 = C\mathbf{e}_1 - C\mathbf{e}_2,$$

$$\tfrac{1}{3}\mathbf{e}_2 = C\mathbf{e}_2 - C\mathbf{e}_3,$$

and so on. Note that though the range of C is dense, it is not closed since otherwise $\operatorname{Ran} C = \ell^2$ and, using the fact that $\ker C = \{0\}$, would mean that C is invertible, which it is not (Theorem 6.1).

We end our discussion of hyponormality with the following observation. Consider the infinite matrix

$$A = \begin{bmatrix} \frac{1}{3} & \frac{1}{4} & \frac{1}{5} & \frac{1}{6} & \frac{1}{7} & \frac{1}{8} \cdots \\ -\frac{2}{3} & \frac{1}{4} & \frac{1}{5} & \frac{1}{6} & \frac{1}{7} & \frac{1}{8} \cdots \\ 0 & -\frac{3}{4} & \frac{1}{5} & \frac{1}{6} & \frac{1}{7} & \frac{1}{8} \cdots \\ 0 & 0 & -\frac{4}{5} & \frac{1}{6} & \frac{1}{7} & \frac{1}{8} \cdots \\ 0 & 0 & 0 & -\frac{5}{6} & \frac{1}{7} & \frac{1}{8} \cdots \\ 0 & 0 & 0 & 0 & -\frac{6}{7} & \frac{1}{8} \cdots \\ \vdots & \vdots & \vdots & \vdots & \vdots & \ddots \end{bmatrix}.$$

An application of Schur's theorem will show that A is a contraction on ℓ^2 and a matrix computation will reveal that

$$C^* = AC.$$

This is a specific case of a more general result which says that for every hyponormal operator T, there is a contraction W such that $T^* = WT$.

8. The Cesàro operator is subnormal

We know from the previous section that the Cesàro operator on ℓ^2 (or H^2) is hyponormal ($C^*C - CC^* \geqslant 0$). One of the highlights in the subject is a result of Kriete and Trutt [**30**] which says that C is subnormal. A bounded operator A on a Hilbert space $\mathcal{H}$ is *subnormal* if there is a Hilbert space $\mathcal{K}$ which contains $\mathcal{H}$ (as a closed subspace) and a bounded normal operator N on $\mathcal{K}$ such that $N\mathcal{H} \subseteq \mathcal{H}$ and $N|_{\mathcal{H}} = A$. The normal operator N is called a *normal extension* of A and is never unique. In short, a subnormal operator is the restriction of a normal operator to an invariant subspace. One can show that a subnormal operator is hyponormal but the converse is not true.

Some immediate examples of subnormal operators come from the following construction. Start with a positive, finite, compactly supported, Borel measure μ on $\mathbb{C}$. The multiplication operator $N_\mu f = zf$ on $L^2(\mu)$ is bounded and normal since $N_\mu^* f = \overline{z} f$ and thus $N_\mu N_\mu^* = N_\mu^* N_\mu$. Furthermore, if $\mathcal{H}^2(\mu)$ denotes the closure of the analytic polynomials in $L^2(\mu)$, i.e., the closed linear span of $\{1, z, z^2, z^3, \ldots\}$, then the multiplication operator $S_\mu f = zf$ is bounded on $\mathcal{H}^2(\mu)$ and is subnormal since $\mathcal{H}^2(\mu)$ is an invariant subspace of N_μ and $N_\mu|_{\mathcal{H}^2(\mu)} = S_\mu$. See [**11**] for a detailed treatment of subnormal operators.

What makes the subnormality of the Cesàro operator C so surprising is that there is no clear normal extension. The Kriete–Trutt result shows that C is unitarily equivalent to S_μ on some $\mathcal{H}^2(\mu)$ space for a measure μ supported in $\overline{\mathbb{D}}$ (the closure of $\mathbb{D}$). This is enough to verify the subnormality of C.

Their construction starts with the functions

$$\varphi_w(z) = (1-z)^{\frac{w}{1-w}}, \quad w \in \mathbb{D}. \tag{8.1}$$

Observe that each function φ_w is analytic on $\mathbb{D}$ and, since the linear fractional transformation

$$w \mapsto \frac{w}{1-w}$$

maps $\mathbb{D}$ to $\{z : \Re z > -\frac{1}{2}\}$, it follows that $\varphi_w \in H^2$ (the sequence of Taylor coefficients belongs to ℓ^2). Since

$$\varphi_{\frac{n}{n+1}}(z) = (1-z)^n, \quad n \in \mathbb{N}_0,$$

we see that the span of $\{\varphi_w : w \in \mathbb{D}\}$ contains the polynomials which constitute a dense set in H^2.

Using the integral form of the adjoint of C from (6.1) we see that

$$C^* \varphi_w = (1-w)\varphi_w. \tag{8.2}$$

The reader will recognize these eigenfunctions from (6.2). For each $f \in H^2$, the function

$$(Kf)(z) := \langle f, \varphi_{\overline{z}} \rangle$$

is analytic on $\mathbb{D}$ and the set

$$\mathcal{K} := \{Kf : f \in H^2\}$$

is a vector space. Since $\{\varphi_w : w \in \mathbb{D}\}$ has dense linear span in H^2, we see that $Kf \equiv 0$ if and only if $f \equiv 0$. This allows us to define a Hilbert space norm on $\mathcal{K}$ as

$$\|Kf\|_{\mathcal{K}} := \|f\|_{H^2}.$$

With this norm, and corresponding inner product, the operator K is unitary from H^2 onto $\mathcal{K}$.

PROPOSITION 8.1. *For all $f \in H^2$, we have*

$$(KCf)(z) = (1-z)(Kf)(z), \quad z \in \mathbb{D}.$$

Thus, C is unitarily equivalent to the operator of multiplication by the function $1-z$ on $\mathcal{K}$.

PROOF. For $f \in H^2$ and $z \in \mathbb{D}$, use (8.2) to see that

$$\begin{aligned}(KCf)(z) &= \langle Cf, \varphi_{\overline{z}} \rangle_{H^2} \\ &= \langle f, C^* \varphi_{\overline{z}} \rangle_{H^2} \\ &= \langle f, (1-\overline{z}) \varphi_{\overline{z}} \rangle_{H^2} \\ &= (1-z) \langle f, \varphi_{\overline{z}} \rangle_{H^2} \\ &= (1-z)(Kf)(z).\end{aligned}$$

□

Therefore, the Cesàro operator is unitarily equivalent to the operator of multiplication by $1-z$ on a Hilbert space $\mathcal{K}$ of analytic functions on $\mathbb{D}$. A discussion of Kriete and Trutt shows that $\mathcal{K}$ contains the polynomials as a dense set and, more importantly (and quite difficult to prove), there is a positive finite Borel measure μ on $\overline{\mathbb{D}}$ such that

$$\int_{\overline{\mathbb{D}}} |p|^2 d\mu = \|p\|_{\mathcal{K}}^2 \quad \text{for all } p \in \mathbb{C}[z]. \tag{8.3}$$

This says that $\mathcal{K} = \mathcal{H}^2(\mu)$ and thus, C is unitarily equivalent to $I - S_\mu$. This last operator is subnormal and hence C is subnormal. The measure μ is quite complicated which makes the space $\mathcal{H}^2(\mu)$ also very complicated.

9. The commutant of the Cesàro operator

We discussed the infinite matrices T which commute with the Cesàro matrix C, i.e., $CT = TC$ in Theorem 3.3. These are the Hausdorff matrices. However, not all of these Hausdorff matrices define bounded operators on ℓ^2. In this section we describe those which define *bounded* operators on ℓ^2.

The commutant of the Cesàro operator on ℓ^2 is defined to be the bounded operators A on ℓ^2 such that $AC = CA$. This set, traditionally denoted by $\{C\}'$, is an algebra that is also closed in the weak operator topology. One can describe $\{C\}'$ via the space $\mathcal{H}^2(\mu)$ used in the previous section to prove that C is subnormal. In [**30**] Kriete and Trutt showed that C is unitarily equivalent to $I - S_\mu$ (multiplication by $1-z$) on $\mathcal{H}^2(\mu)$ for some measure μ on $\overline{\mathbb{D}}$. In [**31**] they went on to show the following.

PROPOSITION 9.1. *Let μ be the measure on $\overline{\mathbb{D}}$ from* (8.3).

(1) *For each $\varphi \in H^\infty$, the algebra of bounded analytic functions on $\mathbb{D}$, the operator $\varphi(S_\mu)$ is a well defined bounded operator on $\mathcal{H}^2(\mu)$ that is equal to $\varphi(S_\mu)f = \varphi f$, $f \in \mathcal{H}^2(\mu)$.*
(2) $\{S_\mu\}' = \{\varphi(S_\mu) : \varphi \in H^\infty\}$.

PROOF. The fact that the operator $f \mapsto \varphi f$ is bounded on $\mathcal{H}^2(\mu)$ for all $\varphi \in H^\infty$ is a detail contained in [**31**]. Moreover, if φ is an analytic function on $\mathbb{D}$ which satisfies $\varphi \mathcal{H}^2(\mu) \subseteq \mathcal{H}^2(\mu)$, then standard results show that $\varphi \in H^\infty$.

Clearly $\varphi(S_\mu)$ is well defined (and is equal to multiplication by φ) for any polynomial φ. If $(\varphi_n)_{n\geqslant 1}$ is a sequence of polynomials that approximate φ in the weak-$*$ topology on H^∞, one can show, as argued in [**31**], that $\varphi_n(S_\mu) \to \varphi(S_\mu)$ in the weak operator topology.

The discussion in the previous paragraph says that $\{\varphi(S_\mu) : \varphi \in H^\infty\} \subseteq \{S_\mu\}'$. Now let $T \in \{S_\mu\}'$. Then for any polynomial φ we have $\varphi(S_\mu)T = T\varphi(S_\mu)$. Apply the previous identity to the constant function 1 to get $T\varphi = \varphi T1$. Now approximate any $f \in \mathcal{H}^2(\mu)$ by a sequence of polynomials in $L^2(\mu)$, and pass to a subsequence if necessary to assume convergence μ-almost everywhere, to see that $Tf = \psi f$, where $\psi = T1$. As discussed earlier in the proof, $\psi \in H^\infty$ and $T = \psi(S_\mu)$. □

As a corollary to this result we see the following result of Shields and Wallen [**43**].

THEOREM 9.2 (Shields–Wallen). *For a bounded operator A on ℓ^2, the following are equivalent.*

(1) $A \in \{C\}'$.
(2) *A belongs to the weak operator closure of* $\{p(C) : p \in \mathbb{C}[z]\}$.
(3) *There is a a bounded analytic function ψ on $\{z : |z-1| < 1\}$ such that* $A = \psi(C)$.

Let us reconnect to the Euler matrices E_λ, $\lambda \in \mathbb{C}$, defined in (3.5). We know they commute with the Cesàro matrix. Which one of these are bounded and thus belong to the commutant of the Cesàro operator on ℓ^2? The answer comes from [**32**].

THEOREM 9.3.

(1) *If $\lambda \in (\frac{1}{2}, 1]$, then E_λ defines a bounded operator on ℓ^2 with* $\|E_\lambda\| = \lambda^{-\frac{1}{2}}$.
(2) *If $\lambda \in (0, \frac{1}{2}]$, then E_λ defines a bounded operator on ℓ^2 with* $\|E_\lambda\| \leqslant (1-\lambda^2)^{-\frac{1}{2}}$.
(3) *If $\lambda \in \mathbb{C} \setminus (0,1]$, then E_λ is unbounded on ℓ^2.*

For $0 < \lambda < 1$, the matrix E_λ defines a bounded operator on ℓ^2 that commutes with the Cesàro operator. By Theorem 9.2, $E_\lambda = F_\lambda(C)$, where F_λ is a bounded analytic function on $\{z : |z-1| < 1\}$. We can compute this functions explicitly. Indeed,

$$F_\lambda(C) = W \begin{bmatrix} F(1) & 0 & 0 & 0 & \cdots \\ * & F(\frac{1}{2}) & 0 & 0 & \cdots \\ * & * & F(\frac{1}{3}) & 0 & \cdots \\ * & * & * & F(\frac{1}{4}) & \cdots \\ \vdots & \vdots & \vdots & \vdots & \ddots \end{bmatrix} W,$$

where $*$ denotes an entry that is not important. The function $F_\lambda(z) = \lambda^{1/z-1}$ is a bounded analytic function on $\{z : |z-1| < 1\}$ with $F_\lambda(\frac{1}{n}) = \lambda^n$.

In general, from Theorem 3.3, the commutant of the Cesàro operator on ℓ^2 is the set of Hausdorff matrices that define bounded operators on ℓ^2.

10. The square root of the Cesàro operator

The Cesáro operator on ℓ^2 has a (bounded) square root (in fact two of them). The first result along these lines is the following. We will use the Taylor series expansion

$$\sqrt{1-z} = 1 - \frac{z}{2} - \frac{z^2}{8} - \frac{z^3}{16} - \frac{5z^4}{128} - \frac{7z^5}{256} \cdots .$$

Notice that the coefficient sequence of $\sqrt{1-z}$ is absolutely summable.

THEOREM 10.1 (Mashreghi–Ptak–Ross, [**35**]). *The following are equivalent for a bounded operator A on ℓ^2.*

(1) $A^2 = C$.

(2)

$$A = \pm\Big(I - \tfrac{1}{2}(I-C) - \tfrac{1}{8}(I-C)^2 - \tfrac{1}{16}(I-C)^3 + \cdots\Big), \tag{10.1}$$

where the previous series converges in operator norm.

PROOF. (2)$\Longrightarrow$(1) follows from multiplication of series. To see that (1)$\Longrightarrow$(2), observe that if A is bounded on ℓ^2 and $A^2 = C$, then $AC = CA$. By Proposition 8.1, this implies that the operator KAK^* commutes with $I - S_\mu$ and thus, by Proposition 9.1, $KAK^* = \varphi(S_\mu)$ for some $\varphi \in H^\infty$. But since KAK^* must also be a bounded square root of $\varphi(S_\mu)$, it must be the case that $\varphi(z) = \sqrt{1-z}$ (or $\varphi(z) = -\sqrt{1-z}$). Now use the fact that the coefficient sequence of $\sqrt{1-z}$ is absolutely summable to see that

$$\varphi(S_\mu) = I - \tfrac{1}{2}S_\mu - \tfrac{1}{8}S_\mu^2 - \tfrac{1}{16}S_\mu^3 - \cdots$$

converges in operator norm. Since $K^*S_\mu K = I - C$ we see that

$$A = \pm\Big(I - \tfrac{1}{2}(I-C) - \tfrac{1}{8}(I-C)^2 - \tfrac{1}{16}(I-C)^3 + \cdots\Big),$$

which completes the proof. □

Theorem 10.1 may seem somewhat unsatisfactory since it does not really yield a formula (in some way) for the two square roots of C. Towards finding a specific formula for the square root of C, recall from (3.4) that $C = WDW$, where

$$W = \begin{bmatrix} 1 & 0 & 0 & 0 & 0 & \cdots \\ 1 & -1 & 0 & 0 & 0 & \cdots \\ 1 & -2 & 1 & 0 & 0 & \cdots \\ 1 & -3 & 3 & -1 & 0 & \cdots \\ 1 & -4 & 6 & -4 & 1 & \cdots \\ \vdots & \vdots & \vdots & \vdots & \vdots & \ddots \end{bmatrix}, \quad D = \begin{bmatrix} 1 & 0 & 0 & 0 & 0 & \cdots \\ 0 & \frac{1}{2} & 0 & 0 & 0 & \cdots \\ 0 & 0 & \frac{1}{3} & 0 & 0 & \cdots \\ 0 & 0 & 0 & \frac{1}{4} & 0 & \cdots \\ 0 & 0 & 0 & 0 & \frac{1}{5} & \cdots \\ \vdots & \vdots & \vdots & \vdots & \vdots & \ddots \end{bmatrix},$$

and $W^2 = I$. It is important to notice here that W does not define a bounded operator on ℓ^2 and the matrix multiplication WDW is merely matrix multiplication, which is justified since the matrices involved are lower triangular.

From here one can immediately see square roots of the Cesàro matrix, namely

$$A^\sigma = W \begin{bmatrix} \pm 1 & 0 & 0 & 0 & \cdots \\ 0 & \pm\sqrt{\frac{1}{2}} & 0 & 0 & \cdots \\ 0 & 0 & \pm\sqrt{\frac{1}{3}} & 0 & \cdots \\ 0 & 0 & 0 & \pm\sqrt{\frac{1}{4}} & \cdots \\ \vdots & \vdots & \vdots & \vdots & \ddots \end{bmatrix} W, \tag{10.2}$$

where the signs along the diagonal of the middle matrix is determined by $\sigma : \mathbb{N} \to \{\pm 1\}$. In fact, A^σ can be computed in closed form as

$$A^\sigma_{ij} = \begin{cases} \binom{i}{j} \sum_{\ell=0}^{i-j} (-1)^\ell \sigma(\ell+j+1) \frac{1}{\sqrt{\ell+j+1}} \binom{i-j}{\ell} & i \geqslant j, \\ 0 & i < j. \end{cases}$$

THEOREM 10.2 (Hupert–Leggett, [**25**]). *For an infinite lower-triangular matrix A, the following are equivalent.*

(1) $A^2 = C$.

(2) $A = A^\sigma$ *for some* $\sigma : \mathbb{N} \to \{\pm 1\}$.

The matrices from (10.1) are all lower triangular and one can see that they are

$$\begin{bmatrix} 1 & 0 & 0 & 0 & 0 & \cdots \\ 1 & -1 & 0 & 0 & 0 & \cdots \\ 1 & -2 & 1 & 0 & 0 & \cdots \\ 1 & -3 & 3 & -1 & 0 & \cdots \\ 1 & -4 & 6 & -4 & 1 & \cdots \\ \vdots & \vdots & \vdots & \vdots & \vdots & \ddots \end{bmatrix} \begin{bmatrix} 1 & 0 & 0 & 0 & \cdots \\ 0 & \sqrt{\frac{1}{2}} & 0 & 0 & \cdots \\ 0 & 0 & \sqrt{\frac{1}{3}} & 0 & \cdots \\ 0 & 0 & 0 & \sqrt{\frac{1}{4}} & \cdots \\ \vdots & \vdots & \vdots & \vdots & \ddots \end{bmatrix} \begin{bmatrix} 1 & 0 & 0 & 0 & 0 & \cdots \\ 1 & -1 & 0 & 0 & 0 & \cdots \\ 1 & -2 & 1 & 0 & 0 & \cdots \\ 1 & -3 & 3 & -1 & 0 & \cdots \\ 1 & -4 & 6 & -4 & 1 & \cdots \\ \vdots & \vdots & \vdots & \vdots & \vdots & \ddots \end{bmatrix}$$

and

$$\begin{bmatrix} 1 & 0 & 0 & 0 & 0 & \cdots \\ 1 & -1 & 0 & 0 & 0 & \cdots \\ 1 & -2 & 1 & 0 & 0 & \cdots \\ 1 & -3 & 3 & -1 & 0 & \cdots \\ 1 & -4 & 6 & -4 & 1 & \cdots \\ \vdots & \vdots & \vdots & \vdots & \vdots & \ddots \end{bmatrix} \begin{bmatrix} -1 & 0 & 0 & 0 & \cdots \\ 0 & -\sqrt{\frac{1}{2}} & 0 & 0 & \cdots \\ 0 & 0 & -\sqrt{\frac{1}{3}} & 0 & \cdots \\ 0 & 0 & 0 & -\sqrt{\frac{1}{4}} & \cdots \\ \vdots & \vdots & \vdots & \vdots & \ddots \end{bmatrix} \begin{bmatrix} 1 & 0 & 0 & 0 & 0 & \cdots \\ 1 & -1 & 0 & 0 & 0 & \cdots \\ 1 & -2 & 1 & 0 & 0 & \cdots \\ 1 & -3 & 3 & -1 & 0 & \cdots \\ 1 & -4 & 6 & -4 & 1 & \cdots \\ \vdots & \vdots & \vdots & \vdots & \vdots & \ddots \end{bmatrix}.$$

Thus, these two matrices are the *bounded* square roots of the Cesàro matrix. The other choices of signs in (10.2) yield square roots of the Cesàro matrix but these do not define bounded operators on ℓ^2. This result was rediscovered and generalized in [**17**].

11. The Cesàro operator and composition operators

There is a relationship between the Cesàro operator on the Hardy space H^2 and composition operators. For any analytic $\varphi : \mathbb{D} \to \mathbb{D}$ (often called an *analytic self map* of the disk), an application of the Littlewood subordination theorem [**15**] says that the composition operator $C_\varphi f = f \circ \varphi$ defines a bounded operator on H^2.

The first connection between C_φ and C appears in a result of Deddens [**13**]. If $0 < \alpha < 1$, consider the composition operator $C_{\alpha+(1-\alpha)z}$. With respect to the standard orthonormal basis $(z^n)_{n \geqslant 0}$ for H^2, the composition operator $C_{\alpha+(1-\alpha)z}$ has the matrix representation

$$\begin{bmatrix} 1 & \alpha & \alpha^2 & \alpha^3 & \cdots \\ 0 & (1-\alpha) & 2\alpha(1-\alpha) & 3\alpha^2(1-\alpha) & \cdots \\ 0 & 0 & (1-\alpha)^2 & 3\alpha(1-\alpha)^2 & \cdots \\ 0 & 0 & 0 & (1-\alpha)^3 & \cdots \\ \vdots & \vdots & \vdots & \vdots & \ddots \end{bmatrix}.$$

Furthermore, by matrix multiplication, $C_{\alpha+(1-\alpha)z}$ commutes with C^* and thus by Theorem 9.2,

$$C_{\alpha+(1-\alpha)z} = F(C^*)$$

for some bounded analytic function on the disk $\{z : |z-1| < 1\}$. In fact, since

$$F(C^*) = \begin{bmatrix} F(1) & * & * & * & * & \cdots \\ 0 & F(\frac{1}{2}) & * & * & * & \cdots \\ 0 & 0 & F(\frac{1}{3}) & * & * & \cdots \\ 0 & 0 & 0 & F(\frac{1}{4}) & * & \cdots \\ 0 & 0 & 0 & 0 & F(\frac{1}{5}) & \cdots \\ \vdots & \vdots & \vdots & \vdots & \vdots & \ddots \end{bmatrix}$$

(the entries with a $*$ are unimportant), one can see that $F(\frac{1}{n}) = (1-\alpha)^{n-1}$ and thus, since the sequence $(\frac{1}{n})_{n \geqslant 1}$ does not form a Blaschke sequence in the disk $\{z : |z-1| < 1\}$, we see that $F(z) = (1-\alpha)^{1/z-1}$. Since C and thus $F(C^*)^*$ is subnormal, it follows that $C^*_{\alpha z+(1-\alpha)}$ is subnormal (often called cosubnormal).

Though we will not get into the details here, the Cesàro operator also connects to a semigroup of composition operators. Though this dates back to [**12, 13**], this connection was used most recently in [**17**] to give some insight into the (complicated) invariant subspace structure for the Cesàro operator. Consider the collection of analytic self maps

$$\varphi_t(z) = e^{-t}z + (1-e^{-t}), \quad t \geqslant 0.$$

One can show that $\varphi_{s+t}(z) = \varphi_s(\varphi_t(z))$ and that this family of mappings form a *holomorphic flow*. Moreover, each of these functions φ_t maps $\mathbb{D}$ onto another disk which is internally tangent to $\mathbb{D}$ at $z = 1$. The family of corresponding composition operators $\{C_{\varphi_t} : t \geqslant 0\}$ forms a strongly continuous semigroup of operators on H^2. Using the fact that one can quickly identify the infinitesimal generator of the semigroup, Gallardo and Partington [**17**] show that

$$(C^* f)(z) = \int_0^\infty e^{-t}(C_{\varphi_t} f)(z)dt.$$

They use this to discuss the invariant subspaces of C in the following Beurling–Lax type of result.

THEOREM 11.1 (Gallardo–Partington, [**17**]). *A closed subspace $\mathcal{M}$ of H^2 is invariant for the Cesàro operator if and only if $\mathcal{M}^\perp$ is invariant for each C_{φ_t}, $t \geqslant 0$.*

12. Invariant subspaces

The invariant subspaces of the Cesàro operator on ℓ^2 seem complicated and do not have a complete description. There are various ways to measure the complexities involved.

As a first order of business, we establish the fact that C is irreducible meaning there are no closed subspaces $\mathcal{M}$ of H^2 for which $C\mathcal{M} \subseteq \mathcal{M}$ and $C^*\mathcal{M} \subseteq \mathcal{M}$. Indeed, if C were reducible, then, via unitary equivalence, the multiplication operator M_z on the Kriete–Trutt space $H^2(\mu)$ would also be reducible. Some basic facts from operator theory would mean there is an orthogonal projection P on $H^2(\mu)$ for which $PM_z = M_z P$. Well known results concerning the commutant of M_z will show that $P = M_\varphi$ for some $\varphi \in H^\infty$. However, from the fact that $P^2 = P$, it follows that $M_{\varphi^2} = M_\varphi$ and thus $\varphi \equiv 0$ or $\varphi \equiv 1$. Thus, $P = 0$ or $P = I$ and so M_z, and hence C, is irreducible.

A paper of Kriete and Trutt [**31**] began to explore some of the complexities of the invariant subspaces of Cesàro operator by means of M_z on $\mathcal{H}^2(\mu)$. For example, if $Z = (z_n)_{n \geqslant 1} \subseteq \mathbb{D}$ and

$$\mathcal{M}(Z) = \{f \in \mathcal{H}^2(\mu) : f(z_n) = 0 \text{ for all } n \geqslant 1\},$$

then $\mathcal{M}(Z)$ is closed in $\mathcal{H}^2(\mu)$ and is an invariant for M_z. In this definition of $\mathcal{M}(Z)$, we impose the condition that the derivatives of order $k-1$ vanish at z_n if the zero z_n is repeated k times in the sequence. Via the functions φ_w from (8.1) and the unitary equivalence between C and $I - M_z$ given in Proposition 8.1, the space $\mathcal{M}(Z)$ gives rise to the C-invariant subspace

$$\Big(\bigvee \{\varphi_{z_n} : n \geqslant 1\}\Big)^{\perp}.$$

Of course there is the issue as to whether $\mathcal{M}(Z) \neq \{0\}$, in other words, Z is a zero set for $\mathcal{H}^2(\mu)$, but this issue seems very far from being resolved. Indeed, as pointed out in [**31**], one can create two sequences Z_1 and Z_2 such that $\mathcal{M}(Z_1) \neq \{0\}$ and $\mathcal{M}(Z_2) \neq \{0\}$ but $\mathcal{M}(Z_1) \cap \mathcal{M}(Z_2) = \{0\}$. This indicates that the zeros of $H^2(\mu)$ can be quite wild and behave more like those of the Bergman space than those of the Hardy and Dirichlet spaces.

Further indications of the complexities of the invariant subspaces of the Cesàro operator come from the paper [**17**] where they show that the invariant subspaces of C are in one-to-one correspondence with the subspaces of $L^2(\mathbb{R}, e^{-2(e^x - 1)}dx)$ which are invariant under the right shift semigroup $\{S_t, t \geqslant 0\}$, where $S_t f(x) = f(x-t)$. Moreover, these right shift invariant subspaces are known to be quite complicated and a complete description of them is unknown.

13. The continuous Cesàro operator

There is a continuous version of the Cesàro operator on $L^2[0,1]$ (inspired by the integral formula in (4.1)) that was explored in [**8**]. Define the operator

$$C_1 : L^2[0,1] \to L^2[0,1], \quad (C_1 f)(x) = \frac{1}{x}\int_0^x f(t)dt.$$

Classical real analysis says that $(C_1 f)(x)$ defines a continuous function on $[0,1]$. Furthermore, an application of Schur's integral test, as was done in [**8**], shows that $f \mapsto C_1 f$, denoted by C_1, is a bounded operator on $L^2[0,1]$. One can also use an integral version of Hardy's inequality, namely

$$\int_0^1 |(C_1 f)(x)|^2 dx \leqslant 4 \int_0^1 |f(t)|^2 dt \quad \text{for all } f \in L^2[0,1],$$

to show that $\|C_1\| \leqslant 2$. In fact, $\|C_1\| = 2$.

Unlike its discrete counterpart on ℓ^2, the continuous Cesàro operator C_1 has eigenvalues. Indeed,

$$(C_1 f)(x) = \lambda f(x) \iff \lambda \frac{d}{dx}(x f(x)) = f(x).$$

The solutions to this differential (Euler) equation are scalar multiplies of the functions

$$f_\lambda(x) = x^{\frac{1}{\lambda} - 1} \quad \text{where } \Re(\frac{1}{\lambda}) > \frac{1}{2}. \tag{13.1}$$

From here one can show that the eigenvalues of C_1 are the open disk $\{z : |z-1| < 1\}$ and, in a similar way that was done for the discrete Cesàro operator, $\sigma(C_1) = \{z : |z-1| \leqslant 1\}$.

There is also an $L^p[0,1]$ version of all of this where one can show that C_1 is a bounded operator on $L^p[0,1]$ when $1 < p < \infty$ with norm equal to $p/(p-1)$. Moreover, the operator C_1 on $L^p[0,1]$ has eigenvalues $\{z : |z - q/2| < q/2\}$ and spectrum the closure of this disk. There is also the resolvent formula [**33**]

$$(\lambda I - C_1)^{-1} = \frac{1}{\lambda} I + \frac{1}{\lambda^2} P_\lambda, \quad |\lambda - 1| > 1,$$

where

$$(P_\lambda f)(x) = \int_0^1 f(xt) t^{-1/\lambda} dt.$$

One can also show that

$$(C_1^* f)(x) = \int_x^1 \frac{f(t)}{t} dt$$

and there is an important connection between C_1^* and the unilateral shift that was discovered in [**8**]. By means of Laplace transforms and Hardy spaces of the right-half plane, they prove the following.

THEOREM 13.1 (Brown–Halmos–Shields). *The operator $I - C_1^*$ is unitarily equivalent to the shift operator $(Sf)(z) = zf(z)$ on the Hardy space H^2.*

As an interesting corollary to this, one can prove the following, which is quite cumbersome to do directly.

COROLLARY 13.2. *The operator $I - C_1^*$ is an isometry on $L^2[0,1]$.*

From Theorem 13.1 one could obtain the invariant subspaces of C_1, or C_1^*, via Beurling's characterization of the invariant subspaces of the shift on H^2 [**15**]. However, the correspondence in [**8**] between the invariant subspaces of the shift on H^2 the the invariant subspaces of $I - C_1^*$ is hidden by unitary operators making the invariant subspaces somewhat difficult to understand. Nevertheless, Nikolskii [**37**, p. 37] was able to use Hardy space theory to describe the invariant subspaces as Theorem 13.3. A direct approach was done recently by Agler and McCarthy in [**1**]. Their description of the invariant subspaces of C_1 goes as follows.

Let $W = \{z : \Re z > -\frac{1}{2}\}$ and for a *finite* subset B of W, let

$$\mathcal{X}(B) := \operatorname{span}\{x^b : b \in B\}.$$

Since

$$C_1 x^b = \frac{1}{b+1} x^b,$$

we see that $\mathcal{X}(B)$ is an invariant subspace of C_1. The condition that $\Re b > -\frac{1}{2}$ assures that $x^b \in L^2[0,1]$. Agler and McCarthy call such $\mathcal{X}(B)$ *finite monomial spaces.* For a sequence $\mathcal{B} := (B_n)_{n \geqslant 1}$ of finite subsets of W define

$$\mathcal{X}_{\mathcal{B}} := \Big\{ f \in L^2[0,1] : \lim_{n \to \infty} \operatorname{dist}(f, \mathcal{X}(B_n)) = 0 \Big\}.$$

One can verify that $\mathcal{X}_{\mathcal{B}}$ is a closed subspace of $L^2[0,1]$ that is invariant for C_1. These are called *monomial spaces.*

THEOREM 13.3. *Let $\mathcal{X}$ be a closed nonzero subspace of $L^2[0,1]$. Then $\mathcal{X}$ is invariant for C_1 if and only if $\mathcal{X}$ is a monomial space.*

There is a version of the continuous Cesàro operator that is defined on $L^2(0,\infty)$ by

$$(C_\infty f)(x) = \frac{1}{x}\int_0^x f(t)dt.$$

An analysis with a version of Hardy's inequality for $L^2(0,\infty)$ shows that $\|C_\infty\| = 2$. Formally, the eigenvectors of C_∞ can be computed as done with C_1 (see (13.1)) but none of them belong to $L^2(0,\infty)$ and thus $\sigma_p(C_\infty) = \varnothing$. There is also the resolvent formula

$$(\lambda I - C_\infty)^{-1} = \frac{1}{\lambda}I + \frac{1}{\lambda^2}P_\lambda, \quad |\lambda - 1| > 1.$$

One also shows that $\sigma(C_\infty)$ is the circle $\{z : |z-1| = 1\}$. Somewhat similar to the situation with C_1, there is the following.

THEOREM 13.4 (Brown–Halmos–Shields, [**8**]). *The operator $I - C_\infty^*$ is unitarily equivalent to the bilateral shift operator $(Mf)(\xi) = \xi f(\xi)$ on $L^2(\mathbb{T})$.*

14. The Cesàro operator in other settings

There are a host of results which document when the Cesàro operator

$$(Cf)(z) = \frac{1}{z}\int_0^z \frac{f(\xi)}{1-\xi}d\xi$$

and its adjoint

$$(Af)(z) = \frac{1}{1-z}\int_z^1 f(\xi)d\xi$$

define bounded linear transformations on various Banach and Fréchet spaces of analytic functions on $\mathbb{D}$. In the Hilbert space H^2 we have $A = C^*$. For a Banach space on analytic functions on $\mathbb{D}$, the adjoint operator acts on the dual space and not the original space. In some cases, one can even compute the spectrum of the operators C and A. We document a selection of these results and then provide some guidance where to discover more about this.

The first set of results deal with the Hardy spaces H^p with $0 < p < \infty$ of analytic function f on $\mathbb{D}$ for which

$$\|f\|_{H^p} := \Big(\sup_{0<r<1}\int_{\mathbb{T}} |f(r\xi)|^p dm(\xi)\Big)^{\frac{1}{p}} < \infty.$$

Note that when $p = 2$, we recover the Hardy space H^2 discussed in previous sections. It is well known [**15**] that when $1 \leqslant p < \infty$, the Hardy space H^p is a Banach space with norm $\|\cdot\|_{H^p}$. When $0 < p < 1$, H^p is a Fréchet space with the metric $\|\cdot\|_{H^p}^p$. As a consequence of [**3**] and [**27**], one can prove that when $1 \leqslant p < \infty$, the Cesàro operator C defines a bounded operator on H^p. In [**45**], Siskakis used semigroups of weighted composition operators to prove the following.

THEOREM 14.1 (Siskakis). *Consider the Cesàro operator C on H^p.*

(1) *If $2 \leqslant p < \infty$, then $\|C\|_{H^p \to H^p} = p$ and the spectrum of C is the closed disk $\{z : |z - \frac{1}{p}| \leqslant \frac{1}{p}\}$.*
(2) *If $1 \leqslant p < 2$, then $p \leqslant \|C\|_{H^p\to H^p} \leqslant 2$ and the spectrum of C contains the closed disk $\{z : |z - \frac{1}{p}| \leqslant \frac{1}{p}\}$.*
(3) *If $1 < p < \infty$, then $\|A\|_{H^p\to H^p} = p/(p-1)$.*

The main driver of Theorem 14.1, and for many other results which were to follow, is the following observation which is interesting in its own right. The set of maps

$$\varphi_t(z) = \frac{e^{-t}z}{(e^{-t}-1)z+1}, \qquad t \geqslant 0,$$

form a semigroup of analytic self maps of $\mathbb{D}$. If we choose the path of integration $t \mapsto \varphi_t(z)$ in the formula

$$(Cf)(z) = \frac{1}{z}\int_0^z \frac{f(\xi)}{1-\xi}d\xi,$$

we obtain

$$\begin{aligned}(Cf)(z) &= \frac{1}{z}\int_0^z \frac{f(\xi)}{1-\xi}d\xi \\ &= -\frac{1}{z}\int_0^\infty \frac{(f\circ\varphi_t)(z)}{1-\varphi_t(z)}\frac{\partial\varphi_t(z)}{\partial t}dt \\ &= \int_0^\infty \frac{\varphi_t(z)}{z} f\circ\varphi_t(z)dt.\end{aligned}$$

Siskakis proved that when $2 \leqslant p < \infty$, the linear transformations

$$(S_t f)(z) = \frac{\varphi_t(z)}{z} f\circ\varphi_t(z)$$

define bounded operators on H^p with $\|S_t\|_{H^p\to H^p} \leqslant e^{-t/p}$. This estimate, along with the previous integration formula, play an important role in estimating $\|C\|_{H^p\to H^p}$. More delicate estimates are needed for the $1 \leqslant p < 2$ case.

One can extend, at least the boundedness result from Theorem 14.1, as follows [**36**].

THEOREM 14.2 (Miao). *For $0 < p < 1$, the Cesàro operator on H^p is bounded.*

Let us make it clear that in this case we are not in a Banach space (but in a Fréchet space), the phrase "C is bounded on H^p" means there is a constant $k_p > 0$ such that

$$\int_{\mathbb{T}} |Cf|^p dm \leqslant k_p \int_{\mathbb{T}} |f|^p dm \quad \text{for all } f \in H^p.$$

Using a similar semigroup analysis, Siskakis discussed the Cesàro operator on the Bergman spaces A^p, $1 \leqslant p \leqslant \infty$, of analytic function f on $\mathbb{D}$ which satisfy

$$\|f\|_{A^p} = \Big(\int_{\mathbb{D}} |f|^p dA\Big)^{\frac{1}{p}} < \infty.$$

In the previous integral, dA denotes planar area measure. The main result from [**46**] is the following.

THEOREM 14.3 (Siskakis). *Consider the Cesàro operator C on A^p.*

(1) *If $4 \leqslant p < \infty$. then $\|C\|_{A^p\to A^p} = \frac{p}{2}$ and the spectrum of C is the closed disk $\{z : |z - \frac{p}{4}| \leqslant \frac{p}{4}\}$.*
(2) *If $1 \leqslant p < 4$, then $\frac{p}{2} \leqslant \|C\|_{A^p\to A^p} \leqslant 2$ and the spectrum of C contains the closed disk $\{z : |z - \frac{p}{4}| \leqslant \frac{p}{4}\}$.*

As one can imagine, the Cesàro operator has been studied for many other spaces of analytic functions on $\mathbb{D}$ and it is impossible to survey them all. We direct the reader to the papers [**18, 28, 38**] which contain useful references to the large literature on this subject.

15. Generalized Cesàro operators

Notice how the Cesàro operator (in integral form) can be written as

$$(Cf)(z) = \frac{1}{z}\int_0^z f(\xi)g'(\xi)d\xi, \quad \text{where } g(z) = \log\Big(\frac{1}{1-z}\Big).$$

For a general $g \in \mathcal{O}(\mathbb{D})$, one can define the linear transformation on $\mathcal{O}(\mathbb{D})$ by

$$(C_g f)(z) = \frac{1}{z}\int_0^z f(\xi)g'(\xi)d\xi.$$

This indicates how one can generalize the classical Cesàro operator on H^2 or other spaces of analytic functions. It was shown in [**39**] (see also [**2**]) that C_g defines a bounded operator on H^2 if and only if g is of bounded mean oscillation. Notice how the logarithm of the function g is of bounded mean oscillation.

Papers [**2, 39**] determine when a generalized Cesàro operator is bounded on H^2. The study of C_g continues with the papers [**34, 38**] which explore the spectral properties of C_g. Other operator theory results such as invariant subspaces (or at least results showing the complexity of them) still need to be developed and this field seems to be wide open and worthy of further exploration.

We end this paper with a segue to another paper on Rhaly matrices that will appear in this volume. This is inspired by the relationship between the Hilbert and Cesàro matrices explored in (4.5). A simple integral substitution shows that the generalized Cesàro operator C_g as a linear transformation on $\mathcal{O}(\mathbb{D})$ can be written in integral form as

$$(C_g f)(z) = \int_0^1 f(tz)g'(tz)dt.$$

With the functions $e_n(z)$ defined by

$$e_n(z) = z^n, \quad n \geqslant 0,$$

observe that if

$$g(z) = \sum_{k=0}^{\infty} g_k z^k,$$

then

$$(C_g e_n)(z) = z^n \sum_{k=1}^{\infty} \frac{k}{k+n} g_k z^{k-1}. \tag{15.1}$$

Thus, the matrix representation of C_g with respect to $(e_n)_{n\geqslant 0}$ is

$$[C_g] = \begin{bmatrix} g_1\frac{1}{0+1} & 0 & 0 & 0 & 0 & \cdots \\ g_2\frac{2}{0+2} & g_1\frac{1}{1+1} & 0 & 0 & 0 & \cdots \\ g_3\frac{3}{0+3} & g_2\frac{2}{1+2} & g_1\frac{1}{2+1} & 0 & 0 & \cdots \\ g_4\frac{4}{0+4} & g_3\frac{3}{1+3} & g_2\frac{2}{2+2} & g_1\frac{1}{3+1} & 0 & \cdots \\ g_5\frac{5}{0+5} & g_4\frac{4}{1+4} & g_3\frac{3}{2+3} & g_2\frac{2}{3+2} & g_1\frac{1}{1+4} & \cdots \\ \vdots & \vdots & \vdots & \vdots & \vdots & \ddots \end{bmatrix}.$$

When

$$g(z) = \log \frac{1}{1-z} = z + \tfrac{1}{2}z^2 + \tfrac{1}{3}z^3 + \cdots,$$

$[C_g]$ becomes the classical Cesàro matrix.

For $g \in \mathcal{O}(\mathbb{D})$, there is the generalized Hilbert operator H_g on $\mathcal{O}(\mathbb{D})$ defined by

$$(H_g f)(z) = \int_0^1 f(t) g'(tz)\, dt.$$

Notice the $f(tz)$ in the integrand of C_g and $f(t)$ (without the z) in the integrand of H_g. A calculation shows that

$$(H_g e_n)(z) = \sum_{k=1}^{\infty} \frac{k}{n+k} g_k z^{k-1} \tag{15.2}$$

and thus, with respect to $(e_n)_{n\geqslant 0}$, the matrix representation of H_g is

$$[H_g] = \begin{bmatrix} g_1 \frac{1}{0+1} & g_1 \frac{1}{1+1} & g_1 \frac{1}{2+1} & g_1 \frac{1}{3+1} & g_1 \frac{1}{4+1} & \cdots \\ g_2 \frac{2}{0+2} & g_2 \frac{2}{1+2} & g_2 \frac{2}{2+2} & g_2 \frac{2}{3+2} & g_2 \frac{2}{4+2} & \cdots \\ g_3 \frac{3}{0+3} & 3 \frac{3}{1+3} & g_3 \frac{3}{2+3} & g_3 \frac{3}{3+3} & g_3 \frac{3}{4+3} & \cdots \\ g_4 \frac{4}{0+4} & g_4 \frac{4}{1+4} & g_4 \frac{4}{2+4} & g_4 \frac{4}{3+4} & g_4 \frac{4}{4+4} & \cdots \\ g_5 \frac{5}{0+5} & g_5 \frac{5}{1+5} & g_5 \frac{5}{2+5} & g_5 \frac{5}{3+5} & g_5 \frac{5}{4+5} & \cdots \\ \vdots & \vdots & \vdots & \vdots & \vdots & \ddots \end{bmatrix}$$

As before with C_g, when

$$g(z) = \log \frac{1}{1-z} = z + \tfrac{1}{2}z^2 + \tfrac{1}{3}z^3 + \cdots,$$

$[H_g]$ becomes the classical Hilbert matrix.

The identities (15.1) and (15.2) show, as before with the Hilbert and Cesàro matrices in (4.5), that one can transform $[H_g]$ into $[C_g]$ by moving the nth column of H_g down n places and filling in the blank spaces with zeros.

References

[1] J. Agler and J. McCarthy, *Beurling's theorem for the Hardy operator on* $L^2[0,1]$, Preprint, arXiv:2206.03462, 2022.

[2] A. Aleman and J. A. Cima, *An integral operator on* H^p *and Hardy's inequality*, J. Anal. Math. **85** (2001), 157–176, DOI 10.1007/BF02788078. MR1869606

[3] R. Bellman, *A note on a theorem of Hardy on Fourier constants*, Bull. Amer. Math. Soc. **50** (1944), 741–744, DOI 10.1090/S0002-9904-1944-08230-X. MR11137

[4] G. Bennett, *Lower bounds for matrices*, Linear Algebra Appl. **82** (1986), 81–98, DOI 10.1016/0024-3795(86)90143-6. MR858964

[5] G. Bennett, *Factorizing the classical inequalities*, Mem. Amer. Math. Soc. **120** (1996), no. 576, viii+130. MR1317938

[6] L. Bouthat and J. Mashreghi, *L-matrices with lacunary coefficients*, Oper. Matrices **15** (2021), no. 3, 1045–1053, DOI 10.7153/oam-2021-15-65. MR4364583

[7] L. Bouthat and J. Mashreghi, *The norm of an infinite L-matrix*, Oper. Matrices **15** (2021), no. 1, 47–58, DOI 10.7153/oam-2021-15-04. MR4237340

[8] A. Brown, P. R. Halmos, and A. L. Shields, *Cesàro operators*, Acta Sci. Math. (Szeged) **26** (1965), 125–137. MR187085

[9] E. Cesàro, *Sur diverses questions d'arithmétique*, Bruxelles: F. Hayez; Opere sclte **1** (1883), 10–362.

[10] E. Cesàro, *Sur la multiplication de séries*, Bull. Sci. Math. **14** (1890), 114–120.

[11] J. B. Conway, *The theory of subnormal operators*, Mathematical Surveys and Monographs, vol. 36, American Mathematical Society, Providence, RI, 1991, DOI 10.1090/surv/036. MR1112128

[12] C. C. Cowen, *Subnormality of the Cesàro operator and a semigroup of composition operators*, Indiana Univ. Math. J. **33** (1984), no. 2, 305–318, DOI 10.1512/iumj.1984.33.33017. MR733903

[13] J. A. Deddens, *Analytic Toeplitz and composition operators*, Canadian J. Math. **24** (1972), 859–865, DOI 10.4153/CJM-1972-085-8. MR310691

[14] P. du Bois-Reymond, *Über die Fourierschen reihen*, Nachr. Kön. Ges. Wiss. Göttingen **21** (1873), 571–582.

[15] P. L. Duren, *Theory of H^p spaces*, Pure and Applied Mathematics, Vol. 38, Academic Press, New York-London, 1970. MR268655

[16] G. Ferraro, *The first modern definition of the sum of a divergent series: an aspect of the rise of 20th century mathematics*, Arch. Hist. Exact Sci. **54** (1999), no. 2, 101–135, DOI 10.1007/s004070050036. MR1705301

[17] E. A. Gallardo-Guti'errez and J. R. Partington, *Insights on the Cesàro operator: shift semigroups and invariant subspaces*, Journal d'Analyse Mathématique (2022), `https://api.semanticscholar.org/CorpusID:249953584`.

[18] V. Glass Miller and T. L. Miller, *The Cesàro operator on the Bergman space $A^2(\mathbb{D})$*, Arch. Math. (Basel) **78** (2002), no. 5, 409–416, DOI 10.1007/s00013-002-8265-6. MR1903676

[19] G. H. Hardy, *Note on a theorem of Hilbert*, Math. Z. **6** (1920), no. 3-4, 314–317, DOI 10.1007/BF01199965. MR1544414

[20] G. H. Hardy, *Divergent Series*, Oxford, at the Clarendon Press, 1949. MR30620

[21] G. H. Hardy, J. E. Littlewood, and G. Pólya, *Inequalities*, Cambridge Mathematical Library, Cambridge University Press, Cambridge, 1988. Reprint of the 1952 edition. MR944909

[22] F. Hausdorff, *Summationsmethoden und Momentfolgen. I* (German), Math. Z. **9** (1921), no. 1-2, 74–109, DOI 10.1007/BF01378337. MR1544453

[23] K. Hoffman, *Banach spaces of analytic functions*, Dover Publications, Inc., New York, 1988. Reprint of the 1962 original. MR1102893

[24] O. Hölder, *Grenzwerthe von Reihen an der Convergenzgrenze* (German), Math. Ann. **20** (1882), no. 4, 535–549, DOI 10.1007/BF01540142. MR1510180

[25] L. E. P. Hupert and A. Leggett, *On the square roots of infinite matrices*, Amer. Math. Monthly **96** (1989), no. 1, 34–38, DOI 10.2307/2323253. MR979593

[26] W. A. Hurwitz and L. L. Silverman, *On the consistency and equivalence of certain definitions of summability*, Trans. Amer. Math. Soc. **18** (1917), no. 1, 1–20, DOI 10.2307/1988924. MR1501058

[27] T. Kawata, *Notes on Fourier series. XII. On Fourier constants*, Proc. Imp. Acad. Tokyo **20** (1944), 218–222. MR14154

[28] E. Kızgut, *The Cesáro operator on weighted Bergman Fréchet and (LB)-spaces of analytic functions*, Filomat **35** (2021), no. 12, 4049–4060. MR4365517

[29] A. Kolmogorov, *Une série e Fourier-Lebesgue divergente presque partout*, Fund. Math **4** (1923), 324–328.

[30] T. L. Kriete III and D. Trutt, *The Cesàro operator in l^2 is subnormal*, Amer. J. Math. **93** (1971), 215–225, DOI 10.2307/2373458. MR281025

[31] T. L. Kriete and D. Trutt, *On the Cesàro operator*, Indiana Univ. Math. J. **24** (1974/75), 197–214, DOI 10.1512/iumj.1974.24.24017. MR350489

[32] M. Lacruz, F. León-Saavedra, S. Petrovic, and O. Zabeti, *Extended eigenvalues for Cesàro operators*, J. Math. Anal. Appl. **429** (2015), no. 2, 623–657, DOI 10.1016/j.jmaa.2015.04.028. MR3342485

[33] G. Leibowitz, *The Cesàro operators and their generalizations: examples in infinite-dimensional linear analysis*, Amer. Math. Monthly **80** (1973), 654–661, DOI 10.2307/2319166. MR322569

[34] A. Limani and B. Malman, *Generalized Cesàro operators: geometry of spectra and quasi-nilpotency*, Int. Math. Res. Not. IMRN **23** (2021), 17695–17707, DOI 10.1093/imrn/rnaa070. MR4349219

[35] J. Mashreghi, M. Ptak, and W. T. Ross, *The square roots of some classical operators*, Studia Math. **269** (2023), no. 1, 83–106, DOI 10.4064/sm210928-19-9. MR4524426

[36] J. Miao, *The Cesàro operator is bounded on H^p for $0 < p < 1$*, Proc. Amer. Math. Soc. **116** (1992), no. 4, 1077–1079, DOI 10.2307/2159491. MR1104399

[37] N. K. Nikol′skiĭ, *Treatise on the shift operator*, Grundlehren der mathematischen Wissenschaften [Fundamental Principles of Mathematical Sciences], vol. 273, Springer-Verlag, Berlin, 1986. Spectral function theory; With an appendix by S. V. Hruščev [S. V. Khrushchëv] and V. V. Peller; Translated from the Russian by Jaak Peetre, DOI 10.1007/978-3-642-70151-1. MR827223

[38] A.-M. Persson, *On the spectrum of the Cesàro operator on spaces of analytic functions*, J. Math. Anal. Appl. **340** (2008), no. 2, 1180–1203, DOI 10.1016/j.jmaa.2007.09.007. MR2390921

[39] Ch. Pommerenke, *Schlichte Funktionen und analytische Funktionen von beschränkter mittlerer Oszillation* (German), Comment. Math. Helv. **52** (1977), no. 4, 591–602, DOI 10.1007/BF02567392. MR454017

[40] J. B. Reade, *On the spectrum of the Cesàro operator*, Bull. London Math. Soc. **17** (1985), no. 3, 263–267, DOI 10.1112/blms/17.3.263. MR806429

[41] H. C. Rhaly Jr., *Posinormal operators*, J. Math. Soc. Japan **46** (1994), no. 4, 587–605, DOI 10.2969/jmsj/04640587. MR1291108

[42] B. E. Rhoades, *Spectra of some Hausdorff operators*, Acta Sci. Math. (Szeged) **32** (1971), 91–100. MR305108

[43] A. L. Shields and L. J. Wallen, *The commutants of certain Hilbert space operators*, Indiana Univ. Math. J. **20** (1970/71), 777–788, DOI 10.1512/iumj.1971.20.20062. MR287352

[44] G. Sinnamon, *Norm of the discrete Cesàro operator minus identity*, Math. Inequal. Appl. **25** (2022), no. 1, 41–48, DOI 10.7153/mia-2022-25-04. MR4366190

[45] A. G. Siskakis, *Composition semigroups and the Cesàro operator on H^p*, J. London Math. Soc. (2) **36** (1987), no. 1, 153–164, DOI 10.1112/jlms/s2-36.1.153. MR897683

[46] A. G. Siskakis, *On the Bergman space norm of the Cesàro operator*, Arch. Math. (Basel) **67** (1996), no. 4, 312–318, DOI 10.1007/BF01197596. MR1407335

[47] A. Zygmund, *Trigonometric series. Vol. I, II*, 3rd ed., Cambridge Mathematical Library, Cambridge University Press, Cambridge, 2002. With a foreword by Robert A. Fefferman. MR1963498

Department of Mathematics and Statistics, University of Richmond, Richmond, VA 23173

Email address: wross@richmond.edu

SELECTED PUBLISHED TITLES IN THIS SERIES

799 **Alberto Condori, Elodie Pozzi, William T. Ross, and Alan A. Sola, Editors,** Recent Progress in Function Theory and Operator Theory, 2024

796 **John Cremona, John Jones, Jennifer Paulhus, Andrew V. Sutherland, and John Voight, Editors,** LuCaNT: LMFDB, Computation, and Number Theory, 2024

795 **Michael A. Jones, David McCune, and Jennifer M. Wilson, Editors,** Mathematical Analyses of Decisions, Voting and Games, 2024

794 **Jean-Pierre Magnot, Editor,** Recent Advances in Diffeologies and Their Applications, 2024

793 **Abba Gumel, Editor,** Mathematical and Computational Modeling of Phenomena Arising in Population Biology and Nonlinear Oscillations, 2024

792 **Shaoming Guo, Zane Kun Li, and Brian Street, Editors,** Recent Developments in Harmonic Analysis and its Applications, 2024

791 **Mee Seong Im, Bach Nguyen, and Arik Wilbert, Editors,** Algebraic and Topological Aspects of Representation Theory, 2024

790 **Peter Crooks and Alexandru I. Suciu, Editors,** Compactifications, Configurations, and Cohomology, 2023

789 **I. S. Krasil′shchik, A. B. Sossinsky, and A. M. Verbovetsky, Editors,** The Diverse World of PDEs, 2023

788 **I. S. Krasil′shchik, A. B. Sossinsky, and A. M. Verbovetsky, Editors,** The Diverse World of PDEs, 2023

787 **Hemen Dutta, Editor,** Mathematical Modelling, 2023

786 **Hemen Dutta, Editor,** Mathematical Modelling, 2023

785 **A. Leroy and S. K. Jain, Editors,** Algebra and Coding Theory, 2023

784 **Dinh-Liem Nguyen, Loc Hoang Nguyen, and Thi-Phong Nguyen, Editors,** Advances in Inverse Problems for Partial Differential Equations, 2023

783 **Alla Detinko, Michael Kapovich, Alex Kontorovich, Peter Sarnak, and Richard Schwartz, Editors,** Computational Aspects of Discrete Subgroups of Lie Groups, 2023

782 **Galina Filipuk, Alberto Lastra, and Sławomir Michalik, Editors,** Recent Trends in Formal and Analytic Solutions of Diff. Equations, 2023

781 **Emanuel Indrei, Diego Marcon, and Levon Nurbekyan, Editors,** Geometric and Functional Inequalities and Recent Topics in Nonlinear PDEs, 2023

780 **Erik Koelink, Stefan Kolb, Nicolai Reshetikhin, and Bart Vlaar, Editors,** Hypergeometry, Integrability and Lie Theory, 2022

779 **Samuele Anni, Valentijn Karemaker, and Elisa Lorenzo García, Editors,** Arithmetic, Geometry, Cryptography, and Coding Theory 2021, 2022

778 **Carlos Galindo, Alejandro Melle Hernández, Julio José Moyano-Fernández, and Wilson A. Zúñiga-Galindo, Editors,** p-Adic Analysis, Arithmetic and Singularities, 2022

777 **Bang-Yen Chen, Nicholas D. Brubaker, Takashi Sakai, Bogdan D. Suceavă, Makiko Sumi Tanaka, Hiroshi Tamaru, and Mihaela B. Vajiac, Editors,** Differential Geometry and Global Analysis, 2022

776 **Aaron Wootton, S. Allen Broughton, and Jennifer Paulhus, Editors,** Automorphisms of Riemann Surfaces, Subgroups of Mapping Class Groups and Related Topics, 2022

775 **Fernando Galaz-García, Cecilia González-Tokman, and Juan Carlos Pardo Millán, Editors,** Mexican Mathematicians in the World, 2021

774 **Randall J. Swift, Alan Krinik, Jennifer M. Switkes, and Jason H. Park, Editors,** Stochastic Processes and Functional Analysis, 2021

For a complete list of titles in this series, visit the
AMS Bookstore at **www.ams.org/bookstore/conmseries/**.